Optimal Sampled-Data Control Systems

Communications and Control Engineering Series

CCES published titles include:

Sampled-Data Control Systems
J. Ackermann

Interactive System Identification
T. Bohlin

The Riccati Equation
S. Bittanti, A.J. Laub and J.C. Willems (Eds.)

Nonlinear Control Systems
A. Isidori

Analysis and Design of Stream Ciphers
R.A. Rueppel

Sliding Modes in Control Optimization
V.I. Utkin

Fundamentals of Robotics
M. Vukobratović

Parametrizations in Control, Estimation and Filtering Problems: Accuracy Aspects
M. Gevers and G. Li

Parallel Algorithms for Optimal Control of Large Scale Linear Systems
Zoran Gajić and Xuemin Shen

Loop Transfer Recovery: Analysis and Design
A. Saberi, B.M. Chen and P. Sannuti

Markov Chains and Stochastic Stability
S.P. Meyn and R.L. Tweedie

Robust Control: Systems with Uncertain Physical Parameters
J. Ackermann in co-operation with A. Bartlett, D. Kaesbauer, W. Sienel and R. Steinhauser

Optimization and Dynamical Systems
U. Helmke and J.B. Moore

Tongwen Chen and Bruce Francis

Optimal Sampled-Data Control Systems

With 222 Figures

Springer

London Berlin Heidelberg New York
Paris Tokyo Hong Kong
Barcelona Budapest

Tongwen Chen, PhD
Department of Electrical and Computer Engineering,
University of Calgary, Calgary, Alberta, Canada T2N 1N4

Bruce Allen Francis, PhD
Department of Electrical and Computer Engineering,
University of Toronto, Toronto, Ontario, Canada M5S 1A4

ISBN 3-540-19949-7 Springer-Verlag Berlin Heidelberg New York

British Library Cataloguing in Publication Data
A catalogue record for this book is available from the British Library

Printed in Great Britain

Typesetting: Camera ready by authors
Printed and bound at the Athenæum Press Ltd., Gateshead
69/3830-543210 Printed on acid-free paper

To Ming and Jingwen, Jessie and Lian

Preface

Many techniques are available for designing linear multivariable analog controllers: pole placement using observer-based controllers, loopshaping, the inverse Nyquist array method, convex optimization in controller parameter space, and so on. One class of techniques is to specify a performance function and then optimize it, and one such performance function is the norm of the closed-loop transfer matrix, suitably weighted. The two most popular norms to optimize are the $\mathcal{H}_2$ and $\mathcal{H}_\infty$ norms. The fact that most new industrial controllers are digital provides strong motivation for adapting or extending these design techniques to digital control systems.

This book is intended as a graduate text in linear sampled-data (SD) control systems. The subject of SD control is a subdomain of digital control; it deals with sampled signals and their discrete-time processing, but not with quantization effects nor with issues of real-time software. SD control systems consist of continuous-time plants to be controlled, discrete-time controllers controlling them, and ideal continuous-to-discrete and discrete-to-continuous transformers.

As a prerequisite, the ideal reader would know multivariable analog control design, especially $\mathcal{H}_2$ and $\mathcal{H}_\infty$ theory—a user's guide to $\mathcal{H}_2$ and $\mathcal{H}_\infty$ theory is presented in Chapter 2. A prior course on digital control at the undergraduate level would also be an asset. Standard facts about state models in continuous and discrete time are collected in the appendix.

Part I (Chapters 2–8) is aimed at first-year graduate students, while Part II (Chapters 9–13) is more advanced. In particular, some of the development in the later chapters is framed in the language of operator theory.

In Part I we present two indirect methods of SD controller design:

- Discretize the plant and design the controller in discrete time.
- Design the controller in continuous time, then discretize it.

These two approaches both involve approximations to the real problem, which involves an analog plant, continuous-time performance specifications, and a SD controller. Part II proposes a direct attack in the continuous-time domain, where SD systems are time-varying (actually, periodic). The main problems addressed are $\mathcal{H}_2$ and $\mathcal{H}_\infty$ optimal SD control. The solutions are presented in forms that can readily be programmed in, for example, MATLAB. MATLAB with the μ-*Tools* toolbox was used for the examples.

Acknowledgements

Graduate courses based on this book are offered at the University of Calgary and the University of Toronto. The first author wishes to thank the following students at the University of Calgary for their careful reading of the drafts: Farhad Ashrafzadeh, Nadra Rafee, Payman Shamsollahi, and Huang Shu. The second author wishes to thank his graduate students at the University of Toronto who collaborated on the research on which this book is based and who made the work so enjoyable: Roger Avedon, Richard Cobden, Geir Dullerud, Freyja Kjartansdottir, Gary Leung, Tony Perry, and Eli Posner. The authors also thank the following people for suggestions, discussions, and collaboration: Abie Feintuch, Bernie Friedland, Toru Fujinaka, Tryphon Georgiou, Tomomichi Hagiwara, Pablo Iglesias, Pramod Khargonekar, Daniel Miller, Li Qiu, Gilbert Strang, and Kemin Zhou.

Various parts of this book in earlier drafts were presented in invited courses: The second author gave a short course at the Centro Internazionale Matematico Estivo, Como, Italy, during June, 1990; he is very grateful to Edoardo Mosca and Luciano Pandolfi for that opportunity. He also gave a course during the fall of 1990 in the Department of Electrical Engineering at the University of Minnesota; he is grateful to the Chairman, Mos Kaveh, and to Allen Tannenbaum for the invitation to do so. Both authors together with Jacob Apkarian gave a short course during May, 1992, at the Fields Institute; the authors thank the Deputy Director, Bill Shadwick, for that invitation. Finally, the second author repeated this short course during June, 1992, in the Department of Automation at Qinghua University, Beijing; for that opportunity he is grateful to the Vice-Chairman, Zheng Da-Zhong.

The authors gratefully acknowledge financial support from the Natural Sciences and Engineering Research Council of Canada.

Contents

II Direct SD Design 207

Chapter 1

Introduction

The signals of interest in control systems—command inputs, tracking errors, actuator outputs, etc.—are usually continuous-time signals and the performance specifications—bandwidth, overshoot, risetime, etc.—are formulated in continuous time. But since digital technology offers many benefits, modern control systems usually employ digital technology for controllers and sometimes sensors. A digital controller performs three functions: It samples and quantizes a continuous-time signal (such as a tracking error) to produce a digital signal; it processes this digital signal using a digital computer; and then it converts the resulting digital signal back into a continuous-time signal. Such a control system thus involves both continuous-time and discrete-time signals, in a continuous-time framework.

Similarly, in many communication systems the input and output signals are continuous-time. For example, in the radio broadcast of a live musical performance, the input to the communication system is the music signal generated by the performers and the output is the audio signal emanating from the radio's speaker. For digital communication, the music signal could be sampled and quantized, giving a digital signal; this latter could be transmitted over a communication channel; and then the received digital signal could be transformed back into continuous time.

Sampled-data systems operate in continuous time, but some continuous-time signals are sampled at certain time instants (usually periodically), yielding discrete-time signals. Sampled-data systems are thus *hybrid* systems, involving both continuous-time and discrete-time signals.

1.1 Sampled-Data Systems

We start with a control system example.

Example 1.1.1 Ball-and-beam. An interesting example is the ball-and-beam setup shown in Figure 1.1. A metal ball rolls along a groove in a

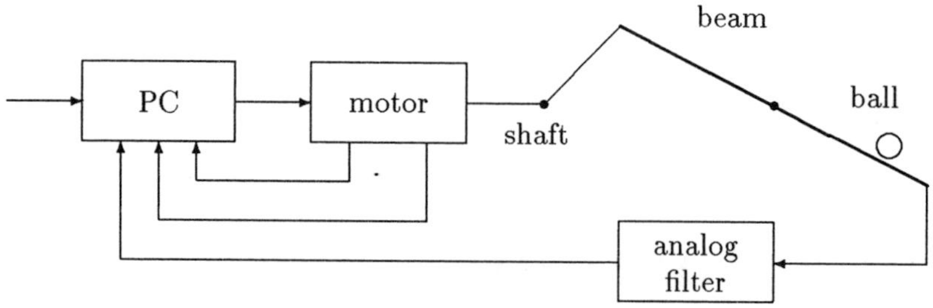

Figure 1.1: Ball-and-beam setup.

balance beam, the angle of tilt of the beam is driven by a motor, and this in turn is driven by a personal computer (PC). The goal is to control the ball's position on the beam.

Look at the motor first. An excitation voltage is applied at its input. This generates a motor torque, τ, by which the angle of tilt of the beam is altered. The distance, d, of the ball from the centre of the beam is sensed (for example, the ball could short a circuit and act as a voltage divider), thus providing a voltage input to the analog filter, a low-pass filter whose purpose is both to provide anti-aliasing and to smooth the measurement of d. Denote by d_f the output of the filter.

Now look at the PC. It has one output—the voltage input to the motor—and three inputs—an externally generated reference voltage (denoted r below) and two other voltages from the motor, one proportional to the angle of the motor's shaft (θ) and the other proportional to its angular velocity ($\dot{\theta}$). The PC is fitted with a card that can do analog-to-digital (A/D) and digital-to-analog (D/A) conversion.

The block diagram is shown in Figure 1.2. The "μ" component denotes the actual processing unit of the PC. Note that a continuous-time signal is drawn by a continuous arrow and a discrete-time signal by a dotted arrow.

This is an example of a digital control system. It is useful to reconfigure the preceding block diagram. First, let us designate four signals:

z = signal to-be-controlled.

Suppose at present that we are concerned only with the tracking requirement that d should be close to r; then we should take

$$z = r - d.$$

The second signal is

$$y \quad = \quad \text{measured signal}$$

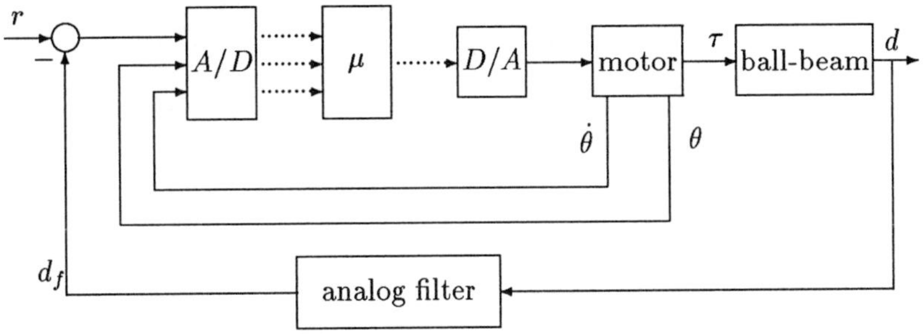

Figure 1.2: Block diagram of ball-and-beam.

$$= \text{ input to digital controller.}$$

In this case we have y a 3-vector:

$$y = \begin{bmatrix} r - d_f \\ \theta \\ \dot{\theta} \end{bmatrix}.$$

The third signal is

$$w = \text{ exogenous input, consisting of reference commands, disturbances, and sensor noise.}$$

Thus w is obtained from all the signals coming from outside of the control system. In this case we have simply

$$w = r.$$

Finally, the fourth signal is

$$\begin{aligned} u &= \text{ control input} \\ &= \text{ output of the digital controller.} \end{aligned}$$

With these four signals we can reconfigure the system as shown in Figure 1.3.

This setup in Figure 1.3 is called the *standard digital control system*. Let us note the following points:

- z, y, w, u are continuous-time signals, perhaps vector-valued, whose amplitudes can be any real numbers. Continuous-time signals will be represented by Roman letters.

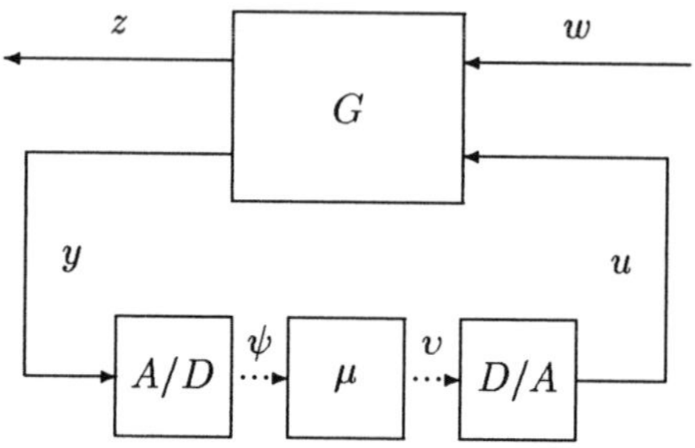

Figure 1.3: Standard digital control system.

- ψ, v are digital signals, possibly vector-valued (like ψ in the example). (A *digital signal* is a discrete-time signal with a quantized amplitude.) Discrete-time signals will be represented by Greek letters.
- G is a dynamical system consisting of physical components (the motor, the ball-beam, the analog filter) and interconnection elements (summing junctions; notice that only one of the two summing junctions above is actually physically present). This system, G, is called the *generalized plant.* It is fixed, that is, designated prior to the control design problem.
- A/D is an analog-to-digital converter, in general, multi-input, multi-output.
- D/A is a digital-to-analog converter, in general, multi-input, multi-output.
- μ is a microprocessor or the central processing unit of a general purpose digital computer.

For purposes of analysis and design, the standard digital control system is idealized. Note that it has the form shown in Figure 1.4, where the controller K consists of the three components

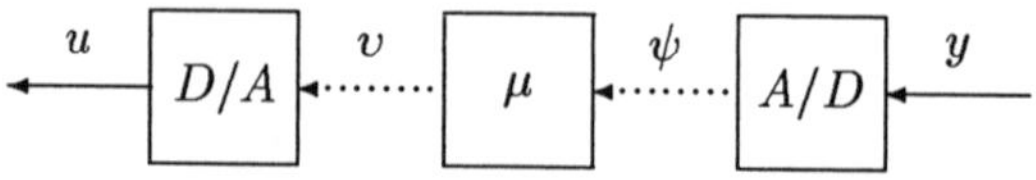

The mathematical idealization of K is

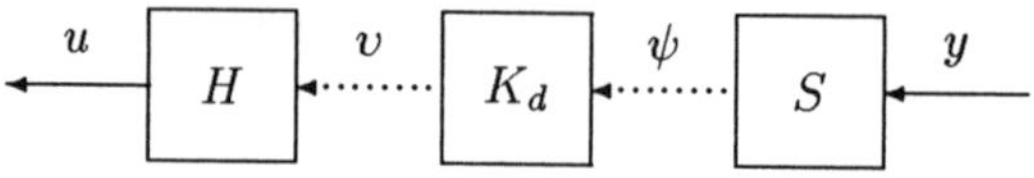

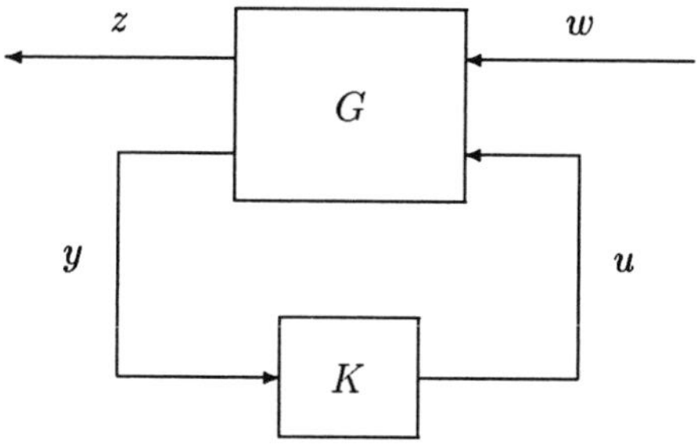

Figure 1.4: Standard control system.

These three components are described as follows:

1. S is the *ideal sampler.* It periodically samples $y(t)$ to yield the discrete-time signal $\psi(k)$. Let h denote the *sampling period.* Thus

   $$\psi(k) := y(kh).$$

   In general, $y(t)$ and $\psi(k)$ are both vectors, of the same dimension. Note that $\psi(k)$ is not quantized in amplitude.

2. K_d is a finite-dimensional (FD), linear time-invariant (LTI), causal, discrete-time system. Its input and output at time k are $\psi(k)$ and $v(k)$. We could use difference equations, state-space equations, or transfer functions to model K_d. (Much more on this later.)

3. H is the *hold operator.* It converts the discrete-time signal v into the continuous-time signal $u(t)$ simply by holding it constant over the sampling intervals. Thus

   $$u(t) = v(k) \quad \text{for} \quad kh \leq t < (k+1)h.$$

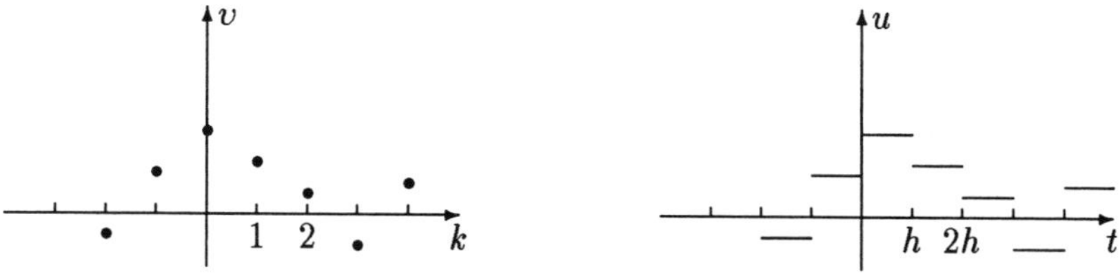

Note that S and H are synchronized, physically by a clock. They are ideal system elements: S instantaneously samples its input; the output of H instantaneously jumps at the sampling instants. (Real A/D and D/A devices are electronic components and obviously don't behave exactly like this.) The full title of H is "zero-order-hold." There are many other ways to convert a discrete-time signal into a continuous-time one.

Let us observe the following terminology:

sampling *period*: h

sampling *instants*: $\ldots, -2h, -h, 0, h, 2h, 3h, \ldots$

sampling *intervals*: $\ldots, [-h, 0), [0, h), [h, 2h), \ldots$

{Recall that $[-h, 0)$ denotes the set of times t with $-h \leq t < 0$.}

Using our idealizations S and H, we obtain the idealized model of the standard digital control system shown in Figure 1.5. This is called the *standard sampled-data (SD) system.* So "sampled-data" refers to a system having both continuous-time and discrete-time signals, whereas "digital" refers to a system having digital signals. In the setup in Figure 1.5 the controller itself is $K = HK_dS$. It is emphasized that K is linear and causal but time-varying (Exercise 1.1). In particular, K does *not* have a transfer function. So transfer function techniques cannot be used directly for analysis or design of the standard SD system. One of the purposes of this book is to develop techniques that *can* be used.

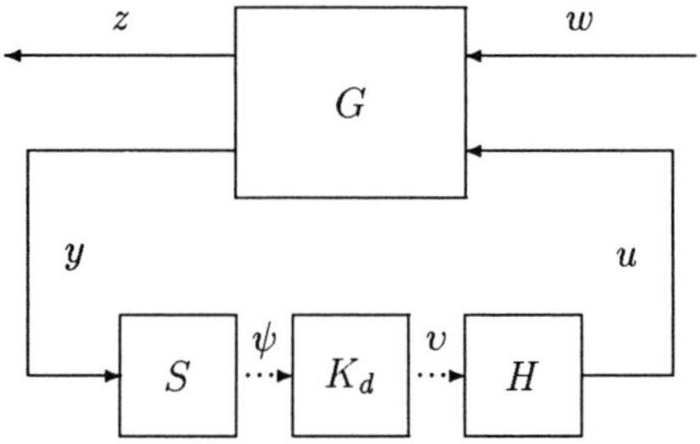

Figure 1.5: Standard sampled-data system.

We conclude this section with some other examples.

Example 1.1.2 Telerobot. A telerobot consists of two robots, a master and a slave, that are in separate locations; for example, the slave robot might be on a space station and the master robot on an earth station. The master robot is manipulated by a human user in order that the slave robot perform some task. Typically, one wants the slave robot to mimic the motion of the master robot when the slave robot is *not* in contact with its environment, and force to be fed back to the human when there *is* contact. The principal obstruction in accomplishing these tasks is time delay in the communication channel.

Figure 1.6 shows a block diagram of a possible configuration of a telerobot with a digital communication channel. The subsystems are

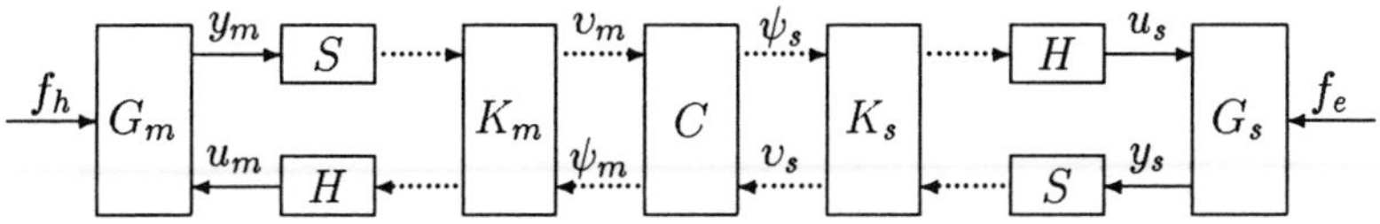

Figure 1.6: Telerobot.

G_m, G_s	master and slave robots
K_m, K_s	master and slave digital controllers
C	digital communication channel

and the signals are

y_m, y_s	master and slave measured signals
u_m, u_s	master and slave control signals
v_m, v_s	signals transmitted from master and slave over the channel
ψ_m, ψ_s	signals received by master and slave
f_h	force human applies to master
f_e	force environment applies to slave.

(Sometimes the dynamics of the human and environment are modelled too.)

Finally we turn to a signal-processing example.

Example 1.1.3 Digital implementation of an analog filter. Let G_a denote an analog filter that has been designed. Suppose we wish to implement G_a digitally, that is, by a system of the form

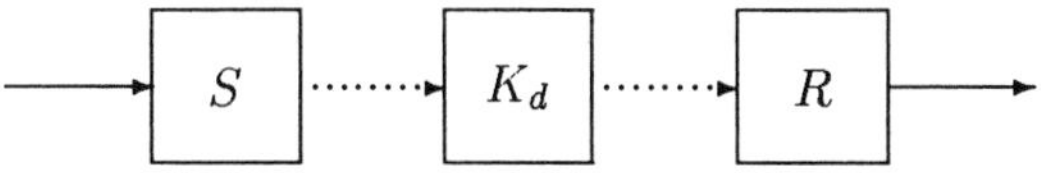

Here K_d is a digital filter, S is the sampler as before, and R is some D/A device: It could be the zero-order hold for realtime applications or it could be the ideal (noncausal) interpolator. We cannot talk about the error between G_a and K_d just as we cannot compare apples and oranges. So it makes sense to look at the *error system*, $G_a - RK_dS$, shown in Figure 1.7. This is a sampled-data system. Clearly, we would consider the discretization to be a

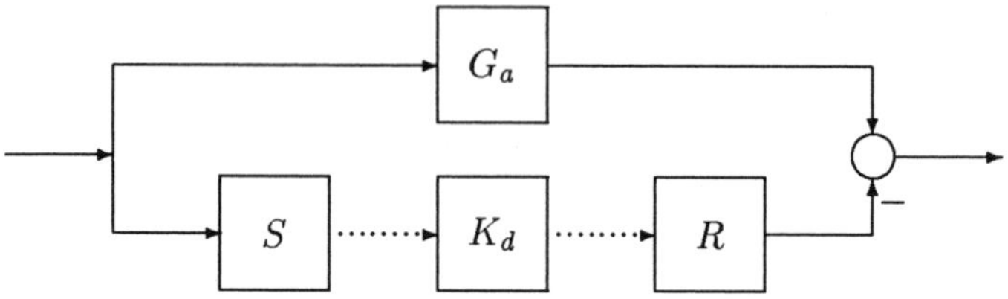

Figure 1.7: The error system.

good one if the error system had small enough gain (suitably defined), ideally, zero gain.

This scenario also arises in digital filter design. To design an IIR (infinite-duration impulse-response) digital filter, common practice is first to design an analog filter and then to transform it into a digital filter for implementation.

1.2 Approaches to SD Controller Design

In order to have something specific to discuss, let us state a *formal synthesis problem* with reference to the standard SD system:

- given the generalized plant G and the sampling period h,
- design a discrete-time controller K_d so that the resulting system has the two properties:
  - internal stability;
  - z is "sufficiently small" for all w in some pre-specified class.

This problem statement is quite vague, but the basic scenario is that G and h are given, while K_d is to be designed. In some applications—such as when you buy an A/D card for a PC with a sampling rate which can be set—h is designable too. Often in an industrial setting, however, the microprocessor is selected upstream from the control engineer and h is then fixed prior to the design of the controller. The term "internal stability" will be defined and studied later, as will what it means for z to be "sufficiently small".

There are essentially three approaches to the SD synthesis problem; the first two are indirect and the third is direct.

Analog design, SD implementation

Let K denote an analog controller designed for G, that is, for satisfactory performance of the system in Figure 1.4. Doing a SD implementation of

K means approximating it by HK_dS for some K_d. The two most common choices of K_d are as follows:

1. K_d is the discretization of K, that is, $K_d = SKH$.

2. The transfer matrix of K_d is obtained from that of K by bilinear transformation.

The advantage of this method is that the design is performed in continuous time, where the performance specifications are most natural. Also, we can expect the analog specifications to be recovered in the limit as $h \to 0$. In practice, however, several technical issues preclude this assumption. First, smaller sampling periods require faster and hence newer and more expensive hardware, so there is a new trade-off between performance and cost in this sense. This is especially relevant in consumer products such as CD players, where a small cost saving per unit results in a large overall saving. Secondly, performing all the control computations may not be feasible if the sampling is too fast. This is relevant in, for example, flexible structures, whose models are high order (think of implementing an LQG controller for a structure with, say, 20 flexible modes). Thirdly, if a plant with slow dynamics is sampled very quickly there will be little difference between successive samples and as result, finite precision arithmetic will demand a large word size to avoid underflow errors. This implies another, different, performance/cost trade-off. And finally, the sampling period is often affected, if not fixed, by other implementation issues unrelated to the control scheme. The microprocessor may be used to perform other functions, constraining the execution time of the digital controller routine. Sensors or actuators may be connected via remote data buses which operate at fixed rates, or may themselves be digital devices.

Discretize the plant; do a discrete-time design

Discretizing the plant means introducing fictitious S and H at z and w as follows:

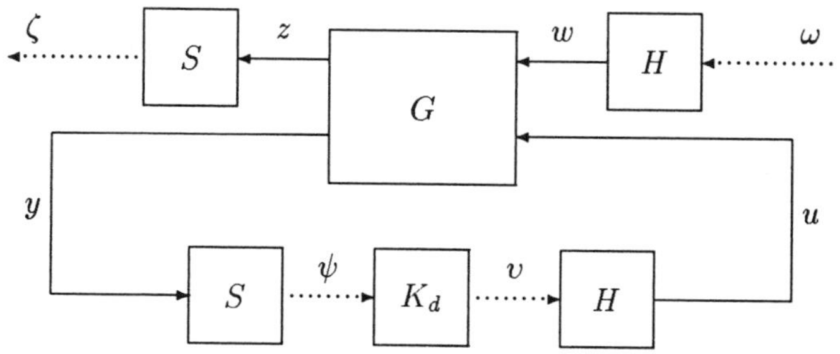

Move the lower S and H around to get

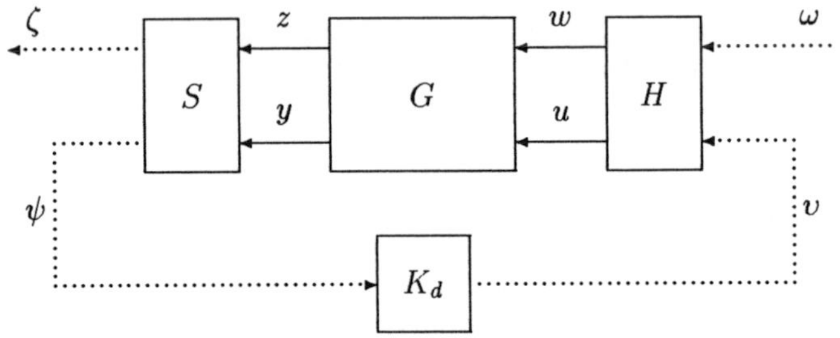

Define $G_d = SGH$, the *discretization* of G. We thus arrive at the purely discrete-time setup

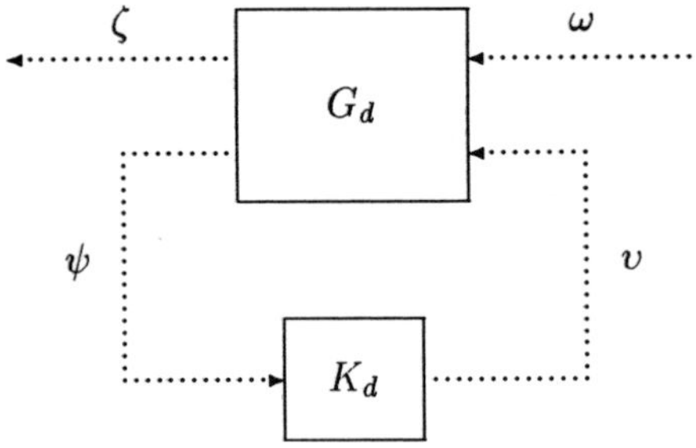

So the next step would be to design K_d for G_d, a purely discrete-time control problem.

The advantage of this approach is its simplicity: We will see that G_d is time-invariant (discrete time) if G is time-invariant (continuous time). But there are disadvantages: The approach completely ignores what is happening between the sampling instants (there might be large intersample amplitudes); continuous-time performance specifications don't always carry over in an obvious way to discrete-time specifications; if h is changed, K_d must be re-designed (G_d depends in a complicated way on h).

Direct SD design

This means design K_d directly for the SD system. The obvious advantage of this method is that it solves the problem with no approximations. The disadvantage is that this approach is harder because the SD system is time-varying.

In this book we will look at all three methods.

1.3 Notation

Good notation is especially important in the study of sampled-data systems. This section summarizes the notation used in this book.

All systems in this book are linear. In the time domain a linear system can be regarded as a linear transformation from one vector space of signals to another (or perhaps the same one). Consider an n-dimensional continuous-time signal, $x(t)$. Assume that t can take on any value in $\mathbb{R}$; then for each t, $x(t) \in \mathbb{R}^n$. Thus x is a function from $\mathbb{R}$ to $\mathbb{R}^n$. Let $\mathcal{L}(\mathbb{R}, \mathbb{R}^n)$ denote the space of all such functions; it is the vector space of all n-dimensional continuous-time signals. Usually n is irrelevant, so we write $\mathcal{L}(\mathbb{R})$. If the starting time for signals were $t = 0$, then $\mathbb{R}_+$, the set of non-negative real numbers, would be the relevant time set and $\mathcal{L}(\mathbb{R}_+, \mathbb{R}^n)$ the associated vector space of signals.

Next, consider an n-dimensional discrete-time signal, $\xi(k)$. Assume that k can take on any value in $\mathbb{Z}$, the set of integers; then ξ is a function from $\mathbb{Z}$ to $\mathbb{R}^n$. Let $\ell(\mathbb{Z}, \mathbb{R}^n)$, or simply $\ell(\mathbb{Z})$, denote the space of all such functions; it is the vector space of all n-dimensional discrete-time signals.

Consider now the standard SD system in Figure 1.5. The four subsystems are linear transformations as follows:

$$\begin{array}{ll} G : \mathcal{L}(\mathbb{R}) \to \mathcal{L}(\mathbb{R}) & K_d : \ell(\mathbb{Z}) \to \ell(\mathbb{Z}) \\ S : \mathcal{L}(\mathbb{R}) \to \ell(\mathbb{Z}) & H : \ell(\mathbb{Z}) \to \mathcal{L}(\mathbb{R}) \end{array}$$

Connecting systems in series amounts to composing the linear transformations; for example, the discretization of G, SGH, is the composition of the three linear transformations:

$$\ell(\mathbb{Z}) \stackrel{H}{\longrightarrow} \mathcal{L}(\mathbb{R}) \stackrel{G}{\longrightarrow} \mathcal{L}(\mathbb{R}) \stackrel{S}{\longrightarrow} \ell(\mathbb{Z})$$

Other points of notation are listed as follows:

- Continuous-time signals are represented by Roman letters, discrete-time signals by Greek letters.
- In a block diagram, continuous-time signals are represented by continuous arrows, discrete-time signals by dotted arrows.
- In continuous time, "dot" denotes time derivative, for example, $\dot{x}(t)$; in discrete time, it denotes forward time advance, for example, $\dot{\xi}(k) := \xi(k+1)$.
- The Laplace transform of a continuous-time signal $x(t)$ is denoted $\hat{x}(s)$.
- If G is a linear time-invariant continuous-time system with impulse-response function $g(t)$, its transfer function (a matrix, in general), the Laplace transform of $g(t)$, is denoted $\hat{g}(s)$. With reference to a state model, the packed notation

$$\left[\begin{array}{c|c} A & B \\ \hline C & D \end{array}\right]$$

denotes the transfer function, $D + C(s - A)^{-1}B$. Here and throughout the book the scalar matrix sI is written simply as s.

- Traditional engineering practice is to use the z-transform in discrete time. However, complex function theory (for example, Cauchy's theorem) deals largely with functions analytic in the unit disc. For this reason it is more convenient to use the λ-transform, where $\lambda = 1/z$. The λ-transform of a discrete-time signal $\xi(k)$ is denoted $\hat{\xi}(\lambda)$, that is,

$$\hat{\xi}(\lambda) = \sum_k \xi(k)\lambda^k.$$

For example, the λ-transform of the unit step starting at time $k = 0$ is

$$1 + \lambda + \lambda^2 + \cdots = \frac{1}{1-\lambda}.$$

- In continuous time the unit impulse and step are denoted respectively by $\delta(t)$ and $1(t)$; in discrete time by $\delta_d(k)$ and $1_d(k)$.

- If G is a linear time-invariant discrete-time system with impulse-response function $g(k)$, its transfer function (a matrix, in general), the λ-transform of $g(k)$, is denoted $\hat{g}(\lambda)$. With reference to a state model, the packed notation

$$\left[\begin{array}{c|c} A & B \\ \hline C & D \end{array}\right]$$

denotes the transfer function, $D + \lambda C(I - \lambda A)^{-1}B$.

- The following abbreviations are used:

LTI	linear time-invariant
FD	finite-dimensional
SISO	single-input, single-output

These are sometimes concatenated, for example, FDLTI.

Finally, consider the standard SD system in Figure 1.5. Since G has two inputs (w, u) and two outputs (z, y), as a linear system it can be partitioned into four components:

$$\begin{aligned} z &= G_{11}w + G_{12}u \\ y &= G_{21}w + G_{22}u. \end{aligned}$$

It can be derived (Exercise 1.10) that the system from w to z equals

$$G_{11} + G_{12}HK_dS(I - G_{22}HK_dS)^{-1}G_{21}, \tag{1.1}$$

or equivalently

$$G_{11} + G_{12}(I - HK_dSG_{22})^{-1}HK_dSG_{21}. \tag{1.2}$$

Exercises

1.1 Look at S and H in tandem:

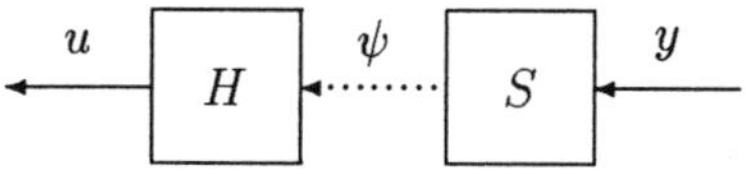

This system, HS, takes a continuous-time signal y into another continuous-time signal u. For example, HS applied to a ramp:

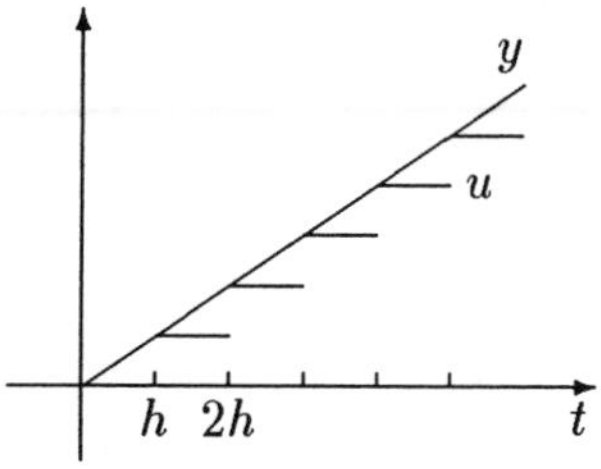

Thus $u = y$ at the sampling instants, but not in between. So $HS \neq I$ (I is the linear system whose output equals its input at every time).

1. Prove that HS is linear.

2. Prove that HS is causal; that is, for every time τ, if $y(t) = 0$ for all $t \leq \tau$, then $u(t) = 0$ for all $t \leq \tau$.

3. Prove that HS is *not* time-invariant.

4. Find a nonzero input y such that $u(t) = 0$ for every t.

1.2 Consider again the sample-and-hold system, HS. This system is linear and causal; and many such systems, including this one, have mathematical models of the form

$$y(t) = \int_{-\infty}^{t} g(t,\tau)u(\tau)\, d\tau.$$

Find the function $g(t,\tau)$. [If this system were time-invariant, then $g(t,\tau)$ would depend only on the difference $t - \tau$.]

1.3 A continuous-time linear system with input $u(t)$ and output $y(t)$ is defined to be *periodic*, of *period* τ, if shifting u by time τ results in shifting y by time τ. (So a time-invariant system is periodic of every period.)

1. Show that HS is periodic, of period h.

2. Consider two sample-and-hold pairs, S_1, H_1 of sampling period h_1 and S_2, H_2 of sampling period h_2. Thus S_1 samples at the times $t = kh_1$ and S_2 samples at the times $t = kh_2$. Connecting them in parallel gives the new system $G = H_1S_1 + H_2S_2$. For what h_1 and h_2 is G periodic?

1.4 Look at S and H in the reverse order:

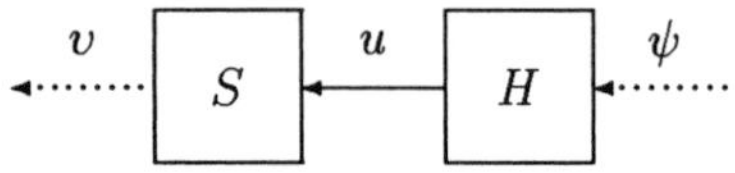

Show that $v(k) = \psi(k)\ \forall k$. Thus $SH = I$, the identity discrete-time system.

1.5 For the ball-and-beam setup in Figure 1.1, find G when the system is configured as in Figure 1.3.

1.6 Reconfigure the following system into the standard SD system:

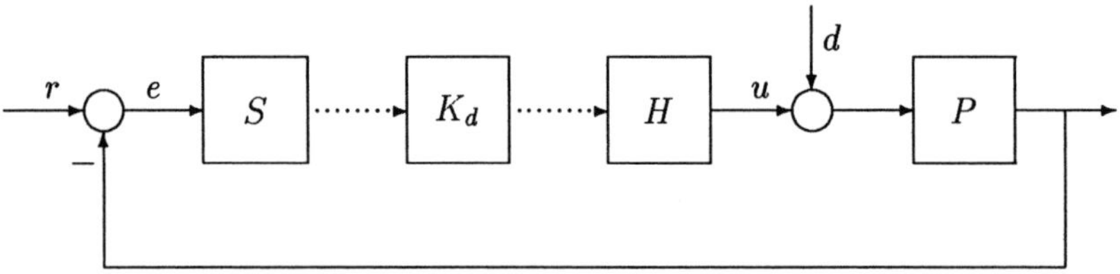

Take

$$z = e, \quad w = \begin{pmatrix} r \\ d \end{pmatrix}, \quad y = e.$$

1.7 In the preceding exercise we took $z = e$, that is, the only signal to-be-controlled was the tracking error. Suppose now that we want in addition to control u. Repeat the previous exercise but with

$$z = \begin{pmatrix} e \\ u \end{pmatrix}.$$

1.8 Figure 1.8 shows two motors controlled by one controller. The motors are identical, with shaft angles θ_1 and θ_2. The left-hand motor is forced by an external torque w. The controller, K, inputs the two shaft positions and their velocities, and outputs two voltages, u_1 and u_2, to the motors. The goal is that the system should act like a telerobot: When a human applies a torque w, the "master" (left-hand) motor should turn appropriately and the "slave" (right-hand) motor should follow it. Take K to be a SD controller and reconfigure the system into the standard SD setup.

1.9 Consider the digital implementation of an analog filter. Reconfigure the error system into the standard SD system.

1.10 Derive equations (1.1) and (1.2).

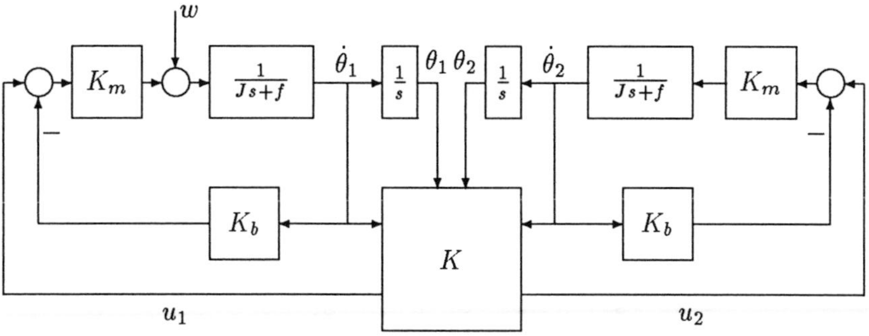

Figure 1.8: Master/slave.

Notes and References

Sampled-data systems originated with the development of automatic-tracking radar systems [55]. Radar signals were pulsed, providing a sampled measurement of position. Important early work on the theory of sampled-data systems was done by Ragazzini and his students at Columbia University in the 1950s; an influential book of that time was [121]. Popular modern introductory books on sampled-data control systems are [9] and [54]. These books only touch on the related topic of real-time software; for this, see [11].

Telerobotics is an interesting subject from the viewpoint of SD control, providing a good platform for testing new SD design techniques. For a general presentation of telerobotics see [123].

Part I

Indirect Design Methods

Chapter 2

Overview of Continuous-Time $\mathcal{H}_2$- and $\mathcal{H}_\infty$-Optimal Control

This chapter gives an overview of the standard $\mathcal{H}_2$ and $\mathcal{H}_\infty$ problems for continuous-time systems. Two examples illustrate how to set problems up as $\mathcal{H}_2$ or $\mathcal{H}_\infty$ problems.

2.1 Norms for Signals and Systems

We begin with the $\mathcal{L}_2$-norm of a signal u in $\mathcal{L}(\mathbb{R}, \mathbb{R}^n)$. For each time t, $u(t)$ is a vector in $\mathbb{R}^n$; denote its Euclidean norm by $||u(t)||$. The $\mathcal{L}_2$-norm of u is then defined to be

$$||u||_2 = \left(\int_{-\infty}^{\infty} ||u(t)||^2 dt \right)^{1/2} .$$

The space $\mathcal{L}_2(\mathbb{R}, \mathbb{R}^n)$, or just $\mathcal{L}_2(\mathbb{R})$ if convenient, consists of all signals for which this norm is finite. For example, the norm is finite if $u(t)$ converges to 0 exponentially as $t \to \pm\infty$. (Caution: $||u||_2 < \infty$ does not imply that $u(t) \to 0$ as $t \to \pm\infty$—think of a counterexample.)

Before defining norms for a transfer matrix, we have to deal with norms for complex matrices. Let R be a $p \times m$ complex matrix, that is, $R \in \mathbb{C}^{p\times m}$. There are many possible definitions for $||R||$; we need two. Let R^* denote the complex-conjugate transpose of R. The matrix R^*R is Hermitian and positive semidefinite. Recall that the *trace* of a square matrix is the sum of the entries on the main diagonal. It is a fact that the trace also equals the sum of the eigenvalues.

The *first definition* for $||R||$ is $[\text{trace}(R^*R)]^{1/2}$.

Example 2.1.1

$$\begin{aligned}
R &= \begin{bmatrix} 2+j & j \\ 1-j & 3-2j \end{bmatrix} \\
R^*R &= \begin{bmatrix} 2-j & 1+j \\ -j & 3+2j \end{bmatrix} \begin{bmatrix} 2+j & j \\ 1-j & 3-2j \end{bmatrix} = \begin{bmatrix} 7 & 6+3j \\ 6-3j & 14 \end{bmatrix} \\
\text{norm} &= (7+14)^{1/2} = \sqrt{21}
\end{aligned}$$

Observe in this example that if r_{ij} denotes the ijth entry in R, then

$$\text{norm} = \left(\sum_i \sum_j |r_{ij}|^2 \right)^{1/2}.$$

This holds in general.

The *singular values* of R are defined as the square roots of the eigenvalues of R^*R. The maximum singular value of R, denoted $\sigma_{\max}(R)$, has the properties required of a norm and is our *second definition* for $\|R\|$.

Example 2.1.2 The singular values of

$$R = \begin{bmatrix} 2+j & j \\ 1-j & 3-2j \end{bmatrix}$$

equal 4.2505, 1.7128. These are computed via the function *svd* in MATLAB. Thus norm = 4.2505.

The importance of this second definition is derived from the following fact. Let $u \in \mathbb{C}^m$ and let $y = Ru$, so $y \in \mathbb{C}^p$. The fact is that

$$\sigma_{\max}(R) = \max\{\|y\| : \|u\| = 1\}.$$

This has the interpretation that if we think of R as a system with input u and output y, then $\sigma_{\max}(R)$ equals the system's gain, that is, maximum output norm over all inputs of unit norm.

Now we can define norms of a stable $p \times m$ transfer matrix $\hat{g}(s)$. Note that for each ω, $\hat{g}(j\omega)$ is a $p \times m$ complex matrix.

$\mathcal{H}_2$-Norm

$$\|\hat{g}\|_2 = \left\{ \frac{1}{2\pi} \int_{-\infty}^{\infty} \text{trace}\,[\hat{g}(j\omega)^* \hat{g}(j\omega)]\, d\omega \right\}^{1/2}$$

Note that the integrand equals the square of the first-definition norm of $\hat{g}(j\omega)$.

$\mathcal{H}_\infty$-Norm

$$\|\hat{g}\|_\infty = \sup_\omega \sigma_{\max}\left[\hat{g}(j\omega)\right]$$

So here we used the second-definition norm of $\hat{g}(j\omega)$.

Concerning these two definitions are two important input-output facts. Let G be a stable, causal, LTI system with input u of dimension m and output y of dimension p:

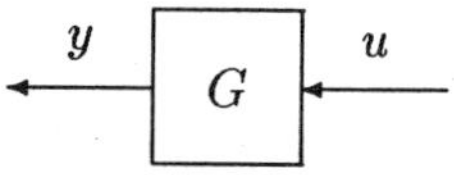

Let e_i, $i = 1, \ldots, m$, denote the standard basis vectors in $\mathbb{R}^m$. Thus, δe_i is an impulse applied to the ith input; $G\delta e_i$ is the corresponding output.

The first fact is that the $\mathcal{H}_2$-norm of the transfer matrix $\hat{g}$ is related to the average $\mathcal{L}_2$-norm of the output when impulses are applied at the input channels.

Theorem 2.1.1 $\|\hat{g}\|_2^2 = \sum_{i=1}^m \|G\delta e_i\|_2^2$

The second fact is that the $\mathcal{H}_\infty$-norm of the transfer matrix $\hat{g}$ is related to the maximum $\mathcal{L}_2$-norm of the output over all inputs of unit norm.

Theorem 2.1.2 $\|\hat{g}\|_\infty = \sup\{\|y\|_2 : \|u\|_2 = 1\}$

Thus the major distinction between $\|\hat{g}_2\|_2$ and $\|\hat{g}\|_\infty$ is that the former is an average system gain for known inputs, while the latter is a worst-case system gain for unknown inputs.

It is useful to be able to compute $\|\hat{g}\|_2$ and $\|\hat{g}\|_\infty$ by state-space methods. Let

$$\hat{g}(s) = \left[\begin{array}{c|c} A & B \\ \hline C & D \end{array}\right],$$

with A stable, that is, all eigenvalues with negative real part. Then $\|\hat{g}\|_2 = \infty$ unless $D = 0$, in which case the following procedure does the job:

Step 1 Solve for L:

$$AL + AL' + BB' = 0.$$

Thus L equals the controllability Gramian.

Step 2 $\|\hat{g}\|_2^2 = \text{trace } CLC'$

The computation of $\|\hat{g}\|_\infty$ using state-space methods is more involved. The formula below involves the Hamiltonian matrix

$$H = \begin{bmatrix} A + B(\gamma^2 - D'D)^{-1}D'C & \gamma B(\gamma^2 - D'D)^{-1}B' \\ -\gamma C'(\gamma^2 - DD')^{-1}C & -[A + B(\gamma^2 - D'D)^{-1}D'C]' \end{bmatrix},$$

where γ is a positive number. The matrices $\gamma^2 - DD'$, $\gamma^2 - D'D$ are invertible provided they are positive definite, equivalently, γ^2 is greater than the largest eigenvalue of DD' (or $D'D$), equivalently, $\gamma > \sigma_{max}(D)$.

Theorem 2.1.3 *Let γ_{max} denote the maximum γ such that H has an eigenvalue on the imaginary axis. Then* $\|\hat{g}\|_\infty = \max\{\sigma_{max}(D), \gamma_{max}\}$.

The theorem suggests the following procedure: Plot, versus γ, the distance from the imaginary axis to the nearest eigenvalue of H; then γ_{max} equals the maximum γ for which the distance equals zero; then $\|\hat{g}\|_\infty = \max\{\sigma_{max}(D), \gamma_{max}\}$. A more efficient procedure is to compute γ_{max} by a bisection search.

2.2 $\mathcal{H}_2$-Optimal Control

Consider the standard setup of Figure 2.1. We must define the concept of

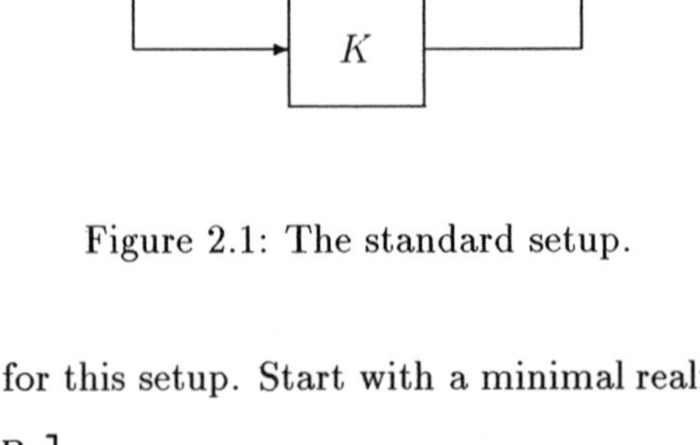

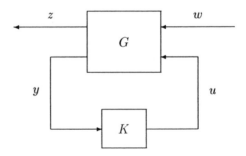

Figure 2.1: The standard setup.

internal stability for this setup. Start with a minimal realization of G:

$$\hat{g}(s) = \left[\begin{array}{c|c} A & B \\ \hline C & D \end{array}\right].$$

The input and output of G are partitioned as

$$\begin{bmatrix} w \\ u \end{bmatrix}, \quad \begin{bmatrix} z \\ y \end{bmatrix}.$$

This induces a corresponding partition of B, C, and D:

$$\begin{bmatrix} B_1 & B_2 \end{bmatrix}, \quad \begin{bmatrix} C_1 \\ C_2 \end{bmatrix}, \quad \begin{bmatrix} D_{11} & D_{12} \\ D_{21} & D_{22} \end{bmatrix}.$$

We shall *assume* that $D_{22} = 0$, that is, the transfer matrix from u to y is strictly proper. This is a condition to guarantee existence of closed-loop transfer matrices. Thus the realization for G has the form

$$\hat{g}(s) = \left[\begin{array}{c|cc} A & B_1 & B_2 \\ \hline C_1 & D_{11} & D_{12} \\ C_2 & D_{21} & 0 \end{array}\right].$$

Also, bring in a minimal realization of K:

$$\hat{k}(s) = \left[\begin{array}{c|c} A_K & B_K \\ \hline C_K & D_K \end{array}\right].$$

Now set $w = 0$ and write the state equations describing the controlled system:

$$\begin{aligned} \dot{x} &= Ax + B_2 u \\ y &= C_2 x \\ \dot{x}_K &= A_K x_K + B_K y \\ u &= C_K x_K + D_K y. \end{aligned}$$

Eliminate u and y:

$$\begin{bmatrix} \dot{x} \\ \dot{x}_K \end{bmatrix} = \begin{bmatrix} A + B_2 D_K C_2 & B_2 C_K \\ B_K C_2 & A_K \end{bmatrix} \begin{bmatrix} x \\ x_K \end{bmatrix}.$$

We call this latter matrix the *closed-loop A-matrix*. It can be checked that its eigenvalues do not depend on the particular minimal realizations chosen for G and K. The closed-loop system is said to be *internally stable* if this closed-loop A-matrix is stable, that is, all its eigenvalues have negative real part. It can be proved that, given G, an internally stabilizing K exists iff (A, B_2) is stabilizable and (C_2, A) is detectable.

Let T_{zw} denote the system from w to z, with transfer matrix $\hat{t}_{zw}(s)$. The $\mathcal{H}_2$-optimal control problem is to compute an internally stabilizing controller K that minimizes $\|\hat{t}_{zw}\|_2$. The following conditions guarantee the existence of an optimal K:

(A1) (A, B_2) is stabilizable and (C_2, A) is detectable;

(A2) the matrices D_{12} and D_{21} have full column and row rank, respectively;

(A3) the matrices

$$\begin{bmatrix} A - j\omega & B_2 \\ C_1 & D_{12} \end{bmatrix}, \quad \begin{bmatrix} A - j\omega & B_1 \\ C_2 & D_{21} \end{bmatrix}$$

have full column and row rank, respectively, $\forall\omega$;

(A4) $D_{11} = 0$.

The first assumption is, as mentioned above, necessary and sufficient for existence of an internally stabilizing controller. In (A2) full column rank of D_{12} means that the control signal u is fully weighted in the output z. This is a sensible assumption, for if, say, some component of u is not weighted, there is no a priori reason for the optimal controller not to try to make this component unbounded. Dually, full row rank of D_{21} means that the exogenous signal w fully corrupts the measured signal y; it's like assuming noise for each sensor. Again, this is sensible, because otherwise the optimal controller may try to differentiate y, that is, the controller may be improper. Assumption (A3) is merely technical—an optimal controller may exist without it. In words, the assumption says there are no imaginary axis zeros in the cross systems from u to z and from w to y. Finally, (A4) guarantees that $\|\hat{t}_{zw}\|_2$ is finite for every internally stabilizing and strictly proper controller (recall that $\hat{t}_{zw}$ must be strictly proper).

The problem is said to be *regular* if assumptions (A1) to (A4) are satisfied. Sometimes when we formulate a problem they are not initially satisfied; for example, we may initially not explicitly model sensor noise. Then we must modify the problem so that the assumptions *are* satisfied. This process is called *regularization.*

Under these assumptions, the MATLAB commands *h2syn* and *h2lqg* compute the optimal controller. The following example illustrates the $\mathcal{H}_2$ design technique.

Example 2.2.1 Bilateral hybrid telerobot. The setup is shown in Figure 2.2. Two robots, a master, G_m, and a slave, G_s, are controlled by one

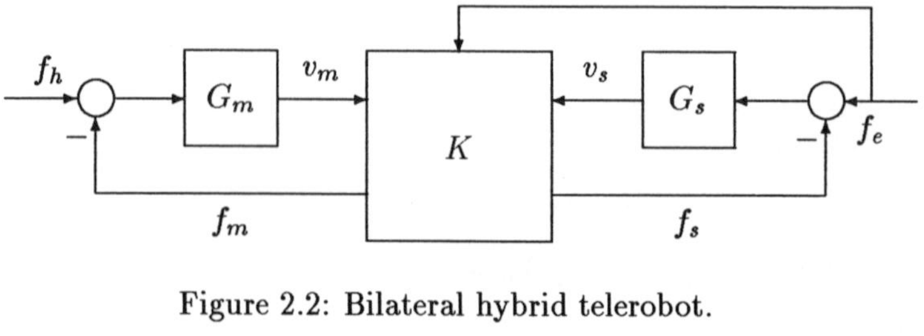

Figure 2.2: Bilateral hybrid telerobot.

controller, K. A human provides a force command, f_h, to the master, while the environment applies a force, f_e, to the slave. The controller measures the two velocities, v_m and v_s, together with f_e via a force sensor. In turn it provides two force commands, f_m and f_s, to the master and slave. Ideally, we want motion following ($v_s = v_m$), a desired master compliance (v_m a desired function of f_h), and force reflection ($f_m = f_e$).

For simplicity of computation we shall take G_m and G_s to be SISO with transfer functions

$$\hat{g}_m(s) = \frac{1}{s}, \quad \hat{g}_s(s) = \frac{1}{10s}.$$

We shall design K for two test inputs, namely, $f_e(t)$ is the finite-width pulse

$$f_e(t) = \begin{cases} 10, & 0 \leq t \leq 0.2 \\ 0, & t > 0.2, \end{cases} \tag{2.1}$$

indicating an abrupt encounter between the slave and a stiff environment, and $f_h(t)$ is the triangular pulse

$$f_h(t) = \begin{cases} 2t, & 0 \leq t \leq 1 \\ -2t+4, & 1 \leq t \leq 2 \\ 0, & t > 2, \end{cases} \tag{2.2}$$

to mimic a ramp-up, ramp-down command.

The generalized error vector is taken to have four components: the velocity error $v_m - v_s$; the compliance error $f_h - v_m$ (for simplicity, the desired compliance is assumed to be $v_m = f_h$); the force-reflection error $f_m - f_e$; and the slave actuator force. The last component is included as part of regularization, that is, to penalize excessive force applied to the slave. Introducing four weights to be decided later, we arrive at the generalized error vector

$$z = \begin{bmatrix} \alpha_v(v_m - v_s) \\ \alpha_c(f_h - v_m) \\ \alpha_f(f_m - f_e) \\ \alpha_s f_s \end{bmatrix}.$$

The Laplace transforms of f_e and f_h are not rational:

$$\hat{f}_e(s) = \frac{10}{s}\left(1 - e^{-0.2s}\right), \quad \hat{f}_h(s) = \frac{2}{s^2}\left(1 - e^{-s}\right)^2.$$

To get a tractable problem, we shall use second- and third-order Padé approximations,

$$e^{-Ts} \approx \left[1 - \frac{Ts}{2} + \frac{(Ts)^2}{12}\right] \Big/ \left[1 + \frac{Ts}{2} + \frac{(Ts)^2}{12}\right]$$

and

$$e^{-Ts} \approx \left[1 - \frac{Ts}{2} + \frac{(Ts)^2}{10} - \frac{(Ts)^3}{120}\right] \Big/ \left[1 + \frac{Ts}{2} + \frac{(Ts)^2}{10} + \frac{(Ts)^3}{120}\right].$$

Using the third-order approximation for $\hat{f}_e(s)$ and the second-order one for $\hat{f}_h(s)$, we get

$$\begin{aligned}\hat{f}_e(s) &\approx 20\left[\frac{0.2}{2}+\frac{0.2^3s^2}{120}\right]\Big/\left[1+\frac{0.2s}{2}+\frac{(0.2s)^2}{10}+\frac{(0.2s)^3}{120}\right]\\ &=: \hat{g}_e(s)\\ \hat{f}_h(s) &\approx 2\Big/\left(1+\frac{s}{2}+\frac{s^2}{12}\right)^2\\ &=: \hat{g}_h(s).\end{aligned}$$

Incorporating these two prefilters into the preceding block diagram leads to Figure 2.3. The two exogenous inputs w_h and w_e are unit impulses. The

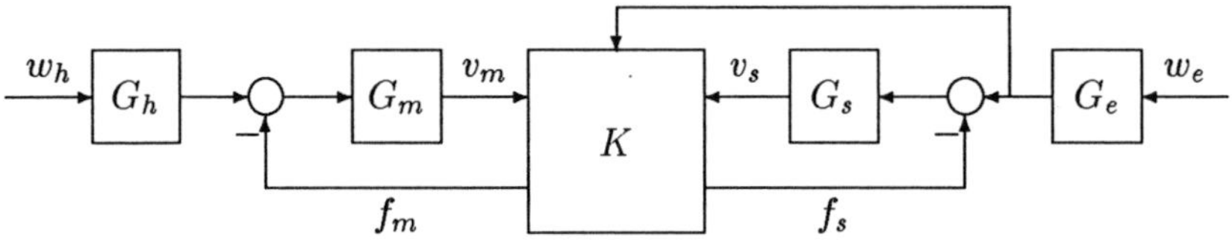

Figure 2.3: Telerobot with prefilters.

vector of exogenous inputs is therefore

$$w = \begin{bmatrix} w_h \\ w_e \end{bmatrix}.$$

Figure 2.4 compares $f_h(t)$ with the impulse response of G_h; and Figure 2.5 is for $f_e(t)$. The error in the second plot is larger because $f_e(t)$ is not continuous.

The control system is shown in Figure 2.6, where z and w are as above and

$$y = \begin{bmatrix} f_e \\ v_s \\ v_m \end{bmatrix}, \quad u = \begin{bmatrix} f_m \\ f_s \end{bmatrix}.$$

Beginning with state models for G_h, G_m, G_s, G_e, namely,

$$\left[\begin{array}{c|c} A_h & B_h \\ \hline C_h & 0 \end{array}\right], \quad \left[\begin{array}{c|c} A_m & B_m \\ \hline C_m & 0 \end{array}\right], \quad \left[\begin{array}{c|c} A_s & B_s \\ \hline C_s & 0 \end{array}\right], \quad \left[\begin{array}{c|c} A_e & B_e \\ \hline C_e & 0 \end{array}\right],$$

with corresponding states x_h, x_m, x_s, x_e, using the interconnections in Figure 2.3, and defining the state

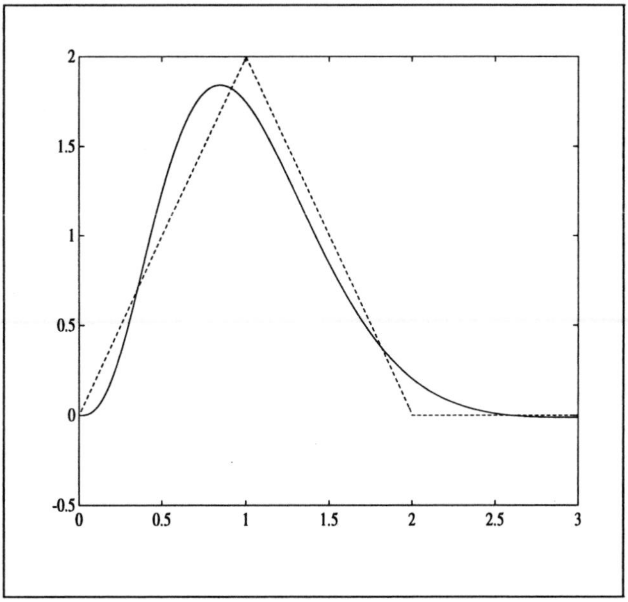

Figure 2.4: $f_h(t)$ (dash) and the impulse response of G_h (solid).

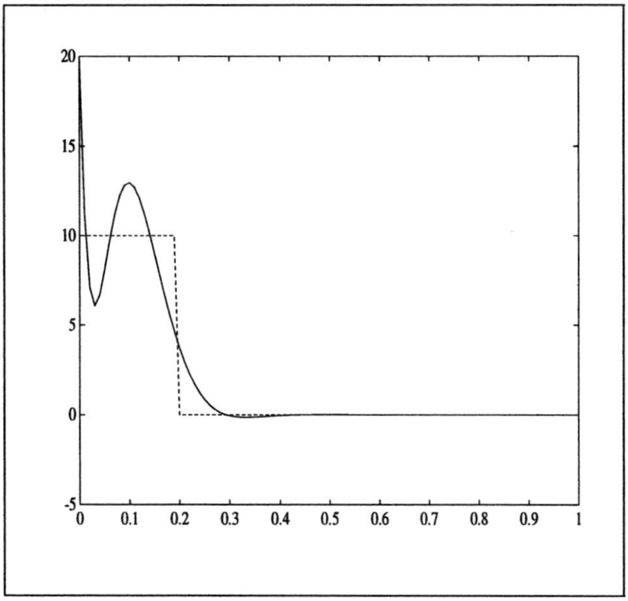

Figure 2.5: $f_e(t)$ (dash) and the impulse response of G_e (solid).

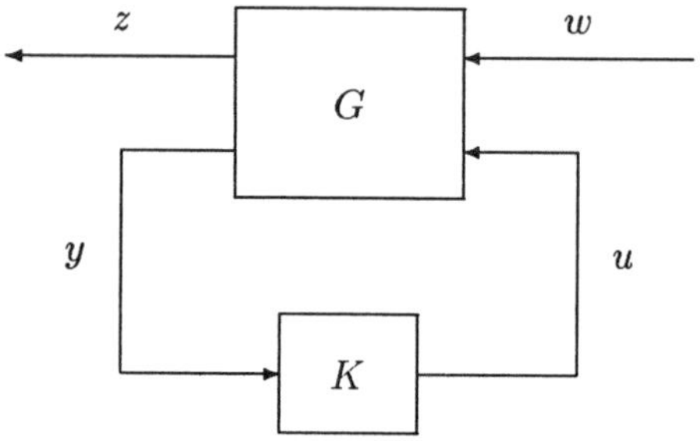

Figure 2.6: Telerobot configured in the standard form.

$$x = \begin{bmatrix} x_m \\ x_s \\ x_e \\ x_h \end{bmatrix}$$

lead to the following state model for G:

$$\left[\begin{array}{c|cc} A & B_1 & B_2 \\ \hline C_1 & 0 & D_{12} \\ C_2 & 0 & 0 \end{array}\right] :=$$

$$\left[\begin{array}{cccc|cccc} A_m & 0 & 0 & B_mC_h & 0 & 0 & -B_m & 0 \\ 0 & A_s & B_sC_e & 0 & 0 & 0 & 0 & -B_s \\ 0 & 0 & A_e & 0 & 0 & B_e & 0 & 0 \\ 0 & 0 & 0 & A_h & B_h & 0 & 0 & 0 \\ \hline \alpha_vC_m & -\alpha_vC_s & 0 & 0 & 0 & 0 & 0 & 0 \\ -\alpha_cC_m & 0 & 0 & \alpha_cC_h & 0 & 0 & 0 & 0 \\ 0 & 0 & -\alpha_fC_e & 0 & 0 & 0 & \alpha_f I & 0 \\ 0 & 0 & 0 & 0 & 0 & 0 & 0 & \alpha_s I \\ & & & & & & & \\ 0 & 0 & C_e & 0 & 0 & 0 & 0 & 0 \\ 0 & C_s & 0 & 0 & 0 & 0 & 0 & 0 \\ C_m & 0 & 0 & 0 & 0 & 0 & 0 & 0 \end{array}\right] . \quad (2.3)$$

For the data at hand, $D_{21} = 0$, so (A2) fails. Evidently, the condition $D_{21} = 0$ reflects the fact that no sensor noise was modelled, that is, perfect measurements of v_m, v_s, f_e were assumed. Let us add sensor noises, say of magnitude ϵ. Then w is augmented to a 5-vector and the state matrices of G change appropriately so that the realization becomes

$$\left[\begin{array}{c|ccc} A & 0 & B_1 & B_2 \\ \hline C_1 & 0 & 0 & D_{12} \\ C_2 & \epsilon I & 0 & 0 \end{array}\right].$$

Some trial-and-error is required to get suitable values for the weights; the following values give reasonable responses:

$$\alpha_v = 10, \quad \alpha_c = 5, \quad \alpha_f = 10, \quad \alpha_s = 0.01, \quad \epsilon = 0.1.$$

The MATLAB functions *h2syn* and *h2lqg* can be used to compute the optimal controller. Figure 2.7 shows plots of $v_s(t)$ and $v_m(t)$ when the system in Figure 2.2 is commanded by $f_h(t)$ (also shown). The velocity tracking and compliance are quite good. Figure 2.8 shows the response of $f_m(t)$

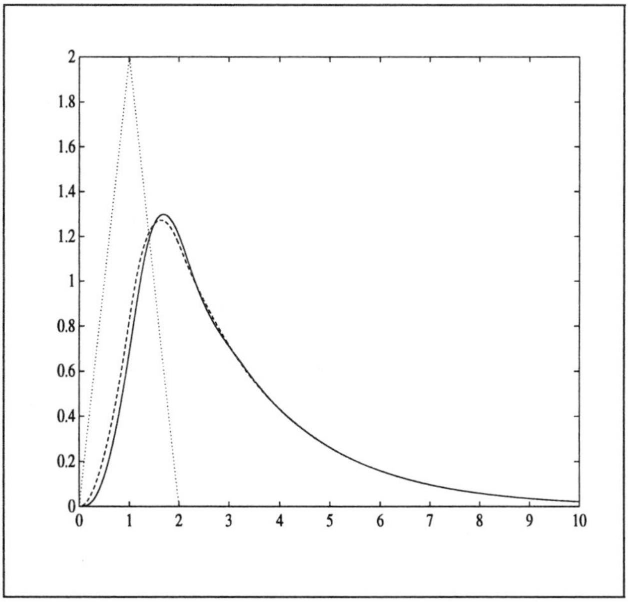

Figure 2.7: Analog design: v_s (solid), v_m (dash), and f_h (dot).

commanded by $f_e(t)$. The force reflection is evident, though there is some oscillation in $f_m(t)$.

2.3 $\mathcal{H}_\infty$-Optimal Control

The $\mathcal{H}_\infty$-optimal control problem is to compute an internally stabilizing controller K that minimizes $\|\hat{t}_{zw}\|_\infty$ for the standard setup of Figure 2.1. This problem is much harder than the $\mathcal{H}_2$ problem. Instead of seeking a controller

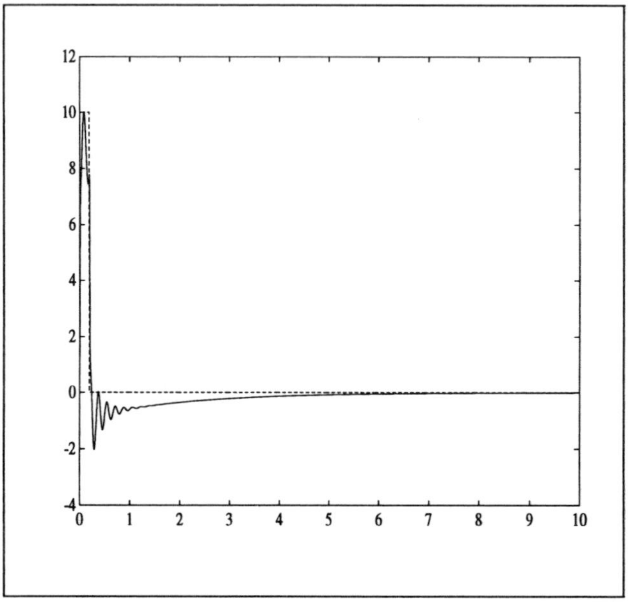

Figure 2.8: Analog design: f_m (solid) and f_e (dash).

that actually minimizes $\|\hat{t}_{zw}\|_\infty$, a simpler problem is to search for a controller that gives $\|\hat{t}_{zw}\|_\infty < \gamma$, where γ is a pre-specified parameter. If γ is too small, a controller will not exist, so we need a test for existence. With this, the following procedure leads to a controller that is close to optimal:

1. Start with a large enough γ so that a controller exists.

2. Test existence for smaller and smaller values of γ until eventually γ is close to the minimum γ for existence.

3. Compute a controller so that $\|\hat{t}_{zw}\|_\infty < \gamma$.

A bisection search can be used.

The MATLAB command *hinfsyn* performs this procedure. The regularity assumptions required are (A1)-(A3), but not (A4), in the preceding section. The following example illustrates how a typical frequency-domain design problem can be formulated as one of $\mathcal{H}_\infty$-optimization.

Example 2.3.1 Figure 2.9 shows a single-loop analog feedback system. The plant is P and the controller K; F is an antialiasing filter for future digital implementation of the controller (it is a good idea to include F at the start of the analog design so that there are no surprises later due to additional phase lag). The basic control specification is to get good tracking

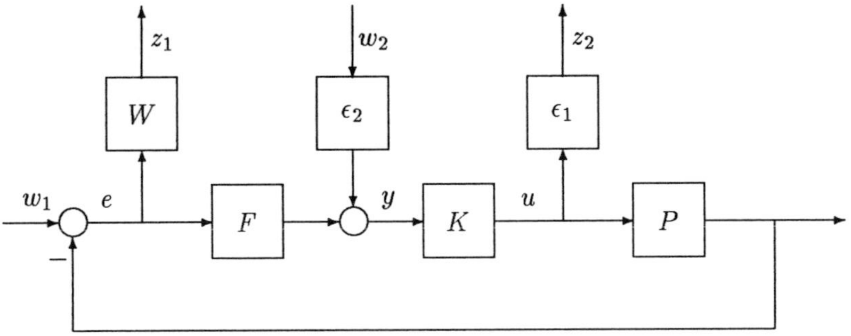

Figure 2.9: Analog feedback system.

over a certain frequency range, say $[0, \omega_1]$; that is, to make the magnitude of the transfer function from w_1 to e small over this frequency range. The weighted tracking error is z_1 in the figure, where the weight W is selected to be a lowpass filter with bandwidth ω_1. We could attempt to minimize the $\mathcal{H}_\infty$-norm from w_1 to z_1, but this problem is not regular. To regularize it, another input, w_2, is added and another signal, z_2, is penalized. The two weights ϵ_1 and ϵ_2 are small positive scalars. The design problem is to minimize the $\mathcal{H}_\infty$-norm

$$\text{from } w = \begin{bmatrix} w_1 \\ w_2 \end{bmatrix} \text{ to } z = \begin{bmatrix} z_1 \\ z_2 \end{bmatrix}.$$

Figure 2.9 can then be converted to Figure 2.1 by stacking the states of P, F, and W to form the state of G.

The plant transfer function is taken to be

$$\hat{p}(s) = \frac{20 - s}{(s + 0.01)(20 + s)}.$$

This can be regarded as an approximation of the time-delay system $\frac{1}{s}e^{-40s}$, an integrator cascaded with a time delay of 40 time units. With a view toward subsequent digital control with $h = 0.5$, the filter F is taken to have bandwidth $\pi/0.5$, the Nyquist frequency ω_N:

$$\hat{f}(s) = \frac{1}{(0.5/\pi)s + 1}.$$

The weight W is then taken to have bandwidth one-fifth the Nyquist frequency:

$$\hat{w}(s) = \left[\frac{1}{(2.5/\pi)s + 1}\right]^2.$$

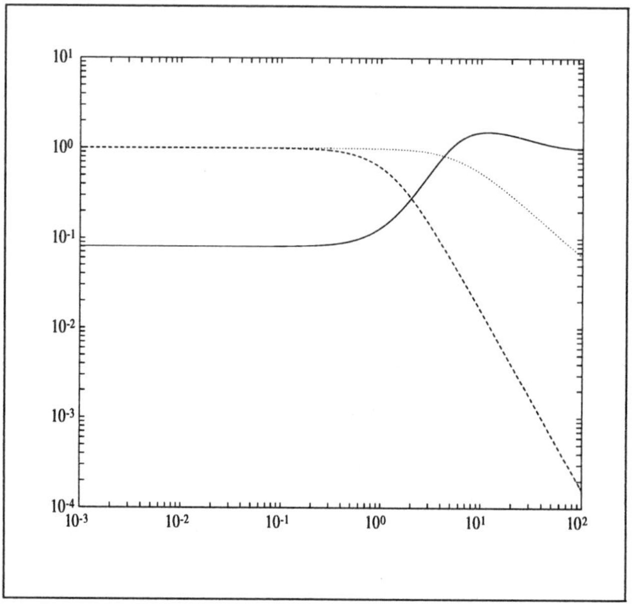

Figure 2.10: Bode magnitude plots: $1/(1+\hat{p}\hat{k}\hat{f})$ (solid), $\hat{w}$ (dash), $\hat{f}$ (dot).

Finally, ϵ_1 and ϵ_2 are both set to 0.01.

Figure 2.10 shows the results of the design using *hinfsyn*. The solid curve is the Bode magnitude plot of the *sensitivity function*, that is, the transfer function from w_1 to e, namely, $1/(1+\hat{p}\hat{k}\hat{f})$. Also shown are the magnitude plots for W (dash) and F (dot). Evidently, the design has achieved some tracking error attenuation over the bandwidth of W. A greater degree of attenuation could be achieved by tuning the weights W, ϵ_1, and ϵ_2.

Notes and References

For a comprehensive treatment of $\mathcal{H}_2$- and $\mathcal{H}_\infty$-optimal control theory see [62], [156].

Chapter 3

Discretization

In this chapter we see how to go from continuous time to discrete time; we look at two discretization techniques: step-invariant transformation and bilinear transformation. There are really two reasons why one might want to discretize a continuous-time system. First, a digital controller sees a discretized plant; for this, the step-invariant transformation is the only choice for discretization if the sampling device is S and the hold device is H. Second, one might want to discretize an analog controller for the purpose of digital implementation; that is, one might want to go from an analog K to a digital K_d. For this second application, the step-invariant transformation is not the only choice. Indeed, there are many other ways to discretize, bilinear transformation being the most common.

3.1 Step-Invariant Transformation

In this section we want to see what happens when we take a continuous-time system G and put H at the input and S at the output:

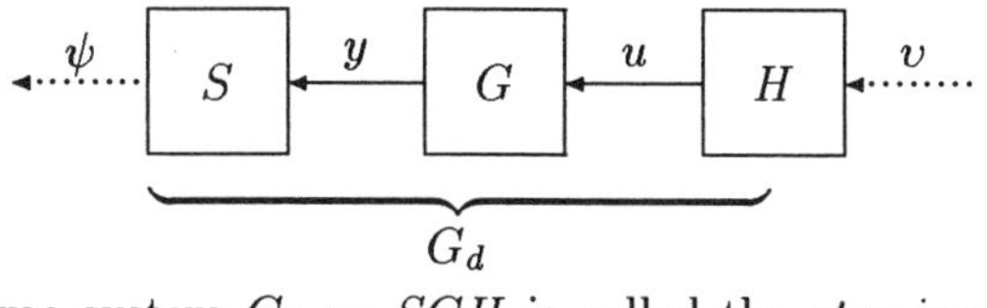

The discrete-time system $G_d := SGH$ is called the *step-invariant transformation* of the continuous-time system G. The reason for this term can be explained as follows. Assume for simplicity that G is a single-input system. When the continuous-time unit step $1(t)$ is applied into G, the sampled output is $SG1$. On the other hand, when the continuous-time unit step $1(t)$ is first sampled, then applied into G_d, the output is G_dS1. It turns out that these two outputs are equal, that is, $G_dS1 = SG1$. The proof is immediate upon noting that $1 = H1_d$, so

$$SG1 = SGH1_d$$

and

$$G_dS1 = G_dSH1_d = G_d1_d \text{ since } SH = I.$$

Now assume that G is FDLTI and let $\hat{g}(s)$ denote its transfer matrix; it is proper and real-rational. Traditional engineering practice is to use the z-transform in discrete time. For our purposes, however, it is more convenient to use the λ-transform, where $\lambda = 1/z$. The transfer matrix for G_d is denoted $\hat{g}_d(\lambda)$. We'll see that $\hat{g}_d(\lambda)$ is real-rational too.

Let us begin with an illustration of how the step-invariant transformation arises in a feedback system.

Example 3.1.1 The block diagram below shows a digital controller and a continuous-time plant.

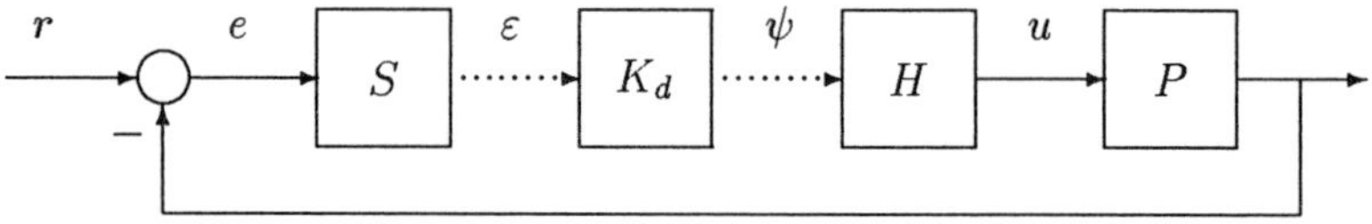

The controller K_d sees a discrete-time input, $\varepsilon(k)$ [the sampled error $e(t)$], and produces a discrete-time output, $\psi(k)$, which is then held to get $u(t)$. Noting that S is linear, we can move it past the summing junction to arrive at the setup

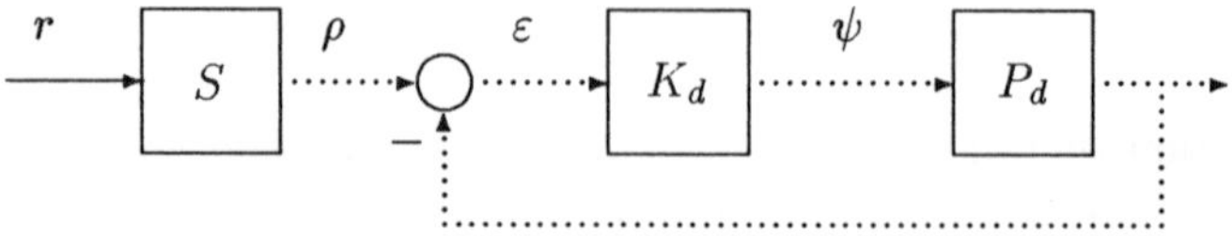

The discretized plant is $P_d = SPH$. The reference input in the second figure is $\rho(k)$, the sampled version of $r(t)$. Notice that the second figure is a purely discrete-time system: It is an exact model of the first figure at the sampling instants.

Now we turn to a procedure for doing step-invariant transformation. Start with state equations for G:

$$\begin{aligned} \dot{x}(t) &= Ax(t) + Bu(t) \\ y(t) &= Cx(t) + Du(t). \end{aligned}$$

It is not necessary for this realization to be minimal. The transfer matrix is

$$\hat{g}(s) = \left[\begin{array}{c|c} A & B \\ \hline C & D \end{array}\right] := D + C(s - A)^{-1}B.$$

For any two times $t_1 < t_2$, the differential equation can be integrated from time t_1 to time t_2:

$$x(t_2) = e^{(t_2 - t_1)A}x(t_1) + \int_{t_1}^{t_2} e^{(t_2 - \tau)A}Bu(\tau)d\tau.$$

Set $t_1 = kh$ and $t_2 = (k+1)h$:

$$x[(k+1)h] = e^{hA}x(kh) + \int_{kh}^{(k+1)h} e^{[(k+1)h - \tau]A}Bu(\tau)d\tau.$$

But $u(\tau)$ is constant over the kth sampling interval; in fact, it equals $v(k)$. Thus

$$x[(k+1)h] = e^{hA}x(kh) + \int_{kh}^{(k+1)h} e^{[(k+1)h - \tau]A}d\tau Bv(k).$$

A change of variables gives

$$\int_{kh}^{(k+1)h} e^{[(k+1)h - \tau]A}d\tau = \int_0^h e^{\tau A}d\tau.$$

Defining the discrete-time state $\xi(k)$ to be the sampled continuous-time state $x(kh)$ and letting "dot" in discrete time denote forward time advance, we get

$$\dot{\xi} = e^{hA}\xi + \int_0^h e^{\tau A}d\tau Bv.$$

Also, sampling the output equation immediately gives

$$\psi = C\xi + Dv.$$

Let us summarize as follows:

Theorem 3.1.1 *The step-invariant transformation maps the state matrices as follows:*

$$(A, B, C, D) \longmapsto (A_d, B_d, C, D)$$

$$A_d := e^{hA}, \quad B_d := \int_0^h e^{\tau A}d\tau B.$$

In particular, only A and B change. If A happens to be nonsingular, we can integrate to get

$$\begin{aligned} \int_0^h e^{\tau A}d\tau &= A^{-1}(e^{hA} - I) \\ &= (e^{hA} - I)A^{-1}. \end{aligned}$$

Usually we'll leave it in integral form. Frequently we shall use the MATLAB-like expression

$$(A_d, B_d) = c2d(A, B, h).$$

The transfer matrix $\hat{g}_d(\lambda)$ can be derived as follows. The λ-*transform* of, for example, the input v is defined to be

$$\hat{v}(\lambda) = v(0) + v(1)\lambda + v(2)\lambda^2 + \cdots.$$

Applying this to the state equations for G_d gives

$$\begin{aligned} \lambda^{-1}\hat{\xi}(\lambda) &= A_d\hat{\xi}(\lambda) + B_d\hat{v}(\lambda) \\ \hat{\psi}(\lambda) &= C\hat{\xi}(\lambda) + D\hat{v}(\lambda). \end{aligned}$$

Solving for $\hat{\xi}(\lambda)$ in the first equation and substituting into the second gives

$$\hat{\psi}(\lambda) = [D + \lambda C(I - \lambda A_d)^{-1}B_d]\hat{v}(\lambda).$$

Thus

$$\hat{g}_d(\lambda) = D + \lambda C(I - \lambda A_d)^{-1}B_d.$$

The transfer matrix on the right-hand side is also written

$$\left[\begin{array}{c|c} A_d & B_d \\ \hline C & D \end{array}\right];$$

this notation is therefore context-dependent.

Now, some examples.

Example 3.1.2 Suppose G is a pure gain D, so G is modelled by

$$y(t) = Du(t).$$

Then G_d is a pure gain too :

$$\psi(k) = Dv(k).$$

In other words,

$$\left[\begin{array}{c|c} 0 & 0 \\ \hline 0 & D \end{array}\right] \longmapsto \left[\begin{array}{c|c} 0 & 0 \\ \hline 0 & D \end{array}\right].$$

Example 3.1.3 Take

$$\hat{g}(s) = \frac{1}{s^3} = \left[\begin{array}{c|c} A & B \\ \hline C & 0 \end{array}\right]$$

$$A = \begin{bmatrix} 0 & 1 & 0 \\ 0 & 0 & 1 \\ 0 & 0 & 0 \end{bmatrix}, \quad B = \begin{bmatrix} 0 \\ 0 \\ 1 \end{bmatrix}$$

$$C = \begin{bmatrix} 1 & 0 & 0 \end{bmatrix}.$$

Observing that

$$\det \begin{bmatrix} \lambda A_d - I & B_d \\ \lambda C & D \end{bmatrix} = \det(\lambda A_d - I)\det[D + \lambda C(I - \lambda A_d)^{-1} B_d],$$

we get that $\hat{g}_d(\lambda)$ equals the ratio

$$\det \begin{bmatrix} \lambda A_d - I & B_d \\ \lambda C & 0 \end{bmatrix} \Big/ \det(\lambda A_d - I).$$

Now A is nilpotent: $A^3 = A^4 = \cdots = 0$. So

$$A_d = I + hA + \frac{h^2}{2!}A^2$$

and

$$B_d = \left(h + \frac{h^2}{2!}A + \frac{h^3}{3!}A^2\right) B.$$

We then compute that

$$\lambda A_d - I = \begin{bmatrix} \lambda - 1 & \lambda h & \lambda h^2/2! \\ 0 & \lambda - 1 & \lambda h \\ 0 & 0 & \lambda - 1 \end{bmatrix},$$

$$B_d = \begin{bmatrix} h^3/3! \\ h^2/2! \\ h \end{bmatrix},$$

and then that

$$\text{denominator } \hat{g}_d = (\lambda - 1)^3, \quad \text{numerator } \hat{g}_d = (-1)^3 \lambda \det M,$$

where

$$M := \begin{bmatrix} 1 & 0 & 0 \\ 0 & 1/h & 0 \\ 0 & 0 & 1/h^2 \end{bmatrix} \begin{bmatrix} \lambda & \lambda/2! & 1/3! \\ \lambda - 1 & \lambda & 1/2! \\ 0 & \lambda - 1 & 1 \end{bmatrix} \begin{bmatrix} h & 0 & 0 \\ 0 & h^2 & 0 \\ 0 & 0 & h^3 \end{bmatrix}.$$

Thus

$$\hat{g}_d(\lambda) = (-1)^3 \frac{h^3 \lambda}{(\lambda - 1)^3} \det \begin{bmatrix} \lambda & \lambda/2! & 1/3! \\ \lambda - 1 & \lambda & 1/2! \\ 0 & \lambda - 1 & 1 \end{bmatrix}.$$

This simplifies to

$$\hat{g}_d(\lambda) = \frac{(-1)^3 h^3}{3!} \frac{\lambda(\lambda^2 + 4\lambda + 1)}{(\lambda - 1)^3}.$$

The zeros of $\hat{g}_d(\lambda)$ are at

$$\lambda = 0, -0.2679, -3.7321.$$

Thus there are two inside the unit disk, indicating a non-minimum-phase characteristic. The first, at $\lambda = 0$, comes from the zero of $\hat{g}(s)$ at $s = \infty$. The second, however, is introduced by the sampling operation.

Example 3.1.4 Take

$$\hat{g}(s) = \frac{1}{s^n}$$

A similar derivation leads to

$$\hat{g}_d(\lambda) = (-1)^n h^n \frac{\lambda \alpha_n(\lambda)}{(\lambda - 1)^n},$$

where α_n is the polynomial

$$\det \begin{bmatrix} \lambda & \lambda/2! & \cdots & \lambda/(n-1)! & 1/n! \\ \lambda - 1 & \lambda & \cdots & \lambda/(n-2)! & 1/(n-1)! \\ \vdots & \vdots & & \vdots & \vdots \\ 0 & 0 & \cdots & \lambda & 1/2! \\ 0 & 0 & \cdots & \lambda - 1 & 1 \end{bmatrix}.$$

The coefficients of $n!\alpha_n(\lambda)$ are as follows:

n	coefficients
2	1 1
3	1 4 1
4	1 11 11 1
5	1 26 66 26 1

For $n \geq 3$ there are zeros in the open unit disk.

In conclusion, for $\hat{g}(s) = 1/s^n$, $\hat{g}_d(\lambda)$ always has a zero at the origin and for $n \geq 3$ it has another zero in $|\lambda| < 1$, indicating non-minimum-phase characteristic in discrete time. Why is this? Look at the step response plots for $\hat{g}(s)$. The integer n is some measure of smoothness at $t = 0+$; as n increases the step response looks more and more sluggish, suggesting a sort of time-delay effect. But a time delay is non-minimum-phase. More on this in Lemma 3.3.3.

3.2 Effect of Sampling

Suppose G is a continuous-time plant with some state realization and we get a realization for $G_d = SGH$ via the step-invariant transformation. What can happen? Can we gain or lose controllability (observability, stabilizability, or detectability)? In this section we answer these questions. We begin with an example.

Example 3.2.1 Again, the setup is

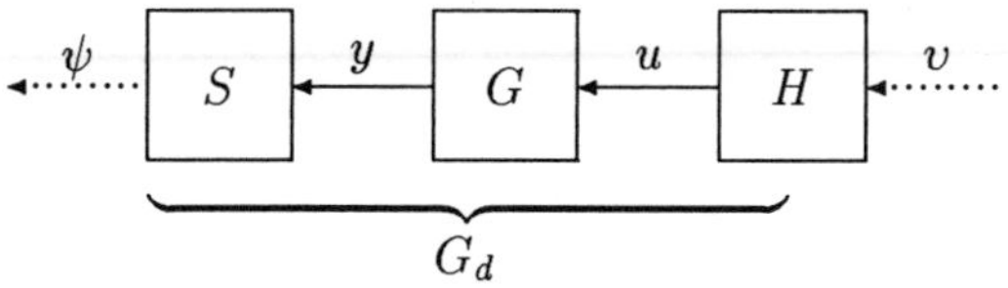

Let ω_s denote the *sampling frequency* (rad/s), that is,

$$\omega_s := \frac{2\pi}{h}.$$

Take G to be an oscillator with

$$\hat{g}(s) = \frac{\omega_s s}{s^2 + \omega_s^2}.$$

Now consider applying the unit step at v. Then u is the unit step in continuous time. Hence

$$\begin{aligned} y(t) &= \text{inverse Laplace transform of } \frac{\omega_s}{s^2+\omega_s^2} \\ &= \sin \omega_s t, \qquad t \geq 0 \end{aligned}$$

and so

$$\psi(k) = 0, \qquad \text{for all } k \geq 0.$$

It follows that G_d must be the zero system, that is, $\hat{g}_d(\lambda) \equiv 0$. So here is a nonzero system G whose discretization is zero. But the discretization of $G = 0$ is also $G_d = 0$. This shows that the mapping $\hat{g}(s) \mapsto \hat{g}_d(\lambda)$ is *not* one-to-one: Two different continuous-time systems can have the same discretization.

Look at the example in terms of a state model:

$$\begin{aligned} \hat{g}(s) &= \left[\begin{array}{c|c} A & B \\ \hline C & D \end{array}\right] = \left[\begin{array}{cc|c} 0 & 1 & 0 \\ -\omega_s^2 & 0 & 1 \\ \hline 0 & \omega_s & 0 \end{array}\right], \qquad \{\text{eigenvalues } A\} = \{\pm j\omega_s\} \\ e^{tA} &= \text{inverse Laplace transform of } (s-A)^{-1} \\ &= \begin{bmatrix} \cos \omega_s t & \frac{1}{\omega_s}\sin \omega_s t \\ -\omega_s \sin \omega_s t & \cos \omega_s t \end{bmatrix} \end{aligned}$$

$$\begin{aligned} A_d &= e^{hA} = \begin{bmatrix} \cos 2\pi & \frac{h}{2\pi}\sin 2\pi \\ -\frac{2\pi}{h}\sin 2\pi & \cos 2\pi \end{bmatrix} = I \\ B_d &= \int_0^h e^{tA} dt B = A^{-1}(e^{hA} - I)B = 0. \end{aligned}$$

Observe that (A, B) is controllable, but (A_d, B_d) is not. So from this point of view, the example shows that sampling can destroy controllability. Clearly the sampling frequency is inappropriate for this system, which is an oscillator at frequency ω_s. This is called pathological sampling.

Now we return to a general discussion. Consider a continuous-time system with state model

$$\left[\begin{array}{c|c} A & B \\ \hline C & D \end{array}\right].$$

Definition 3.2.1 *The sampling frequency* ω_s *is* pathological *(relative to A) if A has two eigenvalues with equal real parts and imaginary parts that differ by an integral multiple of* ω_s. *Otherwise, the sampling frequency is* non-pathological.

Example 3.2.2 Suppose the eigenvalues of A (counting multiplicities) are

$$0, 0, \pm j, 1 \pm 2j.$$

The pathological sampling frequencies can be calculated as follows. The eigenvalues lie on two vertical lines: Re $s = 0$ and Re $s = 1$. Look at the line Re $s = 0$; look at the lowermost eigenvalue, $s = -j$. The distances from it to the other eigenvalues on this line are 1 and 2. Thus the sampling frequency is pathological if $k\omega_s = 1$ or $k\omega_s = 2$ for some positive integer k. Thus the following frequencies are pathological:

$$\left\{\frac{1}{k} : k \geq 1\right\} \cup \left\{\frac{2}{k} : k \geq 1\right\}.$$

By considering the line Re $s = 1$, we get that the following also are pathological frequencies:

$$\left\{\frac{4}{k} : k \geq 1\right\}.$$

Since we have looked at all possible vertical lines, we have counted them all; thus the set of all pathological sampling frequencies is

$$\left\{\frac{1}{k} : k \geq 1\right\} \cup \left\{\frac{2}{k} : k \geq 1\right\} \cup \left\{\frac{4}{k} : k \geq 1\right\}.$$

Since 4 is divisible by 1 and 2, the set of all pathological sampling frequencies is

$$\left\{\frac{4}{k} : k \geq 1\right\}.$$

Notice that this set has an upper bound (4); therefore, ω_s will be non-pathological if it is large enough.

There is another way to describe the effect of pathological sampling, and this needs the *spectral mapping theorem*, which is now briefly described. Let A be a square, real matrix. The eigenvalues of A^2 are equal to the squares of the eigenvalues of A. Another way to say this is that if f is the complex function $f(s) = s^2$, then the eigenvalues of $f(A)$ equal the values of $f(s)$ at the eigenvalues of A:

$$\{\text{eigenvalues of } f(A)\} = \{f(\lambda) : \lambda \text{ is an eigenvalue of } A\}.$$

This equation holds for every complex function f that is analytic at the eigenvalues of A.

From the spectral mapping theorem with $f(s) = e^{hs}$, the eigenvalues of A_d are the points

$$\{e^{h\lambda} : \lambda \text{ is an eigenvalue of } A\}.$$

Also, the function $s \mapsto e^{hs}$ is periodic with period $j\omega_s$. It follows that the sampling frequency is non-pathological iff no two eigenvalues of A are mapped to the same eigenvalue of A_d. Another consequence of non-pathological sampling is that the points $\{\pm j\omega_s, \pm 2j\omega_s, \ldots\}$ cannot be eigenvalues of A. (In the preceding example, the eigenvalues of A are $\{\pm j\omega_s\}$.)

It turns out that controllability and observability are preserved if the sampling frequency is non-pathological.

Theorem 3.2.1 *If the sampling frequency is non-pathological, then*

$$\begin{aligned} (A, B) \textit{ controllable} &\implies (A_d, B_d) \textit{ controllable} \\ (C, A) \textit{ observable} &\implies (C, A_d) \textit{ observable.} \end{aligned}$$

Proof We'll prove just the second implication—the other is similar.

So assume the sampling frequency is non-pathological and (C, A) is observable. To prove that (C, A_d) is observable, we'll show all the eigenvalues of A_d are observable. Now each eigenvalue of A_d has the form $e^{h\lambda}$, where λ is an eigenvalue of A. We must show that

$$\operatorname{rank} \begin{bmatrix} A_d - e^{h\lambda} I \\ C \end{bmatrix} = n \tag{3.1}$$

(i.e., $e^{h\lambda}$ is an observable eigenvalue of A_d) given that

$$\operatorname{rank}\begin{bmatrix} A-\lambda I \\ C \end{bmatrix} = n \tag{3.2}$$

(i.e., λ is an observable eigenvalue of A).

Define the function

$$g(s) = \frac{e^{hs} - e^{h\lambda}}{s-\lambda}.$$

This is analytic everywhere (the "pole" at $s = \lambda$ is cancelled by a "zero" there). Moreover,

$$\begin{aligned} \{\text{zeros of } g\} &= \{s : e^{hs} = e^{h\lambda}, s \neq \lambda\} \\ &= \{s : hs = h\lambda + j2\pi k, k = \pm 1, \pm 2, \ldots\} \\ &= \{s : s = \lambda + jk\omega_s, k = \pm 1, \pm 2, \ldots\}. \end{aligned}$$

By non-pathological sampling, the zeros of g are disjoint from the eigenvalues of A. Now the eigenvalues of the matrix $g(A)$ are precisely the values of g at the eigenvalues of A (this is the spectral mapping theorem, again). Thus 0 is not an eigenvalue of $g(A)$, so $g(A)$ is invertible. Now

$$e^{hs} - e^{h\lambda} = g(s)(s-\lambda).$$

Hence

$$e^{hA} - e^{h\lambda} I = g(A)(A - \lambda I),$$

that is,

$$A_d - e^{h\lambda} I = g(A)(A - \lambda I).$$

Thus

$$\begin{bmatrix} A_d - e^{h\lambda} I \\ C \end{bmatrix} = \begin{bmatrix} g(A) & 0 \\ 0 & I \end{bmatrix} \begin{bmatrix} A - \lambda I \\ C \end{bmatrix}.$$

Since $g(A)$ is invertible, so is

$$\begin{bmatrix} g(A) & 0 \\ 0 & I \end{bmatrix}.$$

Therefore

$$\operatorname{rank}\begin{bmatrix} A_d - e^{h\lambda} I \\ C \end{bmatrix} = \operatorname{rank}\begin{bmatrix} A - \lambda I \\ C \end{bmatrix},$$

so (3.2) implies (3.1). ■

Example 3.2.2 shows we can lose controllability by sampling. What about the converse? Can we gain controllability by sampling? No:

Theorem 3.2.2 *If (A, B) is not controllable, neither is (A_d, B_d).*

Proof If (A, B) is not controllable, there exists an eigenvalue λ of A such that

$$\operatorname{rank}\begin{bmatrix} A - \lambda I & B \end{bmatrix} < n.$$

Then there exists a vector $x \neq 0$ orthogonal to all the columns of this matrix, that is,

$$x^* \begin{bmatrix} A - \lambda I & B \end{bmatrix} = 0.$$

This implies that

$$x^* A = \lambda x^*, \quad x^* B = 0.$$

Now we obtain in succession

$$\begin{aligned} x^* A^2 &= (x^* A) A \\ &= \lambda x^* A \\ &= \lambda^2 x^* \\ x^* A^3 &= \lambda^3 x^* \\ &\text{etc.} \end{aligned}$$

Since

$$e^{hA} = I + hA + \frac{h^2}{2} A^2 + \dots$$

we have

$$\begin{aligned} x^* e^{hA} &= x^* \left(I + hA + \frac{h^2}{2} A^2 + \dots \right) \\ &= x^* + h\lambda x^* + \frac{h^2}{2} \lambda^2 x^* + \dots \\ &= e^{h\lambda} x^*. \end{aligned}$$

Thus

$$\begin{aligned} x^*(A_d - e^{h\lambda} I) &= x^*(e^{hA} - e^{h\lambda} I) \\ &= (e^{h\lambda} - e^{h\lambda}) x^* \\ &= 0 \end{aligned}$$

and

$$\begin{aligned} x^* B_d &= x^* \int_0^h e^{tA} dt B \\ &= \int_0^h e^{t\lambda} dt \underbrace{x^* B}_{=0} \\ &= 0. \end{aligned}$$

Hence

$$x^* \begin{bmatrix} A_d - e^{h\lambda}I & B_d \end{bmatrix} = 0$$

and so

$$\operatorname{rank} \begin{bmatrix} A_d - e^{h\lambda}I & B_d \end{bmatrix} < n.$$

This implies that $e^{h\lambda}$ is an uncontrollable eigenvalue of A_d, proving that (A_d, B_d) is not controllable. ■

Similarly, we can't gain observability by sampling. The results for stabilizability and detectability are entirely analogous; the only change is that in Definition 3.2.1 we only have to consider the eigenvalues in the closed right half-plane.

3.3 Step-Invariant Transformation Continued

In this section we look at how $\hat{g}(s)$ is mapped to $\hat{g}_d(\lambda)$ directly, not through state models. Again, the discretization G_d is defined in Figure 3.1.

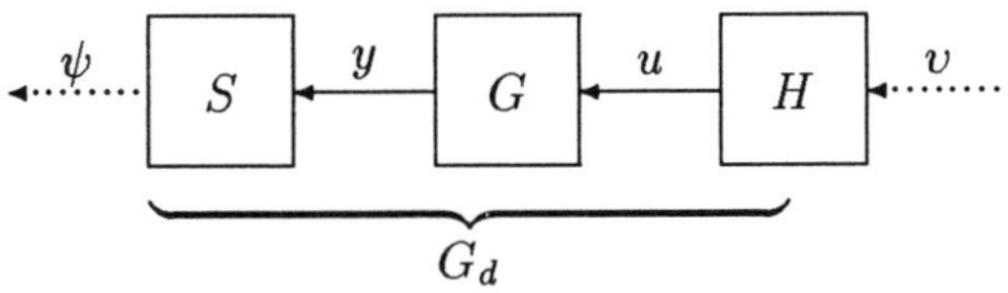

Figure 3.1: Definition of G_d.

We begin with a formula relating the two frequency responses, $\hat{g}(j\omega)$ and $\hat{g}_d\left(e^{j\theta}\right)$. We shall derive this formula in two steps. From now on fix the sampling period h, and let ω_s denote the sampling frequency $(2\pi/h)$ and ω_N the Nyquist frequency (π/h).

The first step is to derive the relationship between the continuous-time Fourier transform of $y(t)$ and the discrete-time Fourier transform of $\psi(k)$, that is, the frequency-domain action of S. The *periodic extension* of a function $\hat{y}(j\omega)$ is

$$\hat{y}_e(j\omega) := \sum_{k=-\infty}^{\infty} \hat{y}(j\omega + jk\omega_s).$$

Note that $\hat{y}_e$ is a periodic function of frequency, of period ω_s:

$$\hat{y}_e[j(\omega + \omega_s)] = \hat{y}_e(j\omega).$$

Lemma 3.3.1 *In the figure*

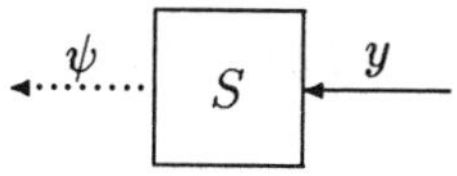

the Fourier transforms of $y(t)$ *and* $\psi(k)$ *are related by the equation*

$$\hat{\psi}\left(e^{-j\omega h}\right) = \frac{1}{h}\hat{y}_e(j\omega). \tag{3.3}$$

Proof It is customary to prove this result using the idea of impulse-train modulation. Define the continuous-time signal $v(t)$:

$$v(t) = y(t) \times \sum_k \delta(t - kh).$$

The impulse train $\sum_k \delta(t - kh)$ is a periodic function of time, of period h; its Fourier series is formally given by

$$\sum_k \delta(t - kh) = \frac{1}{h}\sum_k e^{jk\omega_s t}.$$

Thus we have (formally)

$$\begin{aligned} v(t) &= y(t) \times \frac{1}{h}\sum_k e^{jk\omega_s t} \\ &= \frac{1}{h}\sum_k y(t)e^{jk\omega_s t}. \end{aligned}$$

Taking Fourier transforms, we get

$$\begin{aligned} \hat{v}(j\omega) &= \frac{1}{h}\sum_k \hat{y}(j\omega + jk\omega_s). \\ &= \frac{1}{h}\hat{y}_e(j\omega). \end{aligned} \tag{3.4}$$

On the other hand,

$$\begin{aligned} v(t) &= y(t) \times \sum_k \delta(t - kh) \\ &= \sum_k y(kh)\delta(t - kh) \\ &= \sum_k \psi(k)\delta(t - kh). \end{aligned}$$

Taking Fourier transforms again, we get

$$\hat{v}(j\omega) = \int \left[\sum_k \psi(k)\delta(t - kh)\right] e^{-j\omega t} dt$$

$$
\begin{aligned}
&= \sum_k \psi(k) \int \delta(t-kh) e^{-j\omega t} dt \\
&= \sum_k \psi(k) e^{-j\omega k h} \\
&= \hat{\psi}\left(e^{-j\omega h}\right). \qquad (3.5)
\end{aligned}
$$

Comparing (3.4) and (3.5), we get the desired equation. ■

The second step is to derive the relationship between the discrete-time Fourier transform of $v(k)$ and the continuous-time Fourier transform of $u(t)$, that is, the frequency-domain action of H. We need the system with impulse response

$$r(t) = \frac{1}{h} 1(t) - \frac{1}{h} 1(t-h),$$

that is,

$$r(t) = \begin{cases} 1/h, & 0 \le t < h \\ 0, & \text{elsewhere.} \end{cases} \qquad (3.6)$$

The transfer function is therefore

$$\hat{r}(s) = \frac{1 - e^{-sh}}{sh}. \qquad (3.7)$$

The frequency-response function can be calculated like this:

$$
\begin{aligned}
\hat{r}(j\omega) &= \frac{1 - e^{-j\omega h}}{j\omega h} \\
&= e^{-j\omega \frac{h}{2}} \frac{e^{j\omega \frac{h}{2}} - e^{-j\omega \frac{h}{2}}}{j\omega h} \\
&= e^{-j\omega \frac{h}{2}} \frac{\sin \omega \frac{h}{2}}{w \frac{h}{2}}.
\end{aligned}
$$

Observe that

$$\hat{r}(s) \approx e^{-s\frac{h}{2}} \quad \text{at low frequency.}$$

So at low frequency $\hat{r}$ acts like a time delay of $\frac{h}{2}$.

Lemma 3.3.2 *In the figure*

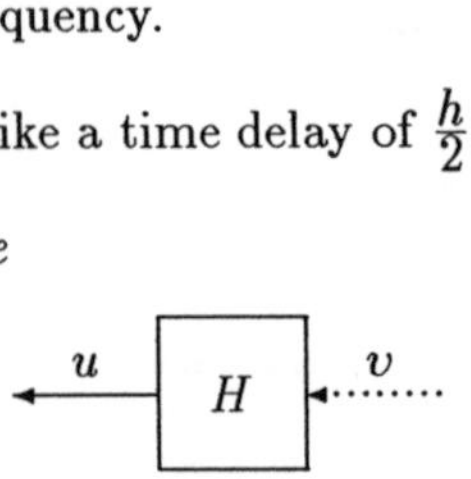

the Fourier transforms of $v(k)$ and $u(t)$ are related by the equation

$$\hat{u}(j\omega) = h\hat{r}(j\omega)\hat{v}\left(e^{-j\omega h}\right). \tag{3.8}$$

Proof From (3.6) and the definition of H, we can write

$$u(t) = h\sum_k v(k)r(t - kh).$$

Taking Fourier transforms, we get

$$\begin{aligned}\hat{u}(j\omega) &= h\sum_k v(k)\hat{r}(j\omega)e^{-j\omega kh} \\ &= h\hat{r}(j\omega)\hat{v}\left(e^{-j\omega h}\right)\end{aligned}$$

∎

Now we put the two preceding lemmas together:

Theorem 3.3.1 *The frequency responses $\hat{g}(j\omega)$ and $\hat{g}_d\left(e^{j\theta}\right)$ are related by*

$$\hat{g}_d\left(e^{-j\omega h}\right) = \sum_{k=-\infty}^{\infty} \hat{g}(j\omega + jk\omega_s)\hat{r}(j\omega + jk\omega_s).$$

Proof In Figure 3.1 let v be the unit impulse. Then

$$\begin{aligned}\hat{u}(j\omega) &= h\hat{r}(j\omega), \quad \text{from Lemma 3.3.2} \\ \hat{y}(j\omega) &= \hat{g}(j\omega)\hat{u}(j\omega) \\ &= \hat{g}(j\omega)h\hat{r}(j\omega) \\ \hat{g}_d\left(e^{-j\omega h}\right) &= \hat{\psi}\left(e^{-j\omega h}\right) \\ &= \frac{1}{h}\sum_{k=-\infty}^{\infty} \hat{y}(j\omega + jk\omega_s), \quad \text{from Lemma 3.3.1.}\end{aligned}$$

∎

The theorem shows that at each frequency ω, $\hat{g}_d\left(e^{-j\omega h}\right)$ depends not only on $\hat{g}(j\omega)$, but also on all the values $\hat{g}(j\omega + jk\omega_s)$ for $k = \pm 1, \pm 2, \ldots$. In the ideal case that $\hat{g}(j\omega)$ is bandlimited to the interval $(-\omega_N, \omega_N)$, then

$$\hat{g}_d\left(e^{-j\omega h}\right) = \hat{g}(j\omega)\hat{r}(j\omega), \quad -\omega_N < \omega < \omega_N.$$

In particular, at low frequencies

$$\hat{g}_d\left(e^{-j\omega h}\right) \approx \hat{g}(j\omega).$$

Example 3.3.1 Figure 3.2 shows the magnitude Bode plot of

$$\hat{g}(s) = \frac{1}{s^2 + 0.1s + 1}$$

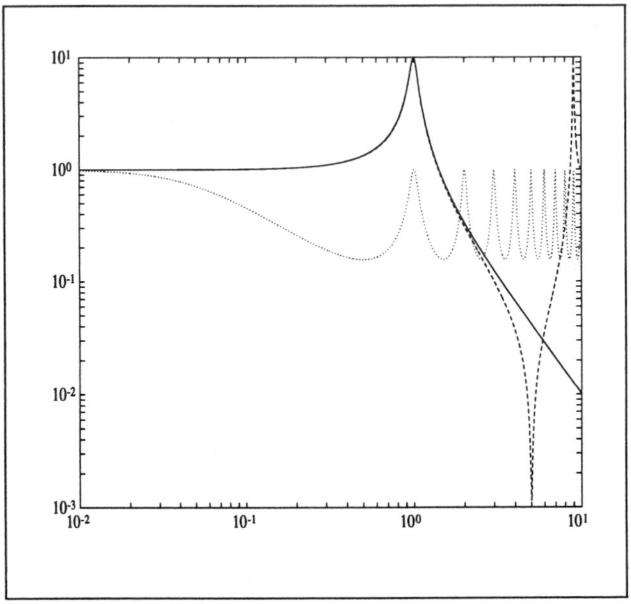

Figure 3.2: Bode magnitude plots: $|\hat{g}(j\omega)|$ (solid); $\left|\hat{g}_d\left(e^{-j\omega h}\right)\right|$ for $\omega_s = 10$ (dash) and $\omega_s = 1$ (dot).

(solid) compared with the graph of $\left|\hat{g}_d\left(e^{-j\omega h}\right)\right|$ for $\omega_s = 10$ (dash) and for $\omega_s = 1$ (dot). For $\omega_s = 10$, $\left|\hat{g}_d\left(e^{-j\omega h}\right)\right| \approx |\hat{g}(j\omega)|$ up to $\omega = 1$, but for $\omega_s = 1$ there is severe distortion except at very small frequencies.

Example 3.3.2 Let us consider digitally implementing the controller designed in Example 2.3.1 by $\mathcal{H}_\infty$-optimization. We had prespecified $h = 0.5$. Setting $K_d = SKH$, we get Figure 3.3. It can be computed that this

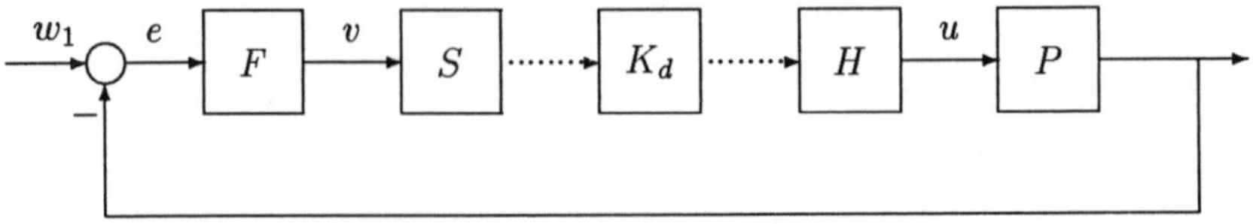

Figure 3.3: Digital implementation of K.

feedback system is, unfortunately, unstable for $h > 0.021$. This shows the disadvantage of this method of design: We can't pre-specify h, do an analog design, implement it by the step-invariant transformation, and expect the resulting digital control system to be stable, let alone achieve the desired performance specifications. We can expect to recover stability and performance only as $h \to 0$.

Let us discretize with $h = 0.01$. To compare the performance of this system with that of the analog system, we should derive the relationship between the Fourier transforms of w_1 and e in Figure 3.3. Recall from Lemmas 3.3.1 and 3.3.2 that the Fourier transforms of $v(t)$ and $u(t)$ are related by the equation

$$\hat{u}(j\omega) = \hat{r}(j\omega)\hat{k}_d\left(e^{-j\omega h}\right)\sum_{k=-\infty}^{\infty}\hat{v}(j\omega + jk\omega_s).$$

If F were an ideal lowpass filter with bandwidth ω_N, then there would be no aliasing and we would have

$$\hat{u}(j\omega) = \hat{r}(j\omega)\hat{k}_d\left(e^{-j\omega h}\right)\hat{v}(j\omega), \quad |\omega| < \omega_N.$$

This would imply the relationship

$$\hat{u}(j\omega) = \hat{r}(j\omega)\hat{k}_d\left(e^{-j\omega h}\right)\hat{f}(j\omega)\hat{e}(j\omega),$$

which in turn would imply that there exists a transfer function from w_1 to e, namely,

$$\frac{1}{1 + \hat{p}(j\omega)\hat{r}(j\omega)\hat{k}_d\left(e^{-j\omega h}\right)\hat{f}(j\omega)},$$

valid over $|\omega| < \omega_N$. Let us denote the latter function by

$$\frac{1}{1 + \hat{p}\hat{r}\hat{k}_d\hat{f}}(j\omega).$$

In summary, the SD system from w_1 to e in Figure 3.3 can be approximated by the transfer function $1/(1 + \hat{p}\hat{r}\hat{k}_d\hat{f})$.

Figure 3.4 shows Bode magnitude plots for $1/(1+\hat{p}\hat{k}\hat{f})$ (solid) and $1/(1+\hat{p}\hat{r}\hat{k}_d\hat{f})$ (dash). There is some deterioration in the digital implementation.

Next we turn to a different formula relating $\hat{g}(s)$ and $\hat{g}_d(\lambda)$. Suppose $\hat{g}(s)$ is strictly proper—the constant $\hat{g}(\infty)$ transfers over directly to $\hat{g}_d$.

Theorem 3.3.2 *Let σ be a real number such that $\frac{1}{s}\hat{g}(s)$ is analytic in Re $s > \sigma$. Then*

$$\hat{g}_d(\lambda) = (1-\lambda)\frac{1}{2\pi j}\int_{\sigma-j\infty}^{\sigma+j\infty}\frac{e^{sh}\lambda}{1-e^{sh}\lambda}\frac{1}{s}\hat{g}(s)ds. \tag{3.9}$$

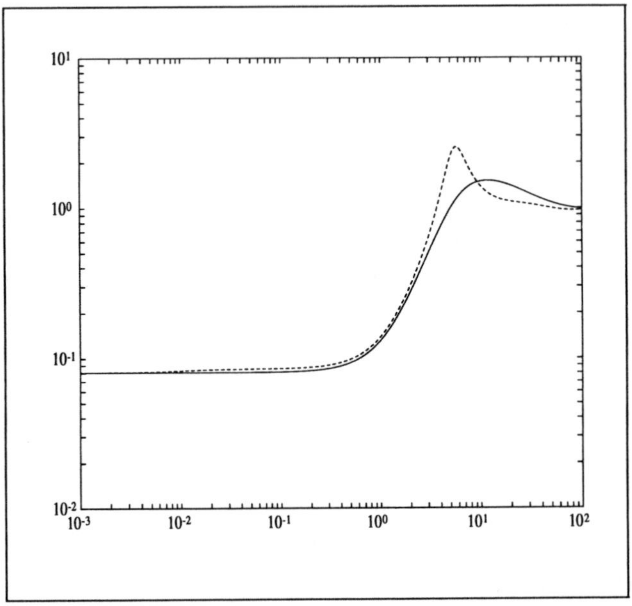

Figure 3.4: Bode magnitude plots: $1/(1+\hat{p}\hat{k}\hat{f})$ (solid) and $1/(1+\hat{p}\hat{r}\hat{k}_d\hat{f})$ (dash) for $h = 0.01$.

Proof Introduce signals as before:

$$
\begin{aligned}
v &= \text{input to } H \\
u &= Hv \\
y &= Gu \\
\psi &= Sy.
\end{aligned}
$$

Let v be one of the standard basis vectors in $\mathbb{R}^{\dim u}$. Set v equal to the unit step times v, that is,

$$\hat{v}(\lambda) = \frac{1}{1-\lambda}v.$$

Then

$$
\begin{aligned}
\hat{u}(s) &= \frac{1}{s}v \\
\hat{y}(s) &= \frac{1}{s}\hat{g}(s)v.
\end{aligned}
$$

The latter equation together with the initial-value theorem shows that $y(0) = 0$. Then

$$y(t) = \text{inverse Laplace of } \frac{1}{s}\hat{g}(s)v$$

$$= \frac{1}{2\pi j}\int_{\sigma-j\infty}^{\sigma+j\infty} \frac{e^{st}}{s}\hat{g}(s)ds\ v.$$

Hence

$$\begin{aligned}\hat{\psi}(\lambda) &= \lambda y(h) + \lambda^2 y(2h) + \cdots \\ &= \frac{1}{2\pi j}\int_{\sigma-j\infty}^{\sigma+j\infty} \frac{1}{s}[e^{sh}\lambda + e^{2sh}\lambda^2 + \cdots]\hat{g}(s)ds\ v.\end{aligned}$$

The series converges for $|e^{sh}\lambda| < 1$, that is, in the disk $|\lambda| < 1/e^{\sigma h}$, and we get

$$\hat{\psi}(\lambda) = \frac{1}{2\pi j}\int_{\sigma-j\infty}^{\sigma+j\infty} \frac{1}{s}\frac{e^{sh}\lambda}{1-e^{sh}\lambda}\hat{g}(s)ds\ v.$$

But

$$\begin{aligned}\hat{\psi}(\lambda) &= \hat{g}_d(\lambda)\hat{v}(\lambda) \\ &= \frac{1}{1-\lambda}\hat{g}_d(\lambda)v.\end{aligned}$$

■

Note that formula (3.9) provides $\hat{g}_d(\lambda)$ analytic initially in the disk $|\lambda| < 1/e^{\sigma h}$. Then do analytic continuation as usual.

We can now consider evaluating the integral in (3.9) by closing the contour either left or right and then using residues. For this we must note that if $|\lambda| < 1/e^{\sigma h}$, then $(1 - e^{sh}\lambda)^{-1}$ is analytic in Re $s < \sigma$.

Close contour to the left

The formula is

$$\hat{g}_d(\lambda) = (1-\lambda)\sum \text{Res}\left[\frac{e^{sh}\lambda}{1-e^{sh}\lambda}\frac{1}{s}\hat{g}(s)\right],$$

where the sum is over residues at poles of $\frac{1}{s}\hat{g}(s)$.

Example 3.3.3 $\hat{g}(s) = \dfrac{1}{s+1}$

$$\text{Res at } \{s = 0\} = \frac{\lambda}{1-\lambda}$$

$$\text{Res at } \{s = -1\} = \frac{\lambda}{\lambda - e^h}$$

$$\hat{g}_d(\lambda) = (1-e^h)\frac{\lambda}{\lambda - e^h}$$

Close contour to the right

Note that the orientation is now negative. First, find the poles of $1/(1-e^{sh}\lambda)$:

$$\begin{aligned} e^{sh}\lambda = 1 \quad &\Longleftrightarrow \quad e^{sh+j2\pi k}\lambda = 1 \\ &\Longleftrightarrow \quad sh = -\ln\lambda - j2\pi k \\ &\Longleftrightarrow \quad s = -\frac{\ln\lambda}{h} - j\frac{2\pi k}{h}. \end{aligned}$$

It is routine to compute that the residue of

$$\frac{e^{sh}\lambda}{1-e^{sh}\lambda}\frac{1}{s}\hat{g}(s)$$

at the pole

$$s_k := -\frac{\ln\lambda}{h} - j\frac{2\pi k}{h}$$

equals

$$-\frac{1}{h}\frac{\hat{g}(s_k)}{s_k}.$$

Thus we get

$$\hat{g}_d(\lambda) = (1-\lambda)\frac{1}{h}\sum_k \frac{\hat{g}(s_k)}{s_k} \tag{3.10}$$

Now we'll see that if the relative degree of $\hat{g}$ is at least 3 and the sampling frequency is sufficiently fast, then $\hat{g}_d$ is non-minimum-phase, even if $\hat{g}$ is not! We emphasize that we're considering the zeros of $\hat{g}_d$ other than the ones at $\lambda = 0$, which arise from the strict properness of $\hat{g}$. The following lemma uses the polynomial $\alpha_l(\lambda)$ introduced in Example 3.1.4.

Lemma 3.3.3 *Suppose G is SISO and the relative degree l of $\hat{g}$ is ≥ 2. Then as $h \longrightarrow 0$, $l-1$ zeros of $\hat{g}_d$ tend to the zeros of $\alpha_l(\lambda)$; hence if the relative degree of $\hat{g}$ is > 2, then $\hat{g}_d$ is non-minimum-phase as $h \longrightarrow 0$.*

Sketch of Proof Noting that the gain of $\hat{g}$ is irrelevant, we can write

$$\hat{g}(s) = \frac{(s-z_1)\cdots(s-z_m)}{(s-p_1)\cdots(s-p_n)},$$

where $l := n - m \geq 2$. Defining

$$\hat{g}_1(s) = \frac{1}{s^l}, \quad \hat{g}_2(s) = \frac{s^l(s-z_1)\cdots(s-z_m)}{(s-p_1)\cdots(s-p_n)},$$

we get $\hat{g} = \hat{g}_1\hat{g}_2$. Notice that $\hat{g}_2(\infty) = 1$.

Fix δ, $0 < \delta \ll 1$, and note that for each k the function

$$\lambda \mapsto \hat{g}_2\left(-\frac{\ln\lambda}{h} - j\frac{2\pi k}{h}\right)$$

converges to 1 uniformly in $|\lambda - 1| \geq \delta$ as $h \longrightarrow 0$. Now apply (3.10) in $|\lambda - 1| \geq \delta$ as $h \longrightarrow 0$:

$$\begin{aligned} \hat{g}_d(\lambda) &= (1-\lambda)\frac{1}{h}\sum \frac{\hat{g}_1(s_k)\hat{g}_2(s_k)}{s_k} \\ &\approx (1-\lambda)\frac{1}{h}\sum \frac{\hat{g}_1(s_k)}{s_k} \\ &= \hat{g}_{1d}(\lambda). \end{aligned}$$

Thus

$$\begin{aligned} \{\text{ zeros of } \hat{g}_d \text{ in } |\lambda - 1| \geq \delta \} &\approx \{\text{ zeros of } \hat{g}_{1d} \text{ in } |\lambda - 1| \geq \delta \} \\ &= \{\text{ zeros of } \lambda\alpha_l(\lambda)\}. \end{aligned}$$

■

3.4 Bilinear Transformation

In this section we look at the other common way of discretization, bilinear transformation (also called Tustin's method).

The method is motivated by considering the trapezoidal approximation of an integrator. Consider an integrator, transfer function $1/s$, input $u(t)$, and output $y(t)$. The trapezoidal approximation of

$$y(kh+h) = y(kh) + \int_{kh}^{kh+h} u(\tau)d\tau$$

is

$$y(kh+h) = y(kh) + \frac{h}{2}[u(kh+h) + u(kh)].$$

The transfer function of the latter equation is

$$\frac{h}{2}\frac{\lambda^{-1}+1}{\lambda^{-1}-1} = \frac{h}{2}\frac{1+\lambda}{1-\lambda}.$$

This motivates the bilinear transformation

$$\frac{1}{s} = \frac{h}{2}\frac{1+\lambda}{1-\lambda},$$

that is,

$$s = \frac{2}{h}\frac{1-\lambda}{1+\lambda}.$$

So a continuous-time transfer matrix $\hat{g}(s)$ is mapped into $\hat{g}_{bt}(\lambda)$, where

$$\hat{g}_{bt}(\lambda) = \hat{g}\left(\frac{2}{h}\frac{1-\lambda}{1+\lambda}\right).$$

It is straightforward to derive that

$$\hat{g}(s) = \left[\begin{array}{c|c} A & B \\ \hline C & D \end{array}\right] \Longrightarrow \hat{g}_{bt}(\lambda) = \left[\begin{array}{c|c} A_{bt} & B_{bt} \\ \hline C_{bt} & D_{bt} \end{array}\right],$$

where

$$\begin{aligned} A_{bt} &= \left(I - \frac{h}{2}A\right)^{-1}\left(I + \frac{h}{2}A\right) \\ B_{bt} &= \frac{h}{2}\left(I - \frac{h}{2}A\right)^{-1} B \\ C_{bt} &= C(I + A_{bt}) \\ D_{bt} &= D + CB_{bt}. \end{aligned}$$

This state formula is valid provided the indicated inverse exists, that is, $2/h$ is not an eigenvalue of A.

Example 3.4.1 If

$$\hat{g}(s) = \frac{1}{s+1},$$

then

$$\hat{g}_{bt}(\lambda) = h\frac{1+\lambda}{(2+h)-(2-h)\lambda}.$$

We can see how the bilinear transformation maps poles and zeros of $\hat{g}(s)$ into those of $\hat{g}_{bt}(\lambda)$: The mapping from s to λ is given by

$$\lambda = \frac{1-\frac{h}{2}s}{1+\frac{h}{2}s}.$$

So for example the right half-plane is mapped into the unit disk. In particular, if $\hat{g}(s)$ has no poles or zeros in the right half-plane, then $\hat{g}_{bt}(\lambda)$ has none in the unit disk.

3.5 Discretization Error

Now we have two ways to discretize a continuous-time system, step-invariant transformation and bilinear transformation. Which is better? Indeed, how can we judge how good a discretization is? This section provides a way to answer this. We'll do the SISO case for simplicity.

Suppose $\hat{g}_a(s)$ is a transfer function of an analog system and $\hat{k}_d(\lambda)$ is a discretization, for example, $\hat{k}_d = \hat{g}_{ad}$ or $\hat{g}_{abt}$. Let G_a and K_d denote the corresponding linear systems. We cannot talk about the error between G_a and K_d just as we cannot compare apples and oranges. Suppose, however, that G_a is an analog controller that has been designed, and K_d is its discretization. Then the actual continuous-time controller that is implemented is HK_dS, as shown in Figure 3.5. So the question that makes sense is, how

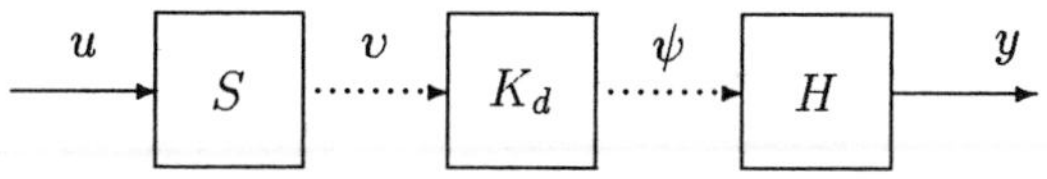

Figure 3.5: Sampled-data implementation of G_a.

well does HK_dS approximate G_a? The *error system*, $G_a - HK_dS$, is shown in Figure 3.6.

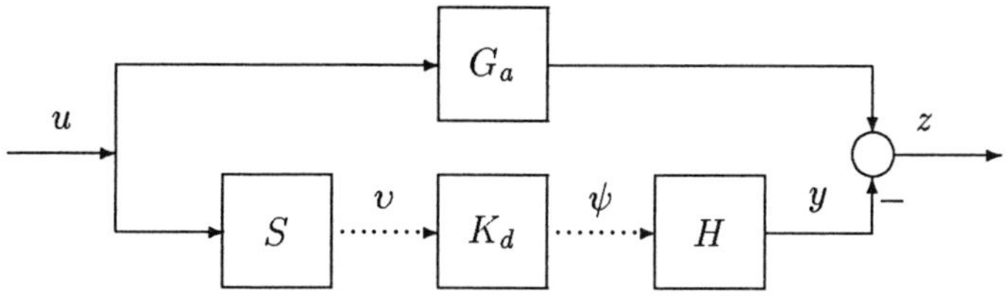

Figure 3.6: The error system.

We are going to study the error system in the frequency domain. For this, we need to know the relationship between the Fourier transforms $\hat{z}(j\omega)$ and $\hat{u}(j\omega)$. (Warning: The error system is time-varying, so it has no transfer function!) Lemmas 3.3.1 and 3.3.2 allow us to write the frequency-domain relationship between $u(t)$ and $y(t)$ in Figure 3.5: From (3.3) and (3.8)

$$\hat{y}(j\omega) = \hat{r}(j\omega)\hat{k}_d\left(e^{-j\omega h}\right)\hat{u}_e(j\omega).$$

Now return to the error system, Figure 3.6. We have

$$\hat{z}(j\omega) = \hat{g}_a(j\omega)\hat{u}(j\omega) - \hat{r}(j\omega)\hat{k}_d\left(e^{-j\omega h}\right)\hat{u}_e(j\omega).$$

Assume $\hat{u}(j\omega)$ is bandlimited to frequencies less than ω_N, that is,

$$\hat{u}(j\omega) = 0 \text{ for } \omega \geq \omega_N.$$

Then

$$\hat{u}_e(j\omega) = \hat{u}(j\omega) \text{ for } \omega < \omega_N,$$

so for $\omega < \omega_N$

$$\hat{z}(j\omega) = \left[\hat{g}_a(j\omega) - \hat{r}(j\omega)\hat{k}_d\left(e^{-j\omega h}\right)\right]\hat{u}(j\omega).$$

This motivates the definition of the *error function*,

$$error(\omega) := \left|\hat{g}_a(j\omega) - \hat{r}(j\omega)\hat{k}_d\left(e^{-j\omega h}\right)\right|,$$

and the *maximum error*,

$$error_{max} := \max_{\omega<\omega_N} error(\omega).$$

Let us recap: G_a is a given continuous-time system and K_d is a discretization of G_a obtained in some way; for inputs that are bandlimited to frequencies less than ω_N, $error_{max}$ is a measure of how closely HK_dS approximates G_a. It is natural to use this measure to compare two different discretizations to see which is "better."

For the bilinear transformation,

$$\begin{aligned}\hat{g}_{abt}\left(e^{-j\omega h}\right) &= \hat{g}_a\left(\frac{2}{h}\frac{1-e^{-j\omega h}}{1+e^{-j\omega h}}\right)\\ &= \hat{g}_a\left(j\frac{2}{h}\tan\frac{\omega h}{2}\right).\end{aligned}$$

Thus

$$error(\omega) := \left|\hat{g}_a(j\omega) - \hat{r}(j\omega)\hat{g}_a\left(j\frac{2}{h}\tan\frac{\omega h}{2}\right)\right|.$$

The error is therefore due to two factors: the presence of the function $\hat{r}$; the "frequency warping"

$$\omega \mapsto \frac{2}{h}\tan\frac{\omega h}{2}.$$

Note that $error(0) = 0$, that is, there is no error at DC. Suppose we would prefer instead that the error were zero at some other frequency, say ω_0. If the transformation is

$$s \mapsto c\frac{1-\lambda}{1+\lambda},$$

then the frequency warping is

$$\omega \mapsto c\tan\frac{\omega h}{2}.$$

So to ensure that ω_0 is mapped to itself, we should take

$$c = \omega_0\left(\tan\frac{\omega h}{2}\right)^{-1}.$$

The mapping

$$s \mapsto \omega_0 \left(\tan \frac{\omega h}{2}\right)^{-1} \frac{1-\lambda}{1+\lambda}$$

is called the *bilinear transformation with prewarping.*

Example 3.5.1 Consider $\hat{g}_a(s)$ to be the elliptic filter with zeros

$$\pm 1.23334j, \ \pm 1.72290j,$$

poles

$$-0.78280, \ -0.07543 \pm 1.05165j, \ -0.379155 \pm 0.875369j,$$

and gain 0.175407. Its magnitude Bode plot is shown in Figure 3.7. Figure 3.8

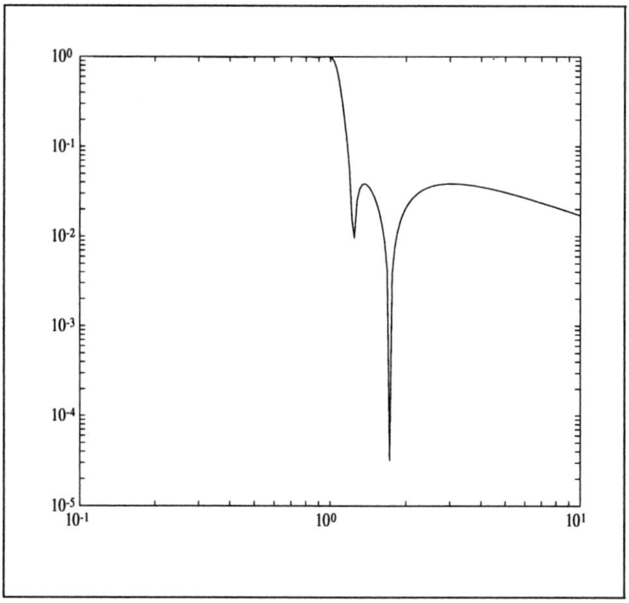

Figure 3.7: Bode Plot of Elliptic Filter.

shows the graph of $error(\omega)$ for the bilinear transformation with $\omega_N = 10$, one decade higher than the cutoff frequency of the filter, and with the prewarping frequency of $\omega_0 = 1$, the cutoff frequency. The error is quite large ($error_{max} = 0.1564$) because ω_N is quite small. This large error suggests that bilinear transformation, a common discretization technique, may be inadequate for some applications and it motivates us to search for a better technique.

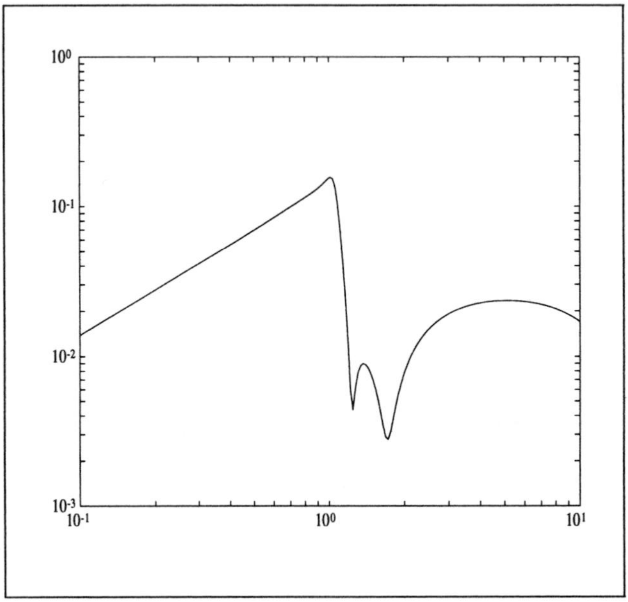

Figure 3.8: Error for Bilinear Transformation with Prewarping of Elliptic Filter.

In certain signal-processing applications, non-causal filters are allowed. Consider Figure 3.5 with $u(t)$ bandlimited to frequencies less than ω_N, but with H replaced by R, the *ideal interpolator* in the Sampling Theorem, that is,

$$y(t) = \sum_k \psi(k) \frac{\sin \omega_N(t - kh)}{\omega_N(t - kh)}.$$

The resulting filter, namely RK_dS, is noncausal. Then the relationship between the Fourier transforms of $u(t)$ and $y(t)$ in Figure 3.5 is known to be

$$\hat{y}(j\omega) = \hat{k}_d\left(\mathrm{e}^{-j\omega h}\right)\hat{u}(j\omega), \quad \omega < \omega_N.$$

It follows that the error in Figure 3.6 is given by

$$\hat{z}(j\omega) = \left[\hat{g}_a(j\omega) - \hat{k}_d\left(\mathrm{e}^{-j\omega h}\right)\right]\hat{u}(j\omega), \quad \omega < \omega_N,$$

so the correct definition of the error function in this case is

$$error(\omega) := \left|\hat{g}_a(j\omega) - \hat{k}_d\left(\mathrm{e}^{-j\omega h}\right)\right|.$$

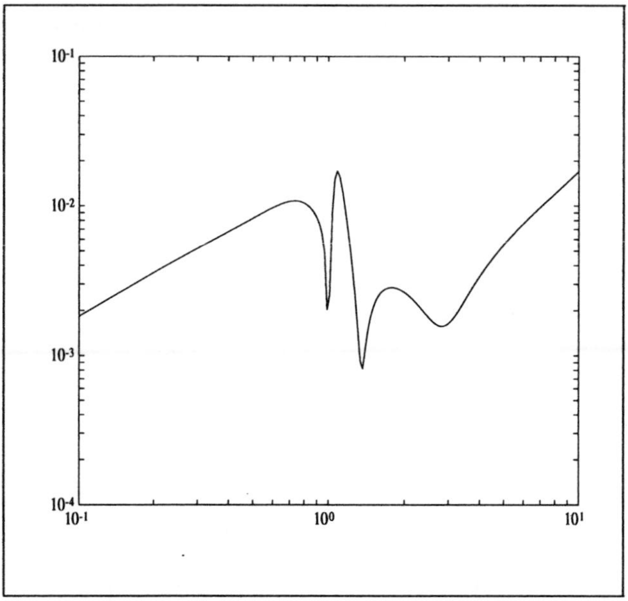

Figure 3.9: Error for Bilinear Transformation with Prewarping of Elliptic Filter, Noncausal Case.

Example 3.5.2 (Continuation of Example 3.5.1) For the same elliptic filter and the same discretization, Figure 3.9 shows the graph of $error(\omega)$ for the noncausal case. The maximum error on the passband is reduced from 0.1564 to only 0.0108. This shows the deteriorating effect of the hold operator.

Exercises

3.1 Consider the continuous-time system G with state model

$$\begin{aligned} \dot{x}(t) &= Ax(t) + Bu(t-h) \\ y(t) &= Cx(t). \end{aligned}$$

Thus the input is delayed by one sampling period. This system can be factored as $G = G_1G_2$, where G_2 is the time-delay system (output equals input delayed by time h) and G_1 is the above continuous-time system but without the time delay. Find the discretized transfer matrix $\hat{g}_d(\lambda)$ by noting the following:

1. $G_d = S \;\; {}_1G_2H$, $\qquad G$

2. $G_2H = HU$, where U is unit time-delay in discrete time.

3.2 Show that if the continuous-time system G has transfer matrix

$$\hat{g}(s) = \left[\begin{array}{c|c} A & B \\ \hline C & 0 \end{array}\right] e^{-\tau s},$$

then SGH has transfer matrix

$$\hat{g}_d(\lambda) = \lambda^l \left[\begin{array}{c|c} A_d & B_d \\ \hline C_d & D_d \end{array}\right],$$

where A_d, B_d are as usual,

$$C_d = Ce^{(lh-\tau)A}, \quad D_d = C\int_0^{lh-\tau} e^{tA}dtB,$$

and l is the integer such that τ lies in the sampling interval $((l-1)h, lh]$.

3.3 Consider the analog control setup

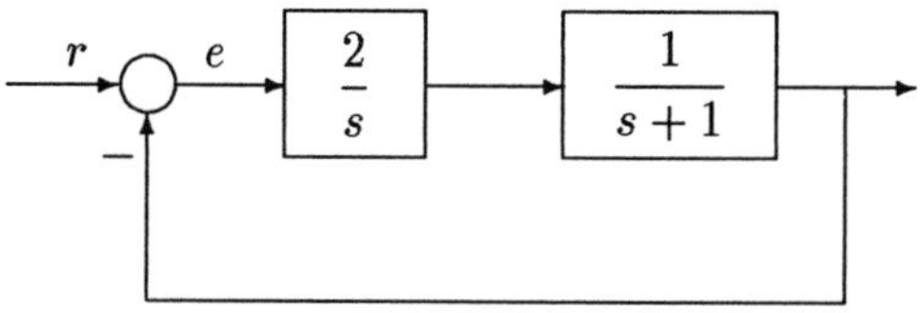

and its sampled-data implementation

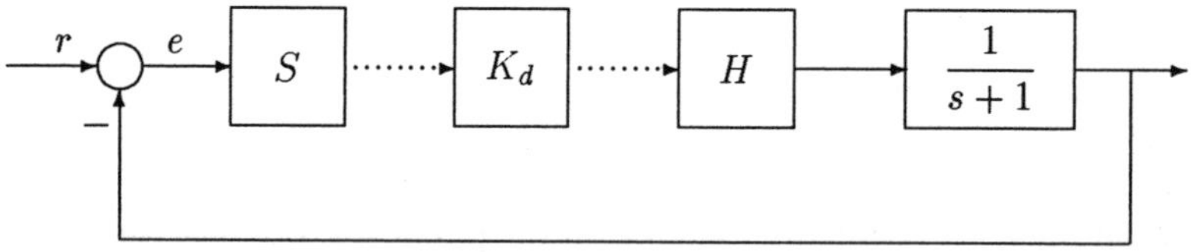

Let K_d be the discretization K_d, via step-invariant transformation, of the analog controller $2/s$.

1. For $h = 1$, find the transfer function $\hat{k}_d(\lambda)$.
2. Let $r(t)$ be the unit step in the second figure and let ε denote the sampled error, that is, $\varepsilon = Se$. Find $\hat{\varepsilon}(\lambda)$.
3. Does $\varepsilon(k) \to 0$ as $k \to \infty$?

Repeat but with $K_d = K_{bt}$.

3.4 Consider a continuous-time linear system with the following A-matrix:

$$\begin{bmatrix} 1 & 2 & 0 & 0 & 0 & 0 & 0 & 0 \\ -2 & 1 & 0 & 0 & 0 & 0 & 0 & 0 \\ 0 & 0 & 0 & 1 & 0 & 0 & 0 & 0 \\ 0 & 0 & 1 & 0 & 0 & 0 & 0 & 0 \\ 0 & 0 & 0 & 0 & 0 & 10 & 0 & 0 \\ 0 & 0 & 0 & 0 & -10 & 0 & 0 & 0 \\ 0 & 0 & 0 & 0 & 0 & 0 & 0 & 0 \\ 0 & 0 & 0 & 0 & 0 & 0 & 0 & 0 \end{bmatrix}$$

What are the pathological sampling frequencies?

3.5 Two pendula, of masses M_1 and M_2 and lengths l_1 and l_2, are coupled by a spring, of stiffness K. The two inputs are the positions u_1 and u_2 of the pivots of the pendula, and the two outputs are their angles y_1 and y_2. The equations of motion are as follows:

$$\begin{aligned} M_1(\ddot{u}_1 - l_1\ddot{y}_1) &= M_1 g y_1 - K(u_1 - l_1 y_1) + K(u_2 - l_2 y_2) \\ M_2(\ddot{u}_2 - l_2\ddot{y}_2) &= M_2 g y_2 + K(u_1 - l_1 y_1) - K(u_2 - l_2 y_2). \end{aligned}$$

1. Derive a state model.

2. Take the following numerical values: $M_1 = 1$ kg, $M_2 = 10$ kg, $l_1 = l_2 = 1$ m, $K = 1$ N/m. Compute the pathological sampling frequencies.

3. Select some non-pathological sampling frequency and compute the transfer matrix of SGH.

3.6 Let G_1 and G_2 be two continuous-time LTI systems. Explain why the following is true: The discretization (via step-invariant transformation) of $G_1 + G_2$ is the sum of the discretizations of G_1 and G_2. Use this fact to compute the transfer function of the discretized G, where the sampling period is h and

$$\hat{g}(s) = \frac{a}{s(s-a)}, \quad a \neq 0.$$

3.7 Let $\hat{g}_1(s)$ and $\hat{g}_2(s)$ be two continuous-time transfer functions, and let $\hat{g}_{1d}(\lambda)$ and $\hat{g}_{2d}(\lambda)$ be their discretizations. True or false: The discretization of the product $\hat{g}_1(s)\hat{g}_2(s)$ equals $\hat{g}_{1d}(\lambda)\hat{g}_{2d}(\lambda)$.

3.8 Consider the setup

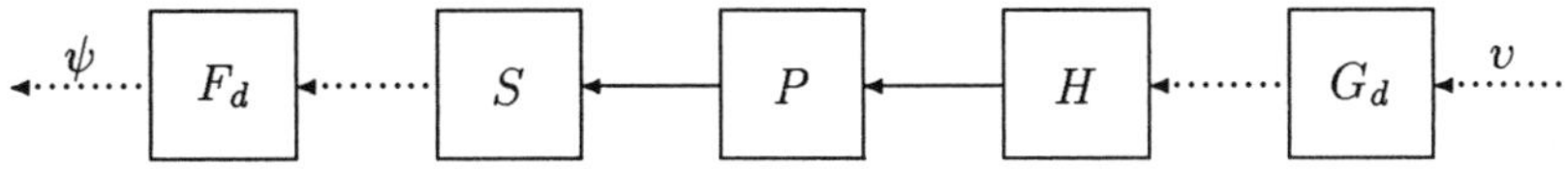

The components are described as follows:

G_d: output equals 0 at odd times; output at time $2k$ equals input at time k

P has transfer function $1/(s+1)$

F_d: output at time k equals input at time $2k$.

Show that the system from v to ψ is LTI and find its transfer function.

3.9 The block diagram shows a modified hold function defined as follows:

$$u(t) = \begin{cases} av(k), & kh \leq t < kh + \frac{h}{2} \\ bv(k), & kh + \frac{h}{2} \leq t < kh + h \end{cases}$$

Here a and b are real scalars. Find the discrete transfer matrix from v to ψ.

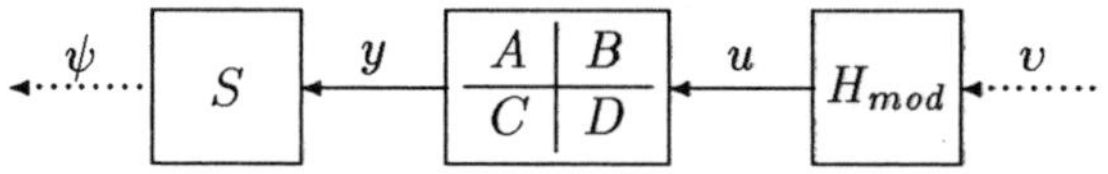

3.10 The following block diagram shows a typical digital control system.

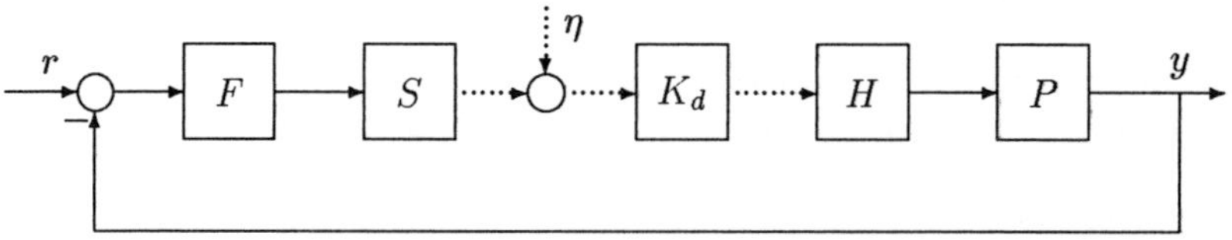

In addition to the usual elements are F, a low-pass (antialiasing) filter, and η, a digital noise signal introduced at the sampler (perhaps to model quantization error). In such a sampled-data system, not all input-output relationships need be time-invariant. Define $\psi = Sy$, the sampled plant output. Is η-to-ψ time-invariant? If so, find its transfer function for

$$\hat{p}(s) = \frac{1}{10s+1}, \quad \hat{f}(s) = \frac{1}{s+1}, \quad \hat{k}_d(\lambda) = \lambda, \quad h = 0.1.$$

Is r-to-y time-invariant?

3.11 A single-input, multi-output continuous-time system has transfer matrix

$$\left[\begin{array}{c|c} A & B \\ \hline C & 0 \end{array}\right].$$

It is forced by a unit-step input and its output is sampled at $h = 0.1$. Find the λ-transform of this sampled output. Repeat with a unit-impulse input.

3.12 Let G be the continuous-time system of delay τ, with $0 < \tau < h$. Find $\hat{g}_d(\lambda)$. (Note that $\hat{g}_d$ is independent of τ, showing that an infinite number of Gs can have the same discretization.)

3.13 Find an example of a pair (A, B) such that (A_d, B_d) is controllable, yet the sampling frequency is pathological. [This shows that non-pathological sampling is not necessary for controllability of (A_d, B_d), only sufficient.] Hint: B must have at least two columns.

3.14 Let G_1 and G_2 be two continuous-time LTI systems. True or false: If a sampling period h is non-pathological relative to both G_1 andG_2, so is it relative to G_1G_2. If your answer is yes, explain; if your answer is no, give a counterexample.

3.15 Suppose $jk\omega_s$ is an eigenvalue of A, where k is an integer. Prove that (A_d, B_d) is not controllable.

3.16 Take

$$\hat{g}_a(s) = \frac{1}{s^2 + 0.2s + 1}.$$

Discretize $\hat{g}_a(s)$ using both the step-invariant transformation and the bilinear transformation, and for both $\omega_N = 10$ and 100. In all four cases plot the discretization error over the range $\omega \le \omega_N$.

3.17 This concerns the discretization problem shown in the setup

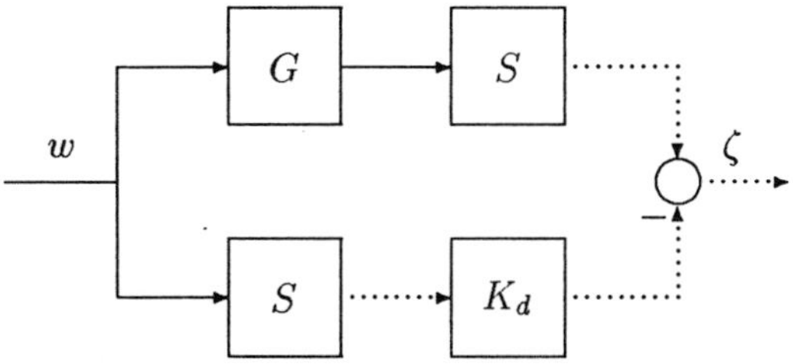

Both G and K_d are SISO. The transfer function $\hat{g}(s)$ is assumed stable and strictly proper. The input $w(t)$ is fixed (traditionally it is a step or ramp). Extending the notion of step-invariant transformation, let us say that the digital system K_d is a *w-invariant discretization of* G if the error ζ is identically zero.

1. Give a condition on w so that K_d is uniquely determined by saying it is a w-invariant discretization. (Answer: $Sw \not\equiv 0$.)

2. Compute $\hat{k}_d(\lambda)$ for

$$w(t) = t1(t), \quad \hat{g}(s) = \frac{1}{s^2 + s + 1}, \qquad h = 1.$$

3. Give an example of a $w(t)$ and a stable $\hat{g}(s)$ for which $\hat{k}_d(\lambda)$ is not stable.

3.18 Consider the discretization setup in the preceding exercise. Both G and K_d are FDLTI, SISO and the input $w(t)$ is the unit ramp $t1(t)$. The goal is, given G and the sampling period h, to design K_d so that the error ζ is identically zero. Give a MATLAB procedure with

input: h, state model (A, B, C, D) for G

output: state model for K_d

Notes and References

Theorem 3.2.1 is due to Kalman, Ho, and Narendra [84]. Lemma 3.3.3 is due to Åström, Hagander, and Sternby [8]. For traditional methods on the design of digital filters, see for example [116] and [147]. In these traditional methods, it is not common to use the error function as the object of merit. Indeed, an approximation error of any sort is seldom considered. For recent works that *do* optimize an approximation error, see [139], [120], and [128]. Having a frequency-domain error measure is important in control systems, since we can then bound the deterioration that results when an analog controller is implemented digitally.

Chapter 4

Discrete-Time Systems: Basic Concepts

The controller in a SD control system sees a discrete-time system, namely, the discretized plant. For this reason it is useful to develop some general techniques for analyzing discrete-time systems.

4.1 Time-Domain Models

We start with scalar-valued signals and move to the vector-valued case in Section 4.4. The discrete time-set is taken to be the integers $\{0, 1, 2, \ldots\}$. So a discrete-time signal is a sequence

$$\{v(0), v(1), v(2), \ldots\},$$

where each $v(k)$ is a real number, the value of the signal at time k. More often it is convenient to write this as an infinite column vector:

$$\begin{bmatrix} v(0) \\ v(1) \\ \vdots \end{bmatrix}.$$

Then we can think of a linear system as a linear transformation on such vectors.

Example 4.1.1 Discretized integrator. Consider the discretization of the pure integrator. The equations are

$$\begin{aligned} \dot{\xi} &= \xi + hv \\ \psi &= \xi. \end{aligned}$$

For zero initial state [i.e., $\xi(0) = 0$] the output is

$$\begin{aligned}
\psi(0) &= 0 \\
\psi(1) &= hv(0) \\
\psi(2) &= hv(0) + hv(1) \\
\psi(3) &= hv(0) + hv(1) + hv(2) \\
&\vdots
\end{aligned}$$

These equations can be written conveniently in vector form:

$$\begin{bmatrix} \psi(0) \\ \psi(1) \\ \psi(2) \\ \psi(3) \\ \vdots \end{bmatrix} = \begin{bmatrix} 0 & 0 & 0 & 0 & \cdots \\ h & 0 & 0 & 0 & \cdots \\ h & h & 0 & 0 & \cdots \\ h & h & h & 0 & \cdots \\ \vdots & \vdots & \vdots & \vdots & \end{bmatrix} \begin{bmatrix} v(0) \\ v(1) \\ v(2) \\ v(3) \\ \vdots \end{bmatrix}.$$

We can regard this infinite matrix as a representation of the system.

More generally, a discrete-time linear system G (we drop the subscript d in this chapter) has an *associated infinite matrix*, denoted $[G]$. The first column is the output signal when the input is the unit impulse,

$$\{1, 0, 0, \ldots\}.$$

The second column is the output when the input is the shifted unit impulse,

$$\{0, 1, 0, 0, \ldots\}.$$

And so on.

Example 4.1.2 Unit delay, U. Suppose the input-output equation is

$$\begin{aligned}
\psi(0) &= 0 \\
\dot{\psi} &= v.
\end{aligned}$$

Thus the output is the one-step delay of the input. The system matrix is

$$[U] = \begin{bmatrix} 0 & 0 & 0 & \cdots \\ 1 & 0 & 0 & \cdots \\ 0 & 1 & 0 & \cdots \\ \vdots & \vdots & \vdots & \end{bmatrix}, \qquad (4.1)$$

that is, all zeros except for 1s on the first subdiagonal.

Example 4.1.3 Moving average model. Here the output at time k is a linear combination of all the inputs:

$$\psi(k) = \sum_{l=0}^{\infty} g(k,l)v(l).$$

The system matrix is

$$\begin{bmatrix} g(0,0) & g(0,1) & \cdots \\ g(1,0) & g(1,1) & \cdots \\ \vdots & \vdots & \end{bmatrix}. \tag{4.2}$$

A special case is the discrete-time convolution equation

$$\psi(k) = \sum_{l=0}^{\infty} g(k-l)v(l)$$

that has system matrix

$$\begin{bmatrix} g(0) & g(-1) & g(-2) & \cdots \\ g(1) & g(0) & g(-1) & \cdots \\ g(2) & g(1) & g(0) & \cdots \\ \vdots & \vdots & \vdots & \end{bmatrix}. \tag{4.3}$$

Another special case is the *truncation projection*, P_k, defined for $k \geq 0$ by

$$\psi(i) = \begin{cases} v(i), & i \leq k \\ 0, & i > k. \end{cases}$$

The corresponding matrix $[P_k]$ has $k+1$ 1s along the main diagonal and zeros elsewhere:

$$[P_k] = \text{diag}\,(1, \ldots, 1, 0, \ldots).$$

Example 4.1.4 State-space system. The system matrix of the state-space model

$$\begin{aligned} \xi(k+1) &= A\xi(k) + Bv(k), \quad \xi(0) = 0 \\ \psi(k) &= C\xi(k) + Dv(k) \end{aligned}$$

is

$$\begin{bmatrix} D & 0 & 0 & 0 & \cdots \\ CB & D & 0 & 0 & \cdots \\ CAB & CB & D & 0 & \cdots \\ CA^2B & CAB & CB & D & \cdots \\ \vdots & \vdots & \vdots & \vdots & \end{bmatrix}. \tag{4.4}$$

Some basic system concepts become transparent when viewed in terms of the system matrix. The idea of causality is that the output at time k depends only on inputs up to time k; in other words, if two inputs are equal up to time k, then the two corresponding outputs should be equal up to time k. The latter description leads to the formal definition: The linear system G is *causal* if

$$(\forall k)(\forall v, \tilde{v}) P_k v = P_k \tilde{v} \Rightarrow P_k G v = P_k G \tilde{v}.$$

It is easy to prove that G is *causal* iff $[G]$ is lower triangular, as in (4.1) and (4.4).

Example 4.1.5 For (4.2) to be lower triangular we need $g(k,l) = 0$ for $k < l$, in which case the moving average model is

$$\psi(k) = \sum_{l=0}^{k} g(k,l) v(l),$$

that is, $\psi(k)$ depends only on $v(0), \ldots, v(k)$. Similarly, the convolution equation represents a causal system iff $g(k) = 0$ for $k < 0$. An example of a non-causal system is the one-step advance, denoted U^*. The defining equation is

$$\psi = \dot{v},$$

and the system matrix is

$$[U^*] = \begin{bmatrix} 0 & 1 & 0 & 0 & \cdots \\ 0 & 0 & 1 & 0 & \cdots \\ 0 & 0 & 0 & 1 & \cdots \\ \vdots & \vdots & \vdots & \vdots & \end{bmatrix}.$$

Check that the advance matrix $[U^*]$ equals the transpose of the delay matrix $[U]$. Verify also that $U^*U = I$ (U^* is a left inverse of U, or U is a right inverse of U^*) and $UU^* = I - P_0$.

Time-invariance means this: If an input $\{v(0), v(1), \ldots\}$ produces the output $\{\psi(0), \psi(1), \ldots\}$, then the input $\{0, v(0), v(1), \ldots\}$ produces an output of the form $\{?, \psi(0), \psi(1), \ldots\}$. The initial value of the latter signal will equal zero if the system is causal. So roughly speaking, time-invariance means that shifting the input shifts the output. The formal definition is that G is *time-invariant* if $U^*GU = G$.

Again, it is easy to prove that G is *time-invariant* iff $[G]$ is constant along diagonals, as in (4.1), (4.3), and (4.4). Such a matrix is called *Toeplitz*.

4.2 Frequency-Domain Models

It is well-known that an LTI system has a simple model in the frequency domain, namely, it is multiplication by a transfer function. We have already used λ-transfer functions, but here is given a more comprehensive presentation.

The λ-*transform* of the signal

$$v = \{v(0), v(1), \ldots\}$$

is defined to be

$$\begin{aligned}\hat{v}(\lambda) &:= v(0) + v(1)\lambda + v(2)\lambda^2 + \cdots \\ &= \sum_{k=0}^{\infty} v(k)\lambda^k.\end{aligned}$$

Example 4.2.1 1. Unit impulse.

$$v = \{1, 0, 0, \ldots\}, \hat{v}(\lambda) = 1$$

2. Unit step.

$$v = \{1, 1, 1, \ldots\}, \hat{v}(\lambda) = 1 + \lambda + \lambda^2 + \cdots = \frac{1}{1-\lambda}, \text{pole at } \lambda = 1$$

3. Geometric series.

$$\begin{aligned}v = \{1, 2, 2^2, 2^3, \ldots\}, \hat{v}(\lambda) &= 1 + 2\lambda + (2\lambda)^2 + \cdots \\ &= \frac{1}{1-2\lambda}, \text{pole at } \lambda = \tfrac{1}{2}\end{aligned}$$

4. Sampled sinusoid.

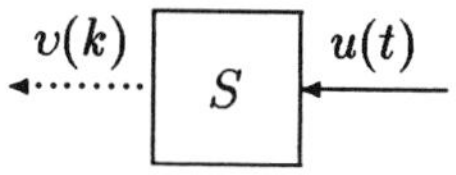

$$u(t) = \sin(\omega t)1(t) \ [1(t) = \text{unit step}]$$

$$\begin{aligned}v &= \{0, \sin\omega h, \sin 2\omega h, \sin 3\omega h, \ldots\} \\ &= \frac{1}{2j}\{0, e^{j\omega h} - e^{-j\omega h}, e^{2j\omega h} - e^{-2j\omega h}, \ldots\}\end{aligned}$$

$$\begin{aligned}\hat{v}(\lambda) &= \frac{\lambda}{2j}\left[e^{j\omega h} - e^{-j\omega h} + (e^{2j\omega h} - e^{-2j\omega h})\lambda + \cdots\right] \\ &= \frac{\lambda}{2j}\left[e^{j\omega h}(1 + e^{j\omega h}\lambda + e^{2j\omega h}\lambda^2 + \cdots)\right.\end{aligned}$$

$$
\begin{aligned}
& - e^{-j\omega h}(1 + e^{-j\omega h}\lambda + e^{-2j\omega h}\lambda^2 + \cdots)] \\
= \; & \frac{\lambda}{2j}\left[\frac{e^{j\omega h}}{1 - e^{j\omega h}\lambda} - \frac{1}{e^{j\omega h} - \lambda}\right] \text{(poles: } \lambda = e^{\pm j\omega h}) \\
= \; & \frac{(\sin\omega h)\lambda}{\lambda^2 - (2\cos\omega h)\lambda + 1}
\end{aligned}
$$

Thus a λ-transform is defined as a power series, and hence is analytic in some disk of sufficiently small radius; in the third example above the disk of analyticity (region of convergence) is $\{\lambda : |\lambda| < \frac{1}{2}\}$.

Now we turn to transfer functions.

Example 4.2.2 Unit delay, U. If the input is

$$v = \{v(0), v(1), v(2), \ldots\},$$

then the output is

$$\psi = \{0, v(0), v(1), \ldots\}.$$

So

$$
\begin{aligned}
\hat{\psi}(\lambda) &= 0 + v(0)\lambda + v(1)\lambda^2 + \cdots \\
&= \lambda[v(0) + v(1)\lambda + \cdots] \\
&= \lambda\hat{v}(\lambda).
\end{aligned}
$$

So the transfer function of U, denoted $\hat{u}(\lambda)$, equals λ.

More generally, we have seen that any causal LTI system G has a matrix of the form

$$
[G] = \begin{bmatrix}
g(0) & 0 & 0 & \cdots \\
g(1) & g(0) & 0 & \cdots \\
g(2) & g(1) & g(0) & \cdots \\
\vdots & \vdots & \vdots &
\end{bmatrix}.
$$

The *impulse response* is the sequence represented by the first column (that is how the first column was defined), and the *transfer function* is the λ-transform of the impulse response:

$$\hat{g}(\lambda) = g(0) + g(1)\lambda + g(2)\lambda^2 + \cdots.$$

Let us summarize the notation:

block diagram :

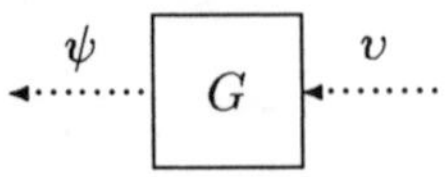

linear transformation equation : $\psi = Gv$

convolution equation : $\psi(k) = \sum_{l=0}^{k} g(k-l)v(l)$

system matrix equation : $\begin{bmatrix} \psi(0) \\ \psi(1) \\ \vdots \end{bmatrix} = \begin{bmatrix} g(0) & 0 & \cdots \\ g(1) & g(0) & \cdots \\ \vdots & \vdots & \end{bmatrix} \begin{bmatrix} v(0) \\ v(1) \\ \vdots \end{bmatrix}$

impulse response : $\{g(0), g(1), \ldots\}$

transfer function : $\hat{g}(\lambda) = \sum_{k=0}^{\infty} g(k)\lambda^k$

Example 4.2.3 State-space model.

$$\begin{aligned} \dot{\xi} &= A\xi + Bv \\ \psi &= C\xi + Dv \end{aligned}$$

The impulse response is

$$\{D, CB, CAB, CA^2B, \ldots\}$$

and therefore the transfer function is

$$\begin{aligned} \hat{g}(\lambda) &= D + CB\lambda + CAB\lambda^2 + \cdots \\ &= D + \lambda C(I + \lambda A + \lambda^2 A^2 + \cdots)B \\ &= D + \lambda C(I - \lambda A)^{-1}B. \end{aligned}$$

Introduce the notation

$$\left[\begin{array}{c|c} A & B \\ \hline C & D \end{array}\right] = D + \lambda C(I - \lambda A)^{-1}B.$$

Example 4.2.4 Discretized double integrator. Take

$$\hat{g}(s) = \frac{1}{s^2}, \quad G_d = SGH.$$

To find $\hat{g}_d(\lambda)$, start with

$$\hat{g}(s) = \left[\begin{array}{c|c} A & B \\ \hline C & D \end{array}\right] = \left[\begin{array}{cc|c} 0 & 1 & 0 \\ 0 & 0 & 1 \\ \hline 1 & 0 & 0 \end{array}\right].$$

Then

$$\hat{g}_d(\lambda) = \left[\begin{array}{c|c} A_d & B_d \\ \hline C & D \end{array}\right],$$

where

$$\begin{aligned} A_d &= e^{hA} = \begin{bmatrix} 1 & h \\ 0 & 1 \end{bmatrix} \\ B_d &= \int_0^h e^{\tau A} d\tau B = \begin{bmatrix} h^2/2 \\ h \end{bmatrix}. \end{aligned}$$

Thus

$$\begin{aligned} \hat{g}_d(\lambda) &= \lambda C(I - \lambda A_d)^{-1} B_d \\ &= \lambda \begin{bmatrix} 1 & 0 \end{bmatrix} \begin{bmatrix} 1-\lambda & -h\lambda \\ 0 & 1-\lambda \end{bmatrix}^{-1} \begin{bmatrix} h^2/2 \\ h \end{bmatrix} \\ &= \frac{\lambda}{(1-\lambda)^2} \begin{bmatrix} 1 & 0 \end{bmatrix} \begin{bmatrix} 1-\lambda & h\lambda \\ 0 & 1-\lambda \end{bmatrix} \begin{bmatrix} h^2/2 \\ h \end{bmatrix} \\ &= \frac{h^2}{2} \frac{\lambda(1+\lambda)}{(1-\lambda)^2}. \end{aligned}$$

The MATLAB functions $tf2ss$ and $ss2tf$ can be used in discrete time, but you have to convert to z.

Example 4.2.5 To get a realization of

$$\frac{\lambda(\lambda-2)}{\lambda^3+\lambda^2-2\lambda+1},$$

first replace λ by z^{-1}, which amounts to reversing the order of the coefficients, that is,

$$\frac{0\lambda^3+\lambda^2-2\lambda+0}{\lambda^3+\lambda^2-2\lambda+1} \longmapsto \frac{0z^3-2z^2+z+0}{z^3-2z^2+z+1},$$

and then call $tf2ss$ on

$$\frac{z(-2z+1)}{z^3-2z^2+z+1}.$$

4.3 Norms

Since the performance of a control system should be measured in terms of the size of various signals of interest (tracking error, plant input, and so on), it is appropriate to define norms for discrete-time signals. We define two norms for the signal $v = \{v(0), v(1), \ldots\}$:

2-Norm $\|v\|_2 = [v(0)^2 + v(1)^2 + \cdots]^{1/2}$
∞-Norm $\|v\|_\infty = \sup_k |v(k)|$

The 2-norm (actually, its square) is associated with energy. For example, suppose $u(t)$ denotes the instantaneous current flowing through a resistor of resistance R in a circuit. The instantaneous power absorbed is $Ru(t)^2$, and the (total) energy absorbed is the integral of this, $\int_0^\infty Ru(t)^2 dt$. Usually we normalize so that $\int_0^\infty u(t)^2 dt$ is interpreted as the *energy* of the signal $u(t)$. In discrete time the corresponding measure is $\sum_0^\infty v(k)^2$.

The ∞-norm is the maximum amplitude of the signal [more precisely, the least upper bound (sup) on the amplitude]. This is an important norm for measuring tracking errors.

Example 4.3.1 1. $v =$ unit impulse: $\|v\|_2 = \|v\|_\infty = 1$

2. $v =$ unit step: $\|v\|_2 = \infty, \|v\|_\infty = 1$

3. $u(t) = e^{-t}1(t)$, $v = Su$

$$\begin{aligned} v &= \{1, e^{-h}, e^{-2h}, \ldots\} \\ \|v\|_2 &= (1 + e^{-2h} + e^{-4h} + \cdots)^{1/2} = \frac{1}{\sqrt{1 - e^{-2h}}} \\ \|v\|_\infty &= 1 \end{aligned}$$

Now we turn to LTI systems, where there are three norms of interest. The first is the 1-norm of the impulse-response function and the second and third are in terms of the transfer function:

1-Norm $\|g\|_1 = |g(0)| + |g(1)| + \cdots$
2-Norm $\|\hat{g}\|_2 = \left(\frac{1}{2\pi} \int_0^{2\pi} |\hat{g}(e^{j\theta})|^2 d\theta\right)^{1/2}$
∞-Norm $\|\hat{g}\|_\infty = \max_\theta |\hat{g}(e^{j\theta})|$

Example 4.3.2

$$\hat{g}(\lambda) = \frac{1}{\lambda - 2}$$

$$\|\hat{g}\|_2 = \left[\frac{1}{2\pi} \int_0^{2\pi} \hat{g}(e^{j\theta}) \overline{\hat{g}(e^{j\theta})} d\theta\right]^{1/2}$$

$$
\begin{aligned}
&= \left[\frac{1}{2\pi}\int_0^{2\pi} \hat{g}(e^{j\theta})\hat{g}(e^{-j\theta})d\theta\right]^{1/2} \\
&= \left[\frac{1}{2\pi j}\oint \hat{g}(\lambda)\hat{g}\left(\frac{1}{\lambda}\right)\frac{1}{\lambda}d\lambda\right]^{1/2} \\
&= \left[\sum \text{residues } \hat{g}(\lambda)\hat{g}\left(\frac{1}{\lambda}\right)\frac{1}{\lambda} \text{ at poles in unit disk}\right]^{1/2} \\
&= \left[\text{residue } \frac{1}{(\lambda-2)(1-2\lambda)} \text{ at } \lambda = \frac{1}{2}\right]^{1/2} \\
&= \frac{1}{\sqrt{3}}
\end{aligned}
$$

$$
\|\hat{g}\|_\infty = \max_\theta \frac{1}{|e^{j\theta}-2|} = \frac{1}{\min|e^{j\theta}-2|} = 1
$$

These norms are related by three important facts pertaining to a stable LTI system G:

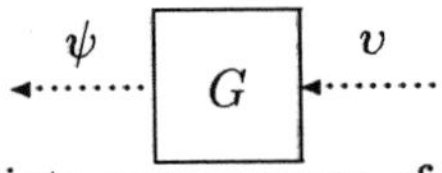

The first fact, an immediate consequence of Parseval's equality, is that if v is the unit impulse, δ_d, then the 2-norm of ψ equals the 2-norm of $\hat{g}$.

Lemma 4.3.1 $\|G\delta_d\|_2 = \|\hat{g}\|_2$

Thus $\|\hat{g}\|_2^2$ equals the energy of the output for a unit impulse input. There is also a random-signal version of this fact: If v is standard white noise [i.e., the sequence $\ldots, v(-1), v(0), v(1), v(2), \ldots$ consists of independent random variables, each of zero mean and unit variance], then the root-mean-square value of $\psi(k)$ equals $\|\hat{g}\|_2$.

The second fact is that the best bound on the ∞-norm of ψ over all inputs of unit ∞-norm equals the 1-norm of g, the impulse-response function:

Lemma 4.3.2 $\sup\{\|\psi\|_\infty : \|v\|_\infty = 1\} = \|g\|_1$

Proof To prove that LHS $\leq$ RHS, let v be any input with $\|v\|_\infty = 1$. From the convolution equation

$$
\psi(k) = \sum_i g(i)v(k-i),
$$

it follows that

$$
\begin{aligned}
|\psi(k)| &\leq \sum_i |g(i)v(k-i)| \\
&\leq \sum_i |g(i)|\|v\|_\infty \\
&= \|g\|_1.
\end{aligned}
$$

So $\|\psi\|_\infty \leq \|g\|_1$.

To prove that LHS $\geq$ RHS, fix k and define the input

$$v(k-i) = |g(i)|/g(i).$$

Then $\|v\|_\infty = 1$ and from the convolution equation

$$\psi(k) = \sum_i |g(i)|,$$

showing that $\|\psi\|_\infty \geq \|g\|_1$. ■

Finally, the third fact is that the best bound on the 2-norm of ψ over all inputs of unit 2-norm equals the ∞-norm of $\hat{g}$:

Lemma 4.3.3 $\sup\{\|\psi\|_2 : \|v\|_2 = 1\} = \|\hat{g}\|_\infty$

Proof That LHS $\leq$ RHS is easy: Let v be any input with $\|v\|_2 = 1$. Then

$$\begin{aligned} \|\psi\|_2 &= \|\hat{\psi}\|_2 \\ &= \|\hat{g}\hat{v}\|_2 \\ &\leq \|\hat{g}\|_\infty \|\hat{v}\|_2 \\ &= \|\hat{g}\|_\infty. \end{aligned}$$

To prove that LHS $\geq$ RHS, let θ_0 be a frequency at which $|\hat{g}|$ attains its maximum, that is,

$$\left|\hat{g}\left(e^{j\theta_0}\right)\right| = \|\hat{g}\|_\infty.$$

Let $\epsilon > 0$ be arbitrary. It suffices to find an input v with $\|v\|_2 \leq 1$ so that $\|\psi\|_2 > \|\hat{g}\|_\infty - \epsilon$. The construction of such v will be sketched.

Since $\left|\hat{g}\left(e^{j\theta}\right)\right|$ is a continuous function of θ, their is a small interval, of width δ, say, centered at θ_0 throughout which

$$\left|\hat{g}\left(e^{j\theta}\right)\right| > \|\hat{g}\|_\infty - \epsilon.$$

This inequality holds also for the interval of width δ centered at $-\theta_0$. Define v by saying that $\left|\hat{v}\left(e^{j\theta}\right)\right|$ is constant on these two intervals; the constant is determined by the normalization condition $\sum_k |v(k)|^2 = 1$, namely, the constant equals $\sqrt{\pi/\delta}$. Thus the graph of $\left|\hat{v}\left(e^{j\theta}\right)\right|$ consists of two narrow rectangles, centered at $\pm\theta_0$. For this input, it is routine to compute that $\|\psi\|_2 > \|\hat{g}\|_\infty - \epsilon$. The only snag is that $v(k)$ is not equal to 0 for $k < 0$. To fix this, pass v through a very long time delay and truncate the output, denoted v_1, by setting it to zero for $k < 0$. Then $v_1 \in \ell_2(\mathbb{Z}_+)$ and $\|v_1\|_2 \leq 1$. If the time delay is long enough, then for this input the output will satisfy $\|\psi\|_2 > \|\hat{g}\|_\infty - \epsilon$. ■

Example 4.3.3 Consider the discrete-time system with

$$\hat{g}(\lambda) = \frac{1}{\lambda - 2}.$$

1. If the input is the unit impulse, then

$$\begin{aligned}(\text{output energy})^{1/2} &= \|\psi\|_2 \\ &= \|\hat{g}\|_2 \\ &= \frac{1}{\sqrt{3}} \text{ (computed before)}\end{aligned}$$

2. If the input satisfies $\|v\|_\infty \leq 1$ (i.e., it lies between ± 1), then $\|\psi\|_\infty \leq \|g\|_1$; furthermore, $\|g\|_1$ is the least upper bound on $\|\psi\|_\infty$. We have

$$\begin{aligned}\hat{g}(\lambda) &= -\frac{1}{2}\frac{1}{1-\frac{1}{2}\lambda} \\ &= -\frac{1}{2}\left(1 + \frac{1}{2}\lambda + \frac{1}{4}\lambda^2 + \cdots\right) \\ g &= \left\{-\frac{1}{2}, -\frac{1}{4}, -\frac{1}{8}, \ldots\right\} \\ \|g\|_1 &= \frac{1}{2} + \frac{1}{4} + \frac{1}{8} + \cdots = 1.\end{aligned}$$

3. If the input satisfies $\|v\|_2 \leq 1$, then $\|\psi\|_2 \leq \|\hat{g}\|_\infty$, and this is the least upper bound. As computed earlier, $\|\hat{g}\|_\infty = 1$.

It is a general fact that

$$\|\hat{g}\|_2 \leq \|\hat{g}\|_\infty$$

(easy proof).

It can be inferred from the second fact that each bounded input (i.e., $\|v\|_\infty < \infty$) produces a bounded output (i.e., $\|\psi\|_\infty < \infty$) iff $\|g\|_1$ is finite, that is, $\sum_{k=0}^{\infty} |g(k)| < \infty$. In the finite-dimensional case (i.e., $\hat{g}(\lambda)$ rational), the following conditions are equivalent:

$$\|g\|_1 < \infty$$

$\hat{g}(\lambda)$ has no poles in $|\lambda| \leq 1$.

Since we will only consider finite-dimensional systems in discrete time, it makes sense to define G to be *stable* if $\hat{g}(\lambda)$ has no poles in $|\lambda| \leq 1$.

Example 4.3.4 State-space model.

$$\hat{g}(\lambda) = \left[\begin{array}{c|c} A & B \\ \hline C & D \end{array}\right]$$

When is G stable? A sufficient condition is that all eigenvalues of A lie in the open unit disk. This is necessary, too, if (A, B) is controllable and (C, A) is observable.

This section concludes with a simple way to compute the 2-norm of $\hat{g}$ using state-space data. Suppose

$$\hat{g}(\lambda) = \left[\begin{array}{c|c} A & B \\ \hline C & D \end{array}\right],$$

with A stable, that is, all eigenvalues in the open unit disk. Remember that we're doing the single-input, single-output (SISO) case, so

$$D : 1 \times 1, \quad C : 1 \times n, \quad B : n \times 1.$$

Let us first look at the matrix equation

$$L = ALA' + BB', \tag{4.5}$$

where prime denotes transpose. This is a linear equation in the unknown matrix L. It is called a (discrete) *Lyapunov equation* and can be solved on MATLAB with the function *dlyap*. The following is a fact: If A is stable, equation (4.5) has a unique, symmetric solution L.

In fact, we can see what the solution is, although not in closed form. Do recursion on (4.5):

$$L = ALA' + BB'$$

$$\begin{aligned} \Longrightarrow L &= A(ALA' + BB')A' + BB' \\ &= A^2LA'^2 + ABB'A' + BB' \end{aligned}$$

$$\Longrightarrow L = A^3LA'^3 + A^2BB'A'^2 + ABB'A' + BB'$$

etc.

so L equals the convergent series of matrices

$$L = BB' + ABB'A' + A^2BB'A'^2 + \cdots.$$

Let us return to computing $\|\hat{g}\|_2$. Parseval's equality in discrete time states that $\|\hat{g}\|_2$ equals the 2-norm of g, the impulse-response function,

$$g = \{D, CB, CAB, CA^2B, \ldots\}.$$

Thus

$$\begin{aligned}
\|\hat{g}\|_2^2 &= \|g\|_2^2 \\
&= D^2 + (CB)^2 + (CAB)^2 + (CA^2B)^2 + \cdots \\
&= D^2 + CBB'C' + CABB'A'C' + CA^2BB'A'^2C' + \cdots \\
&= D^2 + C(BB' + ABB'A' + A^2BB'A'^2 + \cdots)C' \\
&= D^2 + CLC'.
\end{aligned}$$

In summary, a state-space procedure to compute $\|\hat{g}\|_2$ is as follows.

Procedure

Input A realization $\hat{g}(\lambda) = \left[\begin{array}{c|c} A & B \\ \hline C & D \end{array}\right]$, A stable.

Step 1 Solve for L:

$$L = ALA' + BB',$$

Step 2 $\|\hat{g}\|_2 = (D^2 + CLC')^{1/2}$

Example 4.3.5 As a trivial check, let us recompute $\|\hat{g}\|_2$ for

$$\hat{g}(\lambda) = \frac{1}{\lambda - 2}.$$

We have:

$$\begin{aligned}
\hat{g}(\lambda) &= \left[\begin{array}{c|c} 1/2 & 1 \\ \hline -1/4 & -1/2 \end{array}\right] \\
L &= \frac{4}{3} \\
\|\hat{g}\|_2 &= \left(\frac{1}{4} + \frac{1}{4^2} \times \frac{4}{3}\right)^{1/2} = \frac{1}{\sqrt{3}}
\end{aligned}$$

4.4 Multivariable Systems

In this section we summarize the extension of the preceding three sections to the case of vector-valued signals.

A discrete-time signal v is now allowed to be

$$\{v(0), v(1), \ldots\},$$

where each $v(k)$ lies in $\mathbb{R}^n$. Then the column-vector notation has partitioning:

$$v = \begin{bmatrix} v(0) \\ v(1) \\ \vdots \end{bmatrix} \begin{matrix} \}n \\ \}n \\ \vdots \end{matrix}$$

In a multivariable discrete-time linear system G, the input, v, and output, ψ, can be vector-valued:

$$\psi = Gv, \quad \psi(k) \in \mathbb{R}^p, v(k) \in \mathbb{R}^m.$$

Then the matrix $[G]$ has a corresponding partitioning into $p \times m$ blocks. The first column of $[G]$ equals the output signal when a unit impulse is applied at the first input channel, that is,

$$v = \left\{ \begin{bmatrix} 1 \\ 0 \\ \vdots \\ 0 \end{bmatrix}, \begin{bmatrix} 0 \\ 0 \\ \vdots \\ 0 \end{bmatrix}, \begin{bmatrix} 0 \\ 0 \\ \vdots \\ 0 \end{bmatrix}, \ldots \right\};$$

the second column of $[G]$ equals the output signal when a unit impulse is applied at the second input channel, that is,

$$v = \left\{ \begin{bmatrix} 0 \\ 1 \\ 0 \\ \vdots \\ 0 \end{bmatrix}, 0, 0, \ldots \right\};$$

And so on.

Example 4.4.1 State-space model.

$$\begin{aligned} \dot{\xi} &= A\xi + Bv \\ \psi &= C\xi + Dv \end{aligned}$$

$$\begin{aligned} &A : n \times n, \quad &B : n \times m \\ &C : p \times n, \quad &D : p \times m \end{aligned}$$

$$[G] = \begin{bmatrix} D & 0 & 0 & 0 & \cdots \\ CB & D & 0 & 0 & \cdots \\ CAB & CB & D & 0 & \cdots \\ CA^2B & CAB & CB & D & \cdots \\ \vdots & \vdots & \vdots & \vdots & \end{bmatrix}$$

A system G is causal iff $[G]$ is block-lower triangular (as in the preceding example); it is time-invariant iff $[G]$ is constant along block-diagonals (as in the preceding example).

Example 4.4.2 Let us construct a time-varying system as follows. Take

$$\begin{aligned} \dot{\xi}(k) &= a(k)\xi(k) + \upsilon(k), \xi(0) = 0 \\ \psi(k) &= \xi(k), \end{aligned}$$

where $a(k)$ is a periodic coefficient with

$$a(k) = \begin{cases} 0, & k \text{ even} \\ 1, & k \text{ odd.} \end{cases}$$

(Difference equations with periodic coefficients can still be implemented by a finite-memory computer, whereas difference equations with general time-varying coefficients cannot.) The system matrix is

$$\begin{bmatrix} 0 & 0 & 0 & 0 & 0 & 0 & \cdots \\ 1 & 0 & 0 & 0 & 0 & 0 & \cdots \\ 1 & 1 & 0 & 0 & 0 & 0 & \cdots \\ 0 & 0 & 1 & 0 & 0 & 0 & \cdots \\ 0 & 0 & 1 & 1 & 0 & 0 & \cdots \\ 0 & 0 & 0 & 0 & 1 & 0 & \cdots \\ \vdots & \vdots & \vdots & \vdots & \vdots & \vdots & \end{bmatrix},$$

which is not constant along the second subdiagonal. Note, however, that if we redefine the input and output to be

$$\underline{\upsilon}(0) = \begin{bmatrix} \upsilon(0) \\ \upsilon(1) \end{bmatrix}, \quad \underline{\upsilon}(1) = \begin{bmatrix} \upsilon(2) \\ \upsilon(3) \end{bmatrix}, \ldots$$

$$\underline{\psi}(0) = \begin{bmatrix} \psi(0) \\ \psi(1) \end{bmatrix}, \quad \underline{\psi}(1) = \begin{bmatrix} \psi(2) \\ \psi(3) \end{bmatrix}, \ldots$$

then the system *is* time-invariant, because the system matrix is

$$\left[\begin{array}{cc|cc|cc|c} 0 & 0 & 0 & 0 & 0 & 0 & \cdots \\ 1 & 0 & 0 & 0 & 0 & 0 & \cdots \\ \hline 1 & 1 & 0 & 0 & 0 & 0 & \cdots \\ 0 & 0 & 1 & 0 & 0 & 0 & \cdots \\ \hline 0 & 0 & 1 & 1 & 0 & 0 & \cdots \\ 0 & 0 & 0 & 0 & 1 & 0 & \cdots \\ \hline \vdots & \vdots & \vdots & \vdots & \vdots & \vdots & \end{array}\right].$$

Converting the 1-dimensional signal υ into the 2-dimensional signal $\underline{\upsilon}$ is an example of an operation called *lifting*. This technique will be used again in Chapter 8.

The λ-transform of a multivariable signal v is still defined as

$$\hat{v}(\lambda) = v(0) + v(1)\lambda + v(2)\lambda^2 + \cdots.$$

So $\hat{v}(\lambda)$ is a vector, each component being a function of λ. For example, if v is a 3-vector with a unit impulse in the first component, a unit step in the second component, and 0 in the third component, then

$$\hat{v}(\lambda) = \begin{bmatrix} 1 \\ \frac{1}{1-\lambda} \\ 0 \end{bmatrix}.$$

Here's a summary of the notation for an m-input, p-output, causal, LTI system:

block diagram :

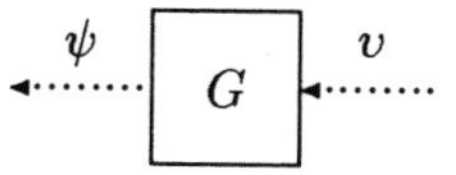

$$v(k) \in \mathbb{R}^m, \psi(k) \in \mathbb{R}^p$$

convolution equation : $\psi(k) = \sum_{l=0}^{k} g(k-l)v(l)$

$g(k) \in \mathbb{R}^{p \times m}$

system matrix equation : $\begin{bmatrix} \psi(0) \\ \psi(1) \\ \vdots \end{bmatrix} = \begin{bmatrix} g(0) & 0 & \cdots \\ g(1) & g(0) & \cdots \\ \vdots & \vdots & \end{bmatrix} \begin{bmatrix} v(0) \\ v(1) \\ \vdots \end{bmatrix}$

impulse response : $\{g(0), g(1), \ldots\}$

transfer function matrix : $\hat{g}(\lambda) = \sum_{k=0}^{\infty} g(k)\lambda^k$

$\hat{g}(\lambda)$ is a $p \times m$ matrix

state-space model : $\hat{g}(\lambda) = D + \lambda C(I - \lambda A)^{-1}B$

$$=: \left[\begin{array}{c|c} A & B \\ \hline C & D \end{array}\right]$$

Norms for multivariable signals and systems is a more complicated topic, so we'll give an abbreviated discussion, namely, the 2-norm and the ∞-norm of a stable $p \times m$ transfer function matrix $\hat{g}(\lambda)$.

2-Norm

$$\|\hat{g}\|_2 = \left\{ \frac{1}{2\pi} \int_0^{2\pi} \operatorname{trace}\left[\hat{g}(e^{j\theta})^* \hat{g}(e^{j\theta})\right] d\theta \right\}^{1/2}$$

∞-Norm

$$\|\hat{g}\|_\infty = \max_\theta \sigma_{\max}\left[\hat{g}(e^{j\theta})\right]$$

Note that the two definitions reduce to the SISO ones when $m = p = 1$.

Concerning these two definitions are two important input-output facts. Let G be a stable LTI system with input v and output ψ:

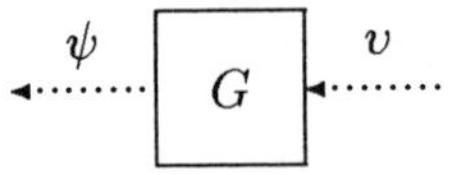

Extending from the one-dimensional case, define the 2-norm of a multivariable signal ϕ to be

$$\|\phi\|_2 = \left(\sum_{k=0}^{\infty} \phi(k)'\phi(k) \right)^{1/2}.$$

Let e_i, $i = 1, \ldots, m$, denote the standard basis vectors in $\mathbb{R}^m$. Thus, $\delta_d e_i$ is an impulse applied to the ith input; $G\delta_d e_i$ is the corresponding output.

The first fact, the MIMO generalization of Lemma 4.3.1, is that the 2-norm of the transfer function $\hat{g}$ is related to the average 2-norm of the output when impulses are applied at the input channels.

Theorem 4.4.1 $\|\hat{g}\|_2^2 = \sum_{i=1}^m \|G\delta_d e_i\|_2^2$

The second fact, the MIMO generalization of Lemma 4.3.3, is that the ∞-norm of the transfer function $\hat{g}$ is related to the maximum 2-norm of the output over all inputs of unit 2-norm.

Theorem 4.4.2 $\|\hat{g}\|_\infty = \sup\{\|\psi\|_2 : \|v\|_2 = 1\}$

Thus the major distinction between $\|\hat{g}_2\|_2$ and $\|\hat{g}\|_\infty$ is that the former is an average system gain for known inputs, while the latter is a worst-case system gain for unknown inputs.

Finally, the state-space procedure to compute the 2-norm is as follows:

Step 1 Solve for L:

$$L = ALA' + BB'.$$

Step 2 $\|\hat{g}\|_2 = [\operatorname{trace}(DD' + CLC')]^{1/2}$

4.5 Function Spaces

In linear algebra we view an n-tuple of real numbers as a point in the vector space $\mathbb{R}^n$. This gives us a position of power: We can say whether two vectors are close to each other, whether they are orthogonal, and so on. So too in the subject of signals and systems it is of value to view a signal or a transfer function as a point in an appropriate space.

Time Domain

You are assumed to know the definitions of vector, normed, and inner-product spaces. Suppose $\mathcal{X}$ and $\mathcal{Y}$ are vector spaces. There is a natural way to add them to get a third vector space, called their *external direct sum*, denoted $\mathcal{X} \oplus \mathcal{Y}$. As a set, $\mathcal{X} \oplus \mathcal{Y}$ consists of all ordered pairs, written

$$\begin{bmatrix} x \\ y \end{bmatrix}.$$

Addition and scalar-multiplication are performed componentwise. If $\mathcal{X}$ and $\mathcal{Y}$ are inner-product spaces, their external direct sum has a natural inner-product, namely

$$\langle \begin{bmatrix} x_1 \\ y_1 \end{bmatrix}, \begin{bmatrix} x_2 \\ y_2 \end{bmatrix} \rangle := < x_1, x_2 > + < y_1, y_2 > .$$

Some notation: $\mathbb{Z}$ is the set of all integers—negative, zero, and positive; $\mathbb{Z}_-$ is the set of negative integers; $\mathbb{Z}_+$ is the set of non-negative integers.

Example 4.5.1 An important example of a vector space is $\ell(\mathbb{Z}_+, \mathbb{R}^n)$, the space of all functions $\mathbb{Z}_+ \to \mathbb{R}^n$, that is, vector-valued signals

$$v = \begin{bmatrix} v(0) \\ v(1) \\ \vdots \end{bmatrix}, \quad v(k) \in \mathbb{R}^n.$$

If n is understood or irrelevant, we may abbreviate by $\ell(\mathbb{Z}_+)$.

An important example of a normed space is $\ell_p(\mathbb{Z}_+, \mathbb{R}^n)$, or just $\ell_p(\mathbb{Z}_+)$. For $1 \le p < \infty$ it is the subspace of $\ell(\mathbb{Z}_+)$ of sequences that are p-power-summable, that is,

$$\|v\|_p := \left(\sum_{k=0}^{\infty} \|v(k)\|^p \right)^{1/p} < \infty.$$

Here $\|v(k)\|$ is some $\mathbb{R}^n$-norm of $v(k)$. It is natural to take the p-norm of the vector $v(k)$—the p-norm of a vector $(x_1, \dots, x_n)$ is defined to be

$$(|x_1|^p + \cdots + |x_n|^p)^{1/p}.$$

For $p = \infty$, $\ell_p(\mathbb{Z}_+)$ is the subspace of $\ell(\mathbb{Z}_+)$ of bounded sequences, that is,

$$\|v\|_\infty := \sup_k \|v(k)\| < \infty.$$

Again, it is natural to take the ∞-norm of the vector $v(k)$; the ∞-norm of the vector $(x_1, \ldots, x_n)$ is defined to be

$$\max_i |x_i|.$$

The special case $\ell_2(\mathbb{Z}_+)$ is an inner-product space with

$$< \psi, v > := \sum_{k=0}^{\infty} \psi(k)' v(k).$$

Two subspaces $\mathcal{V}$, $\mathcal{W}$ of an inner-product space $\mathcal{X}$ are *orthogonal* if $v \perp w$, or $\langle v, w \rangle = 0$, for every v in $\mathcal{V}$ and w in $\mathcal{W}$. Then we write their sum as $\mathcal{V} \oplus \mathcal{W}$. Note that their intersection consists solely of the zero vector. The *orthogonal complement* of $\mathcal{V}$ is

$$\mathcal{V}^\perp := \{x \in \mathcal{X} : x \perp v \ \forall \ v \in \mathcal{V}\}.$$

Example 4.5.2 Take

$$\mathcal{V} := \{v \in \ell_2(\mathbb{Z}_+) : v(0) = 0\}.$$

Then

$$\mathcal{V}^\perp = \{v \in \ell_2(\mathbb{Z}_+) : v(k) = 0 \ \forall \ k > 0\}.$$

The set of real numbers $\mathbb{R}$ enjoys the property that every clustering sequence actually has its limit in $\mathbb{R}$. This is not true of the set of rational numbers: Just think of a sequence of rational numbers converging to, say, π. In this sense $\mathbb{R}$ is "complete." More formally, a sequence $\{x_k\}$ of real numbers is a *Cauchy sequence* provided the numbers cluster together, that is,

$$(\forall \epsilon)(\exists N) k, l > N \Longrightarrow |x_k - x_l| < \epsilon.$$

Then $\mathbb{R}$ is *complete* in the sense that every Cauchy sequence in $\mathbb{R}$ converges to a point in $\mathbb{R}$.

This notion of completeness extends to normed spaces. A *Banach space* is a complete normed space; a *Hilbert space* is a complete inner-product space. It can be proved that $\ell_p(\mathbb{Z}_+)$ is a Banach space for every p, and $\ell_2(\mathbb{Z}_+)$ is a Hilbert space.

Finally, a subspace $\mathcal{V}$ of a normed space $\mathcal{X}$ is *closed* if every sequence in $\mathcal{V}$ which converges in $\mathcal{X}$ actually has its limit in $\mathcal{V}$, that is, $\mathcal{V}$ contains the limits of all its convergent sequences.

Example 4.5.3 Every subspace of $\mathbb{R}^n$ is closed. For an example of a subspace that is not closed, we have to go to an infinite-dimensional space such as $\ell_2(\mathbb{Z}_+)$. In this space, define $\mathcal{V}$ to be the subspace of all sequences that converge to zero in finite time, or in other words, that are nonzero at only finitely many times. This is certainly a subspace (if v_1 and v_2 are in $\mathcal{V}$, so is $v_1 + v_2$; and if v is in $\mathcal{V}$ and c is in $\mathbb{R}$, then $cv \in \mathcal{V}$). To show that $\mathcal{V}$ is not closed, define the following sequence in $\mathcal{V}$:

$$\begin{aligned} v_1 &= \{1, 0, 0, 0, \ldots\} \\ v_2 &= \left\{1, \frac{1}{2}, 0, 0, \ldots\right\} \\ v_3 &= \left\{1, \frac{1}{2}, \frac{1}{2^2}, 0, \ldots\right\} \end{aligned}$$

etc.

This sequence converges in $\ell_2(\mathbb{Z}_+)$ to

$$v = \left\{1, \frac{1}{2}, \frac{1}{2^2}, \frac{1}{2^3}, \ldots\right\},$$

that is $\lim_{n\to\infty} \|v_n - v\|_2 = 0$. But v is not in $\mathcal{V}$.

If $\mathcal{X}$ is an inner-product space and $\mathcal{V}$ is a closed subspace, then $\mathcal{V}$ and $\mathcal{V}^\perp$ sum to all of $\mathcal{X}$, that is, $\mathcal{X} = \mathcal{V} \oplus \mathcal{V}^\perp$.

Frequency Domain

Next we turn to the frequency domain. Let $\mathbb{D}$ denote the open unit disk in the complex plane and $\partial\mathbb{D}$ its boundary, the unit circle. The Lebesgue space $\mathcal{L}_2(\partial\mathbb{D}, \mathbb{C}^{n\times m})$ consists of all square-integrable functions $\partial\mathbb{D} \to \mathbb{C}^{n\times m}$, that is, $n \times m$ matrix-valued functions $\hat{f}(e^{j\theta})$ for which the norm

$$\|\hat{f}\|_2 := \left(\frac{1}{2\pi}\int_0^{2\pi} \text{trace}\left[\hat{f}(e^{j\theta})^*\hat{f}(e^{j\theta})\right] d\theta\right)^{1/2}$$

is finite. Here superscript * means complex-conjugate transpose. This space, abbreviated $\mathcal{L}_2(\partial\mathbb{D})$, is a Hilbert space with inner-product

$$<\hat{f}, \hat{g}> := \frac{1}{2\pi}\int_0^{2\pi} \text{trace}\left[\hat{f}(e^{j\theta})^*\hat{g}(e^{j\theta})\right] d\theta.$$

Example 4.5.4 Consider a polynomial matrix

$$\hat{f}(\lambda) = f(0) + f(1)\lambda + \cdots + f(k)\lambda^k, \quad f(i) \in \mathbb{R}^{n\times m}.$$

Considered as a function on $\partial\mathbb{D}$, that is, for $\lambda = e^{j\theta}$, $\hat{f}$ belongs to $\mathcal{L}_2(\partial\mathbb{D})$. In this way, $\mathcal{L}_2(\partial\mathbb{D})$ contains all polynomials.

Real-rational matrices are ones with real coefficients. The real-rational matrices in $\mathcal{L}_2(\partial\mathbb{D})$ are precisely those with no poles on $\partial\mathbb{D}$. This subset is denoted $\mathcal{RL}_2(\partial\mathbb{D})$, the prefix $\mathcal{R}$ denoting real-rational.

The Hardy space $\mathcal{H}_2(\mathbb{D})$ consists of all complex-valued functions $\hat{f}(\lambda)$ defined and analytic on $\mathbb{D}$ and such that the boundary function $\hat{f}(e^{j\theta})$ belongs to $\mathcal{L}_2(\partial\mathbb{D})$. By identifying $\hat{f}$ and its boundary function we can regard $\mathcal{H}_2(\mathbb{D})$ as a closed subspace of $\mathcal{L}_2(\partial\mathbb{D})$; hence, it has an orthogonal complement, $\mathcal{H}_2(\mathbb{D})^\perp$.

Example 4.5.5 Polynomial matrices are in $\mathcal{H}_2(\mathbb{D})$ too. The real-rational matrices in $\mathcal{H}_2(\mathbb{D})$ are precisely those with no poles in $\mathbb{D}$ or on $\partial\mathbb{D}$; this subset is denoted $\mathcal{RH}_2(\mathbb{D})$. Similarly, $\mathcal{RH}_2(\mathbb{D})^\perp$ consists of the strictly proper real-rational matrices having no poles in $|\lambda| \geq 1$.

Finally, the space $\mathcal{L}_\infty(\partial\mathbb{D})$ consists of matrices $\hat{f}(e^{j\theta})$ whose ∞-norm

$$\|\hat{f}\|_\infty := \sup_\theta \sigma_{\max}[\hat{f}(e^{j\theta})]$$

is finite. The subspace of matrices analytic in $\mathbb{D}$ is $\mathcal{H}_\infty(\mathbb{D})$. Observe that $\mathcal{RH}_2(\mathbb{D}) = \mathcal{RH}_\infty(\mathbb{D})$.

Operators

Let $\mathcal{X}$ and $\mathcal{Y}$ be normed spaces and let $F : \mathcal{X} \to \mathcal{Y}$ be a linear transformation. Then F is *bounded* if

$$(\exists c)(\forall x)\|Fx\| \leq c\|x\|.$$

The least such constant c is called the *norm* of F and is denoted $\|F\|$. A bounded linear transformation is called an *operator*. Alternative expressions for the norm are as follows:

$$\begin{aligned}
\|F\| &= \inf\{c : (\forall x)\|Fx\| \leq c\|x\|\} \\
&= \inf\{c : (\forall x \neq 0)\frac{\|Fx\|}{\|x\|} \leq c\} \\
&= \sup_{x\neq 0} \frac{\|Fx\|}{\|x\|} \\
&= \sup_{\|x\|=1} \|Fx\| \\
&= \sup_{\|x\|\leq 1} \|Fx\|.
\end{aligned}$$

Example 4.5.6 Let G be a stable, multivariable, LTI system. Then it can be regarded as an operator from $\ell_2(\mathbb{Z}_+)$ to itself. Theorem 4.4.2 says that $\|G\| = \|\hat{g}\|_\infty$.

Isomorphism between Time Domain and Frequency Domain

Recall that $\ell_2(\mathbb{Z}_+)$ is the space of square-summable sequences defined for non-negative times. We shall need two additional spaces. Define $\ell_2(\mathbb{Z}_-)$ to be the space of square-summable sequences defined for negative times, $k = \ldots, -2, -1$. A signal v in $\ell_2(\mathbb{Z}_-)$ is written

$$\begin{bmatrix} \vdots \\ v(-2) \\ v(-1) \end{bmatrix}$$

and has the property

$$\sum_{-\infty}^{-1} v(k)'v(k) < \infty.$$

This is a Hilbert space under the inner-product

$$< \psi, v >= \sum_{-\infty}^{-1} \psi(k)'v(k).$$

Also, define $\ell_2(\mathbb{Z})$ to be the external direct sum of $\ell_2(\mathbb{Z}_-)$ and $\ell_2(\mathbb{Z}_+)$:

$$\ell_2(\mathbb{Z}) := \ell_2(\mathbb{Z}_-) \oplus \ell_2(\mathbb{Z}_+).$$

Elements of $\ell_2(\mathbb{Z})$ will be written

$$\begin{bmatrix} \psi \\ v \end{bmatrix}, \quad \psi \in \ell_2(\mathbb{Z}_-), v \in \ell_2(\mathbb{Z}_+).$$

With the inherited inner-product, $\ell_2(\mathbb{Z})$ is a Hilbert space, $\ell_2(\mathbb{Z}_+)$ is a closed subspace, and $\ell_2(\mathbb{Z}_-)$ is its orthogonal complement.

The theorem coming up connects the time-domain Hilbert spaces to the frequency-domain Hilbert spaces. It is compactly stated via the notion of isomorphism. Let $\mathcal{X}$ and $\mathcal{Y}$ be Hilbert spaces. An *isomorphism* from $\mathcal{X}$ to $\mathcal{Y}$ is a linear transformation F having the two properties

It is surjective: $(\forall y)(\exists x) y = Fx$.

It preserves inner-products: $(\forall x_1, x_2) < Fx_1, Fx_2 >=< x_1, x_2 >$.

Such a function automatically has the further properties

It preserves norms: $(\forall x)\|Fx\| = \|x\|$.

It is injective: $(\forall x)Fx = 0 \Longrightarrow x = 0$.

It is bounded.

It has a bounded inverse.

If such an isomorphism exists, then $\mathcal{X}$ and $\mathcal{Y}$ are *isomorphic.*

Theorem 4.5.1 *The λ-transformation is an isomorphism from $\ell_2(\mathbb{Z})$ onto $\mathcal{L}_2(\partial\mathbb{D})$; it maps $\ell_2(\mathbb{Z}_+)$ onto $\mathcal{H}_2(\mathbb{D})$ and $\ell_2(\mathbb{Z}_-)$ onto $\mathcal{H}_2(\mathbb{D})^\perp$.*

This result is a combination of the Riesz-Fischer theorem and Parseval's equality.

4.6 Optimal Discretization of Analog Systems

In Section 3.5 we looked at the problem of doing a digital implementation of an analog system, either a controller or a filter. The bilinear transformation is one method for doing this. In this section we will present a second method that is based on optimization.

The setup is shown in Figure 4.1. Shown are two continuous-time

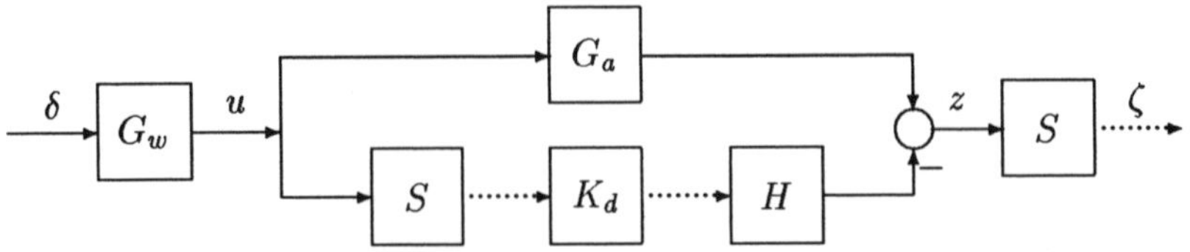

Figure 4.1: The error system.

systems, G_a and G_w, and a discrete-time system, K_d; G_a represents an analog system to be discretized and K_d the discretization. The two systems G_a and HK_dS are compared by applying a common input, u, and observing the error, z, between their outputs. Actually, we choose at this time for simplicity to observe only the sampled error, ζ. The other analog system, G_w, is a weighting filter used as a design parameter in the optimization process. The idea is that G_w should pass only those signals u for which we desire the error to be small. The input to G_w is chosen to be the unit impulse.

The *optimal discretization problem* we pose is as follows: Given G_a and G_w, design K_d to minimize $\|\zeta\|_2$. One strong motivation for choosing the performance measure $\|\zeta\|_2$ is that it makes the problem solvable! We will have to check at the end if a small discretization error is achieved in the sense of Section 3.5.

In the derivation to follow, G_a, G_w, and K_d are assumed to be SISO for simplicity. We assume in addition that $\hat{g}_a(s)$ is stable and proper and that $\hat{g}_w(s)$ is strictly proper. Then $\hat{k}_d(\lambda)$ is constrained to be stable.

It is convenient to define

$$\hat{g}_{w1}(s) = s\hat{g}_w(s), \quad \hat{g}_{w2}(s) = \frac{1}{s},$$

so that $\hat{g}_w = \hat{g}_{w1}\hat{g}_{w2}$. Then in Figure 4.1

$$u = G_{w1}G_{w2}\delta.$$

Now $G_{w2}\delta$ is the unit step, which equals the output of H when its input is the discrete-time unit step, denoted, say σ [$\sigma(k) = 1$ for all $k \geq 0$]. Thus

$$u = G_{w1}H\sigma.$$

The discretized error is therefore

$$\begin{aligned} \zeta &= S(G_a - HK_dS)u \\ &= S(G_a - HK_dS)G_{w1}H\sigma \\ &= (S\ {}_aG_{w1}H - K_dSG_{w1}H)\sigma. \quad G \end{aligned}$$

Define the two discrete-time systems

$$T_{1d} = SG_aG_{w1}H, \quad T_{2d} = SG_{w1}H.$$

Then

$$\zeta = (T_{1d} - K_dT_{2d})\sigma.$$

Since $\|\zeta\|_2 = \|\hat{\zeta}\|_2$, we arrive at the following equivalent optimization problem in $\mathcal{RH}_2(\mathbb{D})$:

$$\min_{\hat{k}_d \in \mathcal{RH}_2(\mathbb{D})} \|(\hat{t}_{1d} - \hat{k}_d\hat{t}_{2d})\hat{\sigma}\|_2.$$

If we are lucky enough that $\hat{t}_{1d}/\hat{t}_{2d}$ is in $\mathcal{RH}_2(\mathbb{D})$, then this is obviously the optimal $\hat{k}_d$. The solution of this optimization problem in the general case is postponed to Section 6.6.

Example 4.6.1 Let $\hat{g}_a(s)$ be the elliptic filter in the example in Section 3.5. Again, take $\omega_N = 10$. Some trial-and-error is required to get a good filter $\hat{g}_w(s)$; the following one gives infinite weight to an error at DC, and has a cutoff frequency of 1 rad/s:

$$\hat{g}_w(s) = \frac{1}{s(s+0.001)(s+1)}.$$

For this data, $\hat{t}_{1d}/\hat{t}_{2d}$ is indeed in $\mathcal{RH}_2(\mathbb{D})$. Figure 4.2 shows the graph of $error(\omega)$ for both the causal (i.e., the filter HK_dS) and noncausal (i.e., the filter RK_dS) cases. The maximum errors on the passband are, respectively,

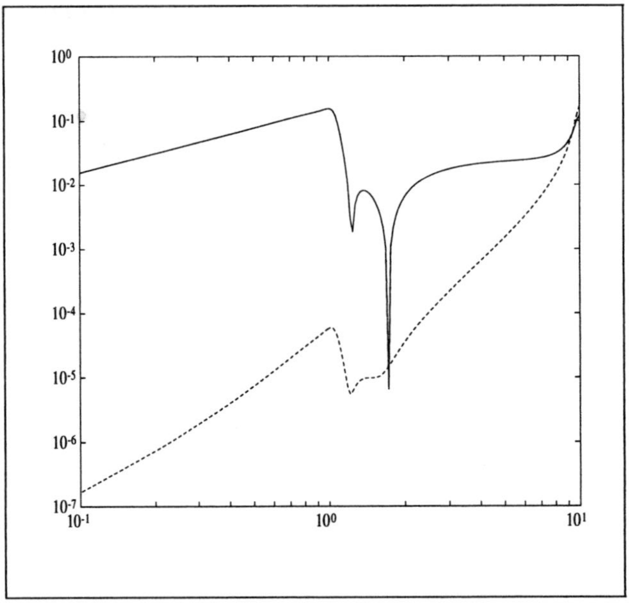

Figure 4.2: Error for Discretization of Elliptic Filter: Causal Case (solid line), Noncausal Case (dash line).

0.1549 and 0.5808×10^{-4}. The latter is a remarkable improvement over the bilinear transformation.

Exercises

4.1 Consider the discrete-time linear system with input v and output ψ, modeled by the difference equation

$$\psi(k) - \psi(k-1) + 3\psi(k-2) = 2v(k) - v(k-2).$$

Find the transfer function from v to ψ. Find a state-space model.

4.2 Consider a linear system G modeled by the difference equation

$$\psi(k+2) + (2k-1)\psi(k+1) + \psi(k) = kv(k+1) + v(k).$$

Find the first five rows of $[G]$.

4.3 Suppose $\psi = Gv$, with

$$\hat{g}(\lambda) = \frac{\lambda}{(2\lambda+3)(\lambda-4)}.$$

Suppose the energy of v is known to equal 0.5. Compute the best upper bound on the energy of ψ.

4.4 For stable SISO transfer functions $\hat{f}(\lambda)$ and $\hat{g}(\lambda)$, prove that

$$\begin{aligned} \|\hat{f}\hat{g}\|_\infty &\leq \|\hat{f}\|_\infty \|\hat{g}\|_\infty, \\ \|\hat{f}\hat{g}\|_2 &\leq \|\hat{f}\|_\infty \|\hat{g}\|_2. \end{aligned}$$

4.5 Consider the LTI system G with

$$[G] = \begin{bmatrix} 0 & 0 & 0 & 0 & \cdots \\ 1 & 0 & 0 & 0 & \cdots \\ \frac{1}{2} & 1 & 0 & 0 & \cdots \\ \frac{1}{3} & \frac{1}{2} & 1 & 0 & \cdots \\ \frac{1}{4} & \frac{1}{3} & \frac{1}{2} & 1 & \cdots \\ \vdots & \vdots & \vdots & \vdots & \end{bmatrix}.$$

Find $\hat{g}(\lambda)$, $\|\hat{g}\|_2$, and $\|\hat{g}\|_\infty$. (This shows that $\mathcal{H}_\infty(\mathbb{D})$ is a proper subset of $\mathcal{H}_2(\mathbb{D})$.)

4.6 For a square matrix A, let $\rho(A)$ denote *spectral radius*—maximum magnitude of all eigenvalues. Is $\rho(A)$ a norm?

4.7 Let $\mathcal{V} := \{v \in \ell_2(\mathbb{Z}_+) : v(0) = 0\}$. Prove that $\mathcal{V}$ is closed.

4.8 A linear system is strictly causal if the output is initially zero and the output at time $k+1$ depends only on inputs up to time k. Give a formal definition of strict causality, and characterize it in terms of the system matrix.

4.9 Consider a SISO discrete-time system G with input $v(k)$ and output $\psi(k)$ related by the equation

$$\psi(k) = \sum_{l=-\infty}^{\infty} \phi(k+l)v(l), \quad -\infty < k < \infty$$

where $\phi(k)$ is a given sequence.

1. Under what conditions on $\phi(k)$ is G time-invariant? Causal? Bounded on $\ell_\infty(\mathbb{Z})$?

2. Find the relationship between the λ-transforms of v and ψ.

4.10 Consider the state-space model G, but where the four matrices $A(k)$, $B(k)$, $C(k)$, $D(k)$ are all periodic, of period N. Then it is not true that $U^*GU = G$, but what is true?

4.11 Show that a linear system G is causal and time-invariant iff it commutes with U.

4.12 A linear system G is *memoryless* if its system matrix is (block) diagonal; thus, the output at time k depends only on the input at time k. Show that G is memoryless and time-invariant iff it commutes with both U and U^*.

4.13 Consider a linear system $G : \ell_2(\mathbb{Z}_+, \mathbb{R}) \rightarrow \ell_2(\mathbb{Z}_+, \mathbb{R})$ which is causal and bounded in the sense that the norm

$$||G|| = \sup_{||v||_2 \leq 1} ||Gv||_2$$

is finite. Since G is time-invariant iff G and U commute, the quantity

$$\tau := ||GU - UG||$$

is a measure of how "time-varying" G is ($GU - UG$ is like the derivative of G).

Take the following specific G: The input $v(k)$ and output $\psi(k)$ satisfy

$$\psi(k) = \begin{cases} v(k), & k \text{ even} \\ 2v(k), & k \text{ odd.} \end{cases}$$

Compute τ.

4.14 For this exercise only, we shall redefine some notation. Consider sequences defined for all time, $k \in \mathbb{Z}$. Define the backward and forward shifts U and U^* on $\ell(\mathbb{Z})$ in the obvious way. Finally, let G be a linear transformation on $\ell(\mathbb{Z})$. Show that the following three conditions are equivalent:

$U^*GU = G$

G commutes with U

G commutes with both U and U^*

4.15 In this problem and the next, all functions are scalar-valued. Fix a point a in the open unit disk. Find a function $\hat{f}$ in $\mathcal{H}_2(\mathbb{D})$ such that for every $\hat{g}$ in $\mathcal{H}_2(\mathbb{D})$

$$< \hat{f}, \hat{g} >= \hat{g}(a).$$

(Hint: Cauchy's integral formula.)

4.16 Define $\hat{f}(\lambda) = 2\lambda - 1$ and $\mathcal{V} = \hat{f}\mathcal{H}_2(\mathbb{D})$; that is, $\mathcal{V}$ equals the set of all functions $\hat{f}\hat{g}$, as $\hat{g}$ ranges over $\mathcal{H}_2(\mathbb{D})$. Show that $\mathcal{V}$ is a closed subspace of $\mathcal{H}_2(\mathbb{D})$. Show that the dimension of $\mathcal{V}^\perp$, the orthogonal complement in $\mathcal{H}_2(\mathbb{D})$, equals 1. Find a basis for $\mathcal{V}^\perp$.

4.17 Find the projections in $\mathcal{H}_2(\mathbb{D})$ and $\mathcal{H}_2(\mathbb{D})^\perp$ of

$$\begin{bmatrix} \frac{\lambda^3(\lambda+3)}{2\lambda^2-5\lambda+2} \\ \frac{1}{\lambda^2-2\lambda} \end{bmatrix}.$$

4.18 Compute the $\mathcal{H}_\infty(\mathbb{D})$-norm of

$$\begin{bmatrix} \frac{\lambda^2+1}{\lambda+2} & 1 \\ \frac{\lambda}{\lambda^2-\lambda-6} & \lambda \end{bmatrix}.$$

4.19 Prove that if an $n \times m$ matrix $\hat{f}$ is in $\mathcal{H}_\infty(\mathbb{D})$, then

$$\|\hat{f}\|_2 \leq \sqrt{m}\|\hat{f}\|_\infty,$$

so that $\hat{f} \in \mathcal{H}_2(\mathbb{D})$.

4.20 Compute the $\mathcal{H}_2(\mathbb{D})$-norm of

$$\begin{bmatrix} \frac{\lambda^2+1}{\lambda+2} & 1 \\ \frac{\lambda}{\lambda^2-\lambda-6} & \lambda \end{bmatrix}.$$

4.21 Consider the system

$$\begin{aligned} \dot{\xi} &= A\xi + Bv \\ \psi &= C\xi + Dv. \end{aligned}$$

Apply an input with $v(k) = 0$ for k odd. Defining $\phi(k) = \psi(2k)$ and $\nu(k) = v(2k)$, find the transfer matrix from ν to ϕ.

4.22 Consider a discrete-time transfer matrix $\hat{g}$ with minimal realization

$$\hat{g}(\lambda) = \left[\begin{array}{c|c} A & B \\ \hline C & D \end{array}\right].$$

Assume A has the property that no two eigenvalues outside the closed unit disk add up to zero. Show that if

$$\left[\begin{array}{c|c} A^2 & AB \\ \hline C & D \end{array}\right] \in \mathcal{H}_\infty(\mathbb{D}),$$

then $\rho(A) < 1$.

4.23 Write a MATLAB program to compute the optimal K_d for the elliptic filter in Example 4.6.1. Reproduce the plots in Figure 4.2.

Notes and References

Regarding linear systems as linear transformations on function spaces puts the subject into the mathematical domain of *linear operator theory*, a very rich source of results. For the use of norms in characterizing performance and proofs of the facts in Sections 4.3 and 4.4, see [21] and [39]. General references for Section 4.5 are [33], [58], and [67]. The optimization problem, technique of solution, and example in Section 4.6 are due to Smith [128].

Chapter 5

Discrete-Time Feedback Systems

This chapter collects some useful material in linear control theory: observer-based controllers; feedback stability; parametrization of all stabilizing controllers; tracking step inputs.

5.1 Connecting Subsystems

One frequently wants to connect subsystems having state models, and then to obtain a state model for the resulting system. The purpose of this short section is to develop some formulas for the connection of two subsystems. The systems can be continuous-time or discrete-time—the formulas are the same. For convenience the block diagrams are drawn as continuous-time systems (solid lines).

The first formula is

Parallel connection

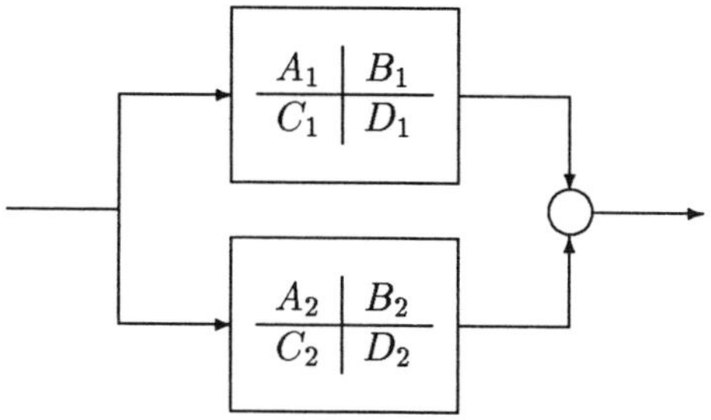

$$\left[\begin{array}{cc|c} A_1 & 0 & B_1 \\ 0 & A_2 & B_2 \\ \hline C_1 & C_2 & D_1 + D_2 \end{array}\right]$$

The figure shows the connection of two systems in parallel and state models for each component; following it is a state model for the combined system. The result is easy to derive, as follows. Start with state models for the two subsystems:

$$\begin{aligned} \dot{x}_1 &= A_1 x_1 + B_1 u \\ y_1 &= C_1 x_1 + D_1 u \\ \dot{x}_2 &= A_2 x_2 + B_2 u \\ y_2 &= C_2 x_2 + D_2 u. \end{aligned}$$

The equation at the summing junction is

$$y = y_1 + y_2.$$

Therefore the combined system is modelled by

$$\begin{aligned} \begin{bmatrix} \dot{x}_1 \\ \dot{x}_2 \end{bmatrix} &= \begin{bmatrix} A_1 & 0 \\ 0 & A_2 \end{bmatrix} \begin{bmatrix} x_1 \\ x_2 \end{bmatrix} + \begin{bmatrix} B_1 \\ B_2 \end{bmatrix} u \\ y &= \begin{bmatrix} C_1 & C_2 \end{bmatrix} \begin{bmatrix} x_1 \\ x_2 \end{bmatrix} + (D_1 + D_2) u. \end{aligned}$$

The other formulas are as follows:

- *Series connection*

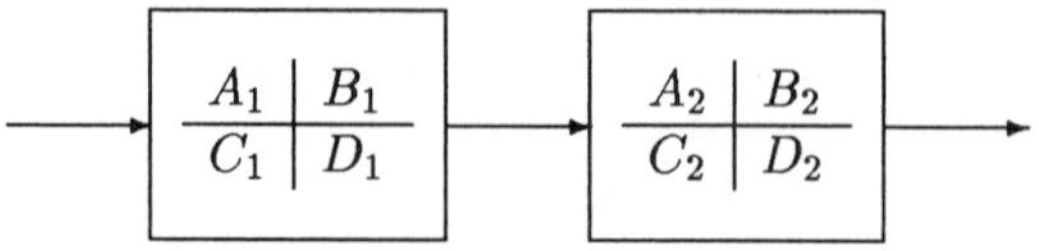

$$\left[\begin{array}{cc|c} A_1 & 0 & B_1 \\ B_2 C_1 & A_2 & B_2 D_1 \\ \hline D_2 C_1 & C_2 & D_2 D_1 \end{array}\right] = \left[\begin{array}{cc|c} A_2 & B_2 C_1 & B_2 D_1 \\ 0 & A_1 & B_1 \\ \hline C_2 & D_2 C_1 & D_2 D_1 \end{array}\right]$$

- *Feedback connection, no. 1*

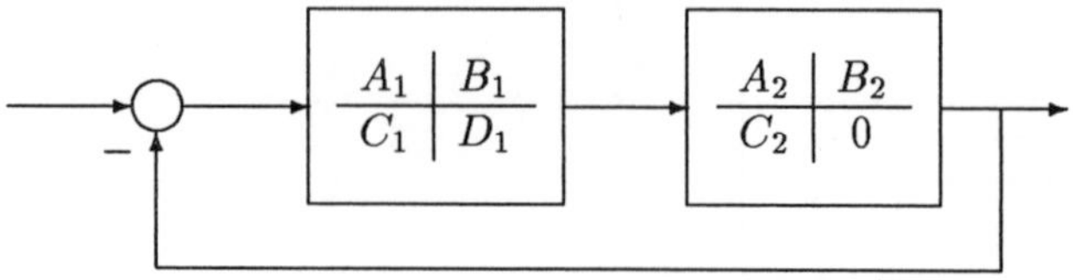

$$\left[\begin{array}{cc|c} A_1 & -B_1C_2 & B_1 \\ B_2C_1 & A_2 - B_2D_1C_2 & B_2D_1 \\ \hline 0 & C_2 & 0 \end{array}\right]$$

- *Feedback connection, no. 2*

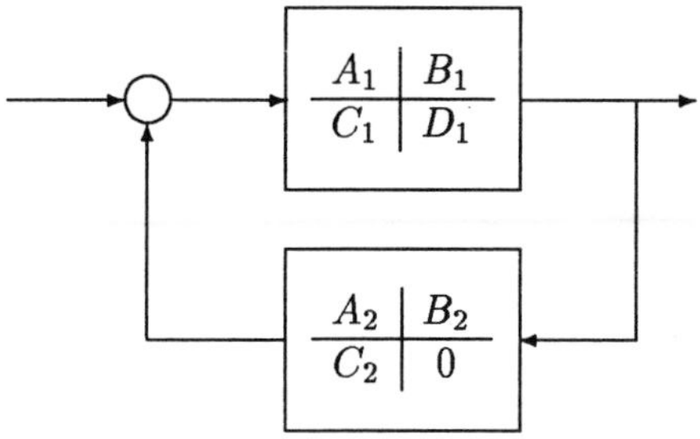

$$\left[\begin{array}{cc|c} A_1 & B_1C_2 & B_1 \\ B_2C_1 & A_2 + B_2D_1C_2 & B_2D_1 \\ \hline C_1 & D_1C_2 & D_1 \end{array}\right]$$

- *General interconnection*

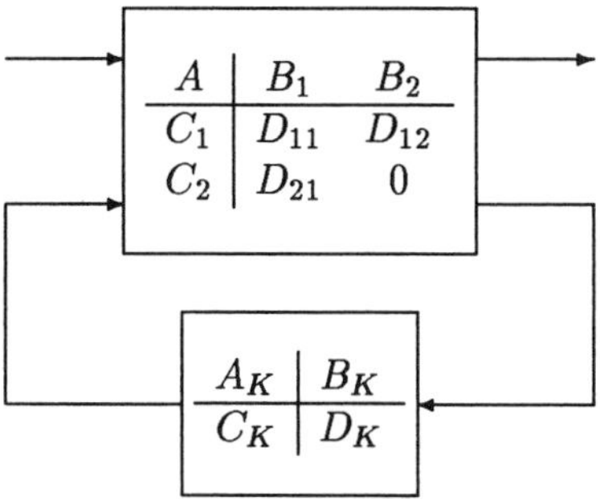

$$\left[\begin{array}{cc|c} A + B_2D_KC_2 & B_2C_K & B_1 + B_2D_KD_{21} \\ B_KC_2 & A_K & B_KD_{21} \\ \hline C_1 + D_{12}D_KC_2 & D_{12}C_K & D_{11} + D_{12}D_KD_{21} \end{array}\right]$$

5.2 Observer-Based Controllers

This section presents basic results on observers and observer-based controllers. Consider the plant model

$$\begin{aligned} \dot{\xi} &= A\xi + Bv \\ \psi &= C\xi + Dv. \end{aligned}$$

(Remember that $\dot{\xi}$ in discrete time means unit time advance.) An *observer*—more properly, an asymptotic state estimator—is another system, with inputs v, ψ and output $\tilde{\xi}$, having the property that the state error vector $\xi(k)-\tilde{\xi}(k)$ converges to 0 as $k \to \infty$. The general form of an observer is

$$\begin{aligned} \dot{\xi}_o &= A_o\xi_o + B_{o1}v + B_{o2}\psi \\ \tilde{\xi} &= C_o\xi_o + D_{o1}v + D_{o2}\psi. \end{aligned}$$

So the two systems are hooked up as in Figure 5.1. The observer is to provide

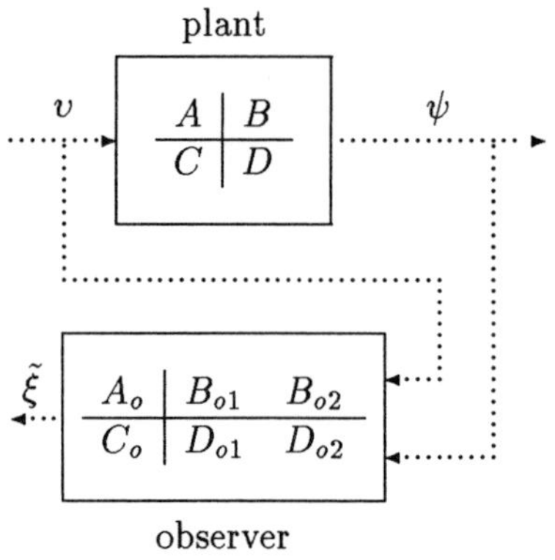

Figure 5.1: Plant and observer.

state estimation in the following sense: For every initial state $\xi(0)$ and $\xi_o(0)$ and every input v, we should have

$$\xi(k) - \tilde{\xi}(k) \to 0 \quad \text{as} \quad k \to \infty.$$

Let us define the estimation error, $\varepsilon := \xi - \tilde{\xi}$, and then view the plant-observer combination as a system with input v and output ε. We have

$$\begin{aligned} \dot{\xi} &= A\xi + Bv \\ \dot{\xi}_o &= A_o\xi_o + B_{o1}v + B_{o2}(C\xi + Dv) \\ \varepsilon &= \xi - C_o\xi_o - D_{o1}v - D_{o2}(C\xi + Dv). \end{aligned}$$

So Figure 5.1 becomes Figure 5.2. Again, we want $\varepsilon(k) \to 0$ as $k \to \infty$ for every initial state and every input. In Figure 5.2, (A, B, C, D) is given and $(A_o, B_{o1}, B_{o2}, C_o, D_{o1}, D_{o2})$ is to be designed. There is a simple solution. It turns out that an observer exists if (C, A) is detectable. To get an observer, first select L so that $A + LC$ is stable, and then take the observer equations to be

$$\begin{aligned} \dot{\xi}_o &= A\xi_o + Bv + L(C\xi_o + Dv - \psi) \\ \tilde{\xi} &= \xi_o, \end{aligned}$$

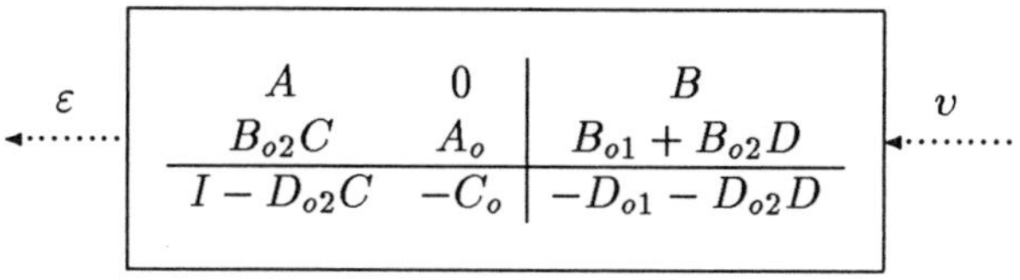
ε
v
A 0 B
B_{o2}C A_o B_{o1} + B_{o2}D
I − D_{o2}C −C_o −D_{o1} − D_{o2}D

Figure 5.2: Plant-observer combination.

that is,

$$\left[\begin{array}{c|cc} A_o & B_{o1} & B_{o2} \\ \hline C_o & D_{o1} & D_{o2} \end{array}\right] = \left[\begin{array}{c|cc} A+LC & B+LD & -L \\ \hline I & 0 & 0 \end{array}\right]. \tag{5.1}$$

Notice that the observer has this special structure:

$$\underbrace{\dot{\xi}_o = A\xi_o + Bv}_{\text{simulation of plant}} \quad + L\underbrace{[\quad \underbrace{C\xi_o + Dv}_{\text{estimate of output}} \quad - \psi]}_{\text{output error}}.$$

To verify that this is an observer, write out the corresponding equations:

$$\begin{aligned} \dot{\xi} &= A\xi + Bv \\ \dot{\xi}_o &= (A+LC)\xi_o - LC\xi + Bv \\ \varepsilon &= \xi - \xi_o. \end{aligned}$$

Subtract the first two:

$$\dot{\xi} - \dot{\xi}_o = (A+LC)(\xi - \xi_o).$$

Thus

$$\dot{\varepsilon} = (A+LC)\varepsilon,$$

so the transfer matrix from v to ε equals zero. Also, $\varepsilon(k) \to 0$ for every $\xi(0)$, $\xi_o(0)$, and v.

Let us summarize:

Theorem 5.2.1 *For the setup in Figure 5.3, if (C, A) is detectable and L is chosen so that $A + LC$ is stable, then $\xi(k) - \xi_o(k) \to 0$ for every $\xi(0)$, $\xi_o(0)$, v.*

Only detectability of (C, A) was required for the preceding construction, but if we want fast convergence of $\xi(k) - \xi_o(k)$, then we should place the eigenvalues of $A + LC$ well inside $\mathbb{D}$. To reassign all eigs, of course we need observability, not just detectability.

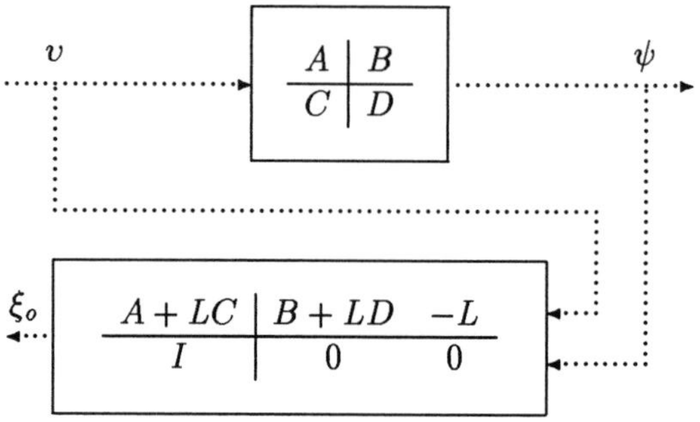

Figure 5.3: Plant and observer.

We are poised now to think of a two-stage controller: The first stage would be an observer to generate an estimate of the plant's state; the second stage would be to feed back this estimate as though it were the state. Let us see this when the goal is to stabilize an unstable plant.

Let the plant be

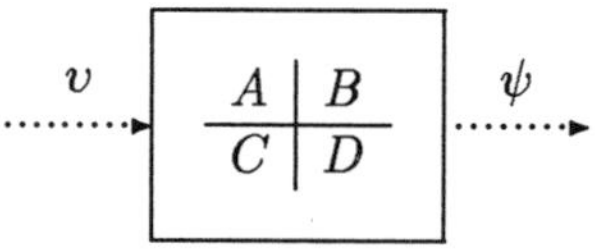

with (A, B) stabilizable and (C, A) detectable. Choosing L so that $A + LC$ is stable, hook up the observer as in Figure 5.3. Finally, close the loop via $v = F\xi_o$, where F is chosen to stabilize $A+BF$. In this way we get Figure 5.4. This can be simplified by writing the equations from ψ to v:

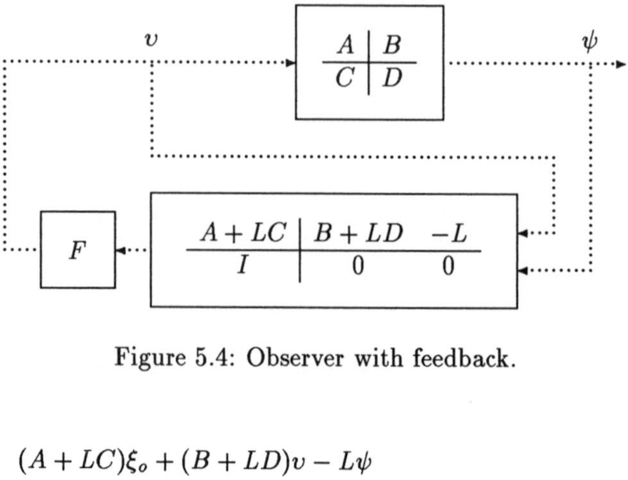

Figure 5.4: Observer with feedback.

$$\dot{\xi}_o = (A + LC)\xi_o + (B + LD)v - L\psi$$

$$v = F\xi_o.$$

Thus

$$\begin{aligned} \dot{\xi}_o &= (A + BF + LC + LDF)\xi_o - L\psi \\ v &= F\xi_o. \end{aligned}$$

Thus the configuration reduces to that of Figure 5.5.

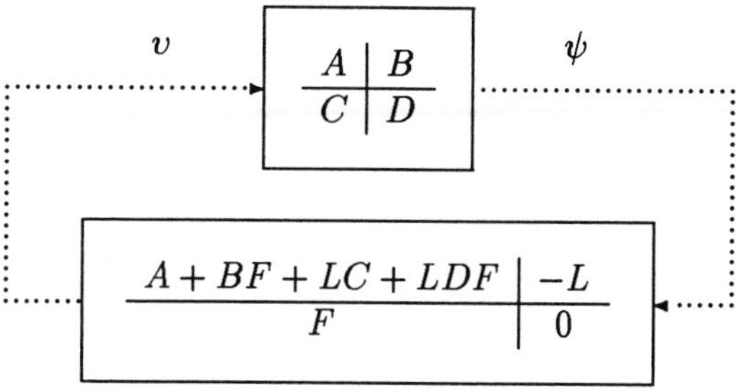

Figure 5.5: Observer-based controller.

This system is autonomous—it has no inputs. But it is internally stable in the following sense: For every initial value of the states, $\xi(0)$ and $\xi_o(0)$, the states $\xi(k)$ and $\xi_o(k)$ converge to zero as $k \to \infty$.

Proof Write the state equations corresponding to Figure 5.5:

$$\begin{aligned} \dot{\xi} &= A\xi + Bv \\ &= A\xi + BF\xi_o \\ \dot{\xi}_o &= (A + BF + LC + LDF)\xi_o - L\psi \\ &= (A + BF + LC + LDF)\xi_o - L(C\xi + Dv) \\ &= (A + BF + LC + LDF)\xi_o - L(C\xi + DF\xi_o) \\ &= (A + BF + LC)\xi_o - LC\xi. \end{aligned}$$

Thus

$$\begin{bmatrix} \dot{\xi} \\ \dot{\xi}_o \end{bmatrix} = \underline{A} \begin{bmatrix} \xi \\ \xi_o \end{bmatrix},$$

where

$$\underline{A} := \begin{bmatrix} A & BF \\ -LC & A + BF + LC \end{bmatrix}. \tag{5.2}$$

So stability is equivalent to the condition that $\underline{A}$ be stable. To show that $\underline{A}$ is indeed stable, define

$$T := \begin{bmatrix} I & 0 \\ I & I \end{bmatrix}.$$

A simple computation gives

$$T^{-1}\underline{A}T = \begin{bmatrix} A+BF & BF \\ 0 & A+LC \end{bmatrix}.$$

Thus the eigs of $\underline{A}$ are the eigs of $A+BF$ together with the eigs of $A+LC$. But these latter two matrices are stable by design. ■

Let us put the configuration into the more conventional form of the unity-feedback system of Figure 5.6. Suppose P is a given plant and K is a con-

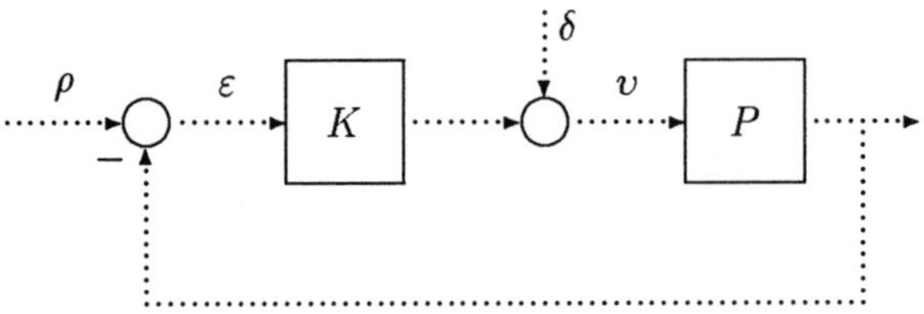

Figure 5.6: Unity-feedback system.

troller to be designed. Beginning with a stabilizable, detectable realization of P,

$$\hat{p}(\lambda) = \left[\begin{array}{c|c} A & B \\ \hline C & D \end{array}\right],$$

choose F and L to stabilize $A+BF$ and $A+LC$, and then take the transfer matrix of K to be

$$\hat{k}(\lambda) = \left[\begin{array}{c|c} A+BF+LC+LDF & L \\ \hline F & 0 \end{array}\right].$$

[It's $+L$ instead of $-L$ because we switched to negative feedback in Figure 5.6.] In this way we have Figure 5.7. Using one of the formulas from

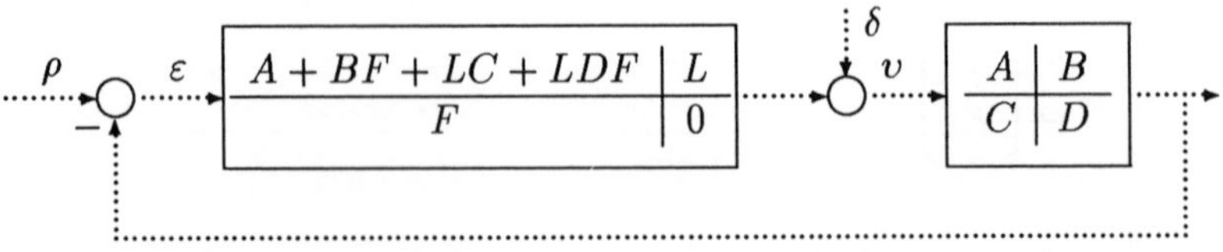

Figure 5.7: Observer-based controller.

the preceding section, we get that the transfer matrix from $\begin{bmatrix} \rho \\ \delta \end{bmatrix}$ to $\begin{bmatrix} \varepsilon \\ v \end{bmatrix}$ equals

$$\left[\begin{array}{cc|cc} A & BF & 0 & B \\ -LC & A+BF+LC & L & -LD \\ \hline -C & -DF & I & -D \\ 0 & F & 0 & I \end{array}\right].$$

The A-matrix here is precisely $\underline{A}$ in (5.2). Thus the transfer matrix from $\begin{bmatrix} \rho \\ \delta \end{bmatrix}$ to $\begin{bmatrix} \varepsilon \\ \upsilon \end{bmatrix}$ is stable, so the feedback system is input-output stable in this sense as well.

5.3 Stabilization

The preceding section showed how to stabilize a plant using an observer-based controller. That is just one way. In this section we look at stabilization more generally.

Let us start with the multivariable unity-feedback setup in Figure 5.6. We shall *assume* that P is strictly causal [$\hat{p}(0) = 0$] and K is causal [$\hat{k}(0)$ finite]. Suppose we start with minimal realizations of P and K:

$$\hat{p}(\lambda) = \left[\begin{array}{c|c} A_P & B_P \\ \hline C_P & 0 \end{array}\right], \quad \hat{k}(\lambda) = \left[\begin{array}{c|c} A_K & B_K \\ \hline C_K & D_K \end{array}\right].$$

Let ξ_P, ξ_K denote the state vectors. *Internal stability* means that

$$\xi_P(k), \xi_K(k) \to 0 \text{ as } k \to \infty \text{ for every } \xi_P(0), \xi_K(0).$$

The A-matrix of the autonomous system (i.e., $\rho = 0$, $\delta = 0$) is

$$\underline{A} = \begin{bmatrix} A_P - B_P D_K C_P & B_P C_K \\ -B_K C_P & A_K \end{bmatrix},$$

so internal stability is equivalent to the condition that $\underline{A}$ has all its eigenvalues inside $\mathbb{D}$.

In Figure 5.6 the transfer matrix from $\begin{bmatrix} \rho \\ \delta \end{bmatrix}$ to $\begin{bmatrix} \varepsilon \\ \upsilon \end{bmatrix}$ equals

$$\left[\begin{array}{cc|cc} A_P - B_P D_K C_P & B_P C_K & B_P D_K & B_P \\ -B_K C_P & A_K & B_K & 0 \\ \hline -C_P & 0 & I & 0 \\ -D_K C_P & C_K & D_K & I \end{array}\right].$$

Again, the A-matrix here is precisely $\underline{A}$, so internal stability implies that this input-output transfer matrix is stable as well.

Now we extend the discussion to the general case in Figure 5.8. Let us first observe what kind of control structure is depicted in this figure. Since G has two inputs (ω and υ) and two outputs (ζ and ψ), it can be partitioned as

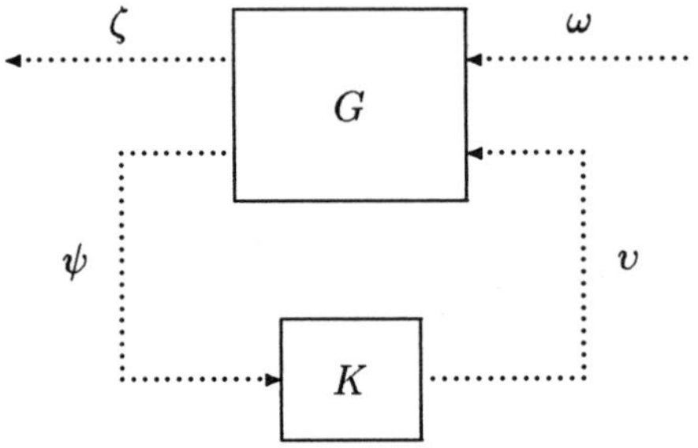

Figure 5.8: General discrete-time system.

$$G = \begin{bmatrix} G_{11} & G_{12} \\ G_{21} & G_{22} \end{bmatrix},$$

that is, the inputs and outputs are related by the equations

$$\begin{aligned} \zeta &= G_{11}\omega + G_{12}\upsilon \\ \psi &= G_{21}\omega + G_{22}\upsilon. \end{aligned}$$

This together with $\upsilon = K\psi$ allows us to derive the input-output system from ω to ζ:

$$\zeta = [G_{11} + G_{12}K(I - G_{22}K)^{-1}G_{21}]\omega$$

(provided the inverse exists). As a function of K, this input-output system is called a *linear-fractional transformation* (LFT) and is denoted by $\mathcal{F}(G, K)$.

To define precisely what internal stability means in Figure 5.8, start with stabilizable, detectable state realizations for G and K:

$$\hat{g}(\lambda) = \left[\begin{array}{c|c} A & B \\ \hline C & D \end{array}\right], \quad \hat{k}(\lambda) = \left[\begin{array}{c|c} A_K & B_K \\ \hline C_K & D_K \end{array}\right].$$

Let ξ_G, ξ_K denote the state vectors. *Internal stability* means that

$$\xi_G(k), \xi_K(k) \rightarrow 0 \text{ as } k \rightarrow \infty \text{ for every } \xi_G(0), \xi_K(0) \text{ and with } \omega = 0.$$

Partition B, C, D corresponding to the two inputs (ω and υ) and two outputs (ζ and ψ) of G:

$$\hat{g}(\lambda) = \left[\begin{array}{c|cc} A & B_1 & B_2 \\ \hline C_1 & D_{11} & D_{12} \\ C_2 & D_{21} & 0 \end{array}\right].$$

For simplicity it has been assumed that $D_{22} = 0$, that is, the block of G from υ to ψ, denoted G_{22}, is strictly causal. It is routine to derive that the A-matrix of the autonomous system (i.e., $\omega = 0$) is

$$\underline{A} = \begin{bmatrix} A + B_2 D_K C_2 & B_2 C_K \\ B_K C_2 & A_K \end{bmatrix}. \tag{5.3}$$

So internal stability is equivalent to the condition that $\underline{A}$ has all its eigenvalues inside $\mathbb{D}$.

Let us note several points:

1. For $\underline{A}$ to have all its eigenvalues inside $\mathbb{D}$, it is necessary that (A, B_2) be stabilizable and that (C_2, A) be detectable.

2. If (A, B_2) is stabilizable and (C_2, A) is detectable, then there does indeed exist an internally stabilizing K. Matrix $\underline{A}$ in (5.3) has exactly the same form as $\underline{A}$ in (5.2), so K could be an observer-based controller. In particular, it suffices that K internally stabilize G_{22} as in Figure 5.9.

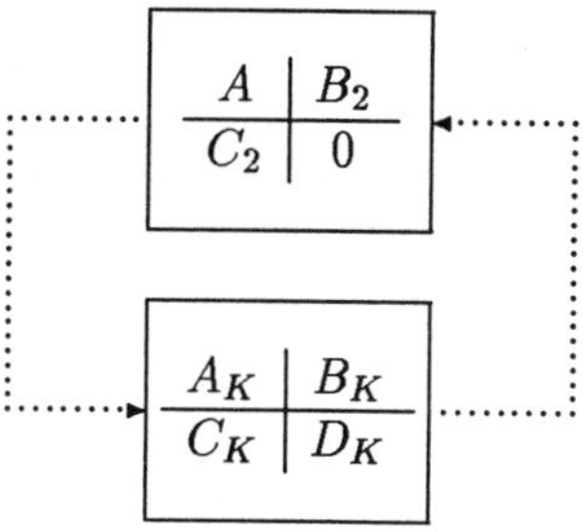

Figure 5.9: Controller with G_{22}.

3. When K is connected in Figure 5.8, the transfer matrix from ω to ζ equals

$$\left[\begin{array}{cc|c} A + B_2 D_K C_2 & B_2 C_K & B_1 + B_2 D_K D_{21} \\ B_K C_2 & A_K & B_K D_{21} \\ \hline C_1 + D_{12} D_K C_2 & D_{12} C_K & D_{11} + D_{12} D_K D_{21} \end{array}\right]$$

   The A-matrix here is precisely $\underline{A}$, so this transfer matrix is stable.

Let us summarize:

Theorem 5.3.1 *For the setup in Figure 5.8, start with a minimal realization of G. An internally stabilizing controller exists iff (A, B_2) is stabilizable and (C_2, A) is detectable. When these conditions hold, it suffices to choose K to internally stabilize G_{22} in Figure 5.9. An internally stabilizing controller achieves stability of the transfer matrix from ω to ζ.*

5.4 All Stabilizing Controllers

We continue the study of internal stability for the general setup of Figure 5.8 with

$$\hat{g}(\lambda) = \left[\begin{array}{c|cc} A & B_1 & B_2 \\ \hline C_1 & D_{11} & D_{12} \\ C_2 & D_{21} & 0 \end{array}\right]. \tag{5.4}$$

We wish to parametrize all Ks that achieve internal stability for G. As shown in the preceding section, this reduces to internally stabilizing G_{22} under the assumption that (A, B_2) is stabilizable and (C_2, A) is detectable.

Parametrization

Start with some matrices F and L such that $A+B_2F$ and $A+LC_2$ are stable. All stabilizing controllers for G can be parametrized by a linear fractional transformation of some arbitrary but stable system, as follows.

Theorem 5.4.1 *The set of all (FDLTI and causal) Ks achieving internal stability in Figure 5.8 is parametrized by the formula*

$$\hat{k} = \mathcal{F}(\hat{\jmath}, \hat{q}), \quad \hat{q} \in \mathcal{RH}_\infty(\mathbb{D}), \tag{5.5}$$

where

$$\hat{\jmath}(\lambda) = \left[\begin{array}{c|cc} A + B_2F + LC_2 & -L & -B_2 \\ \hline F & 0 & -I \\ -C_2 & I & 0 \end{array}\right]. \tag{5.6}$$

Note that the controller K in (5.5) is an LFT of Q, which is FDLTI, causal, and stable. Such a controller is represented as the input-output transfer matrix of the block diagram

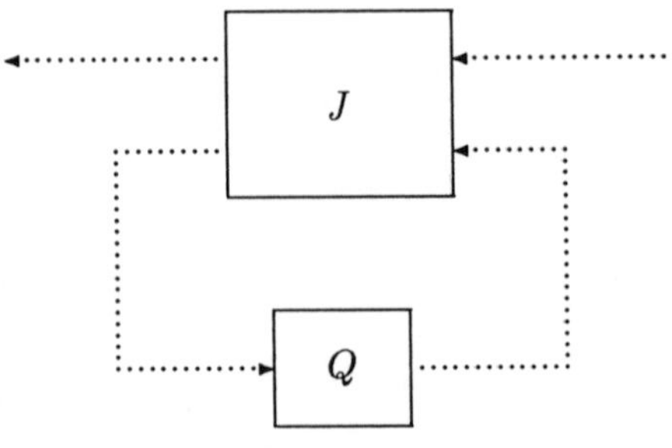

Here J is partitioned as usual. As a special case, if $Q = 0$, the controller K reduces to J_{11} with the transfer matrix

$$\left[\begin{array}{c|c} A + B_2F + LC_2 & -L \\ \hline F & 0 \end{array}\right];$$

this is simply an observer-based controller for G_{22} (see Figure 5.5).

In this theorem, we have taken $D_{22} = 0$ for simplicity; the result can be generalized to the case $D_{22} \neq 0$—see Exercise 5.9 for the details.

Proof of Theorem 5.4.1 Rewrite (5.5) as

$$K = J_{11} + J_{12}Q(I - J_{22}Q)^{-1}J_{21}. \tag{5.7}$$

First, we claim that the mapping $Q \mapsto K$ is a bijection (one-to-one and onto) from the set of FDLTI, causal systems to itself. Since J is FDLTI and causal with J_{22} strictly causal, it follows easily that K is FDLTI and causal if Q is. Conversely, assume K in (5.7) is FDLTI and causal. Since J_{12}^{-1} and J_{21}^{-1} exist as FDLTI and causal systems, solve equation (5.7) for Q:

$$Q = J_{12}^{-1}(K - J_{11})J_{21}^{-1}\left[I + J_{12}^{-1}(K - J_{11})J_{21}^{-1}J_{22}\right]^{-1}.$$

Hence Q is also FDLTI and causal. Now it suffices to show that K in (5.7) internally stabilizes G_{22} iff Q is stable.

Assume first that K in (5.7) internally stabilizes G_{22}. Let ξ_G denote the state of G corresponding to the realization in (5.4); let ξ_J denote the state of J for the realization in (5.6); and let ξ_Q denote the state of Q for any minimal realization. It can be verified that the induced realization for $\mathcal{F}(J,Q)$ is stabilzable and detectable, so $\xi_K := \begin{bmatrix} \xi_J \\ \xi_Q \end{bmatrix}$ is a satisfactory state for K. Connecting the parametrized controller to G_{22}, we have Figure 5.10. In this figure, by internal stability

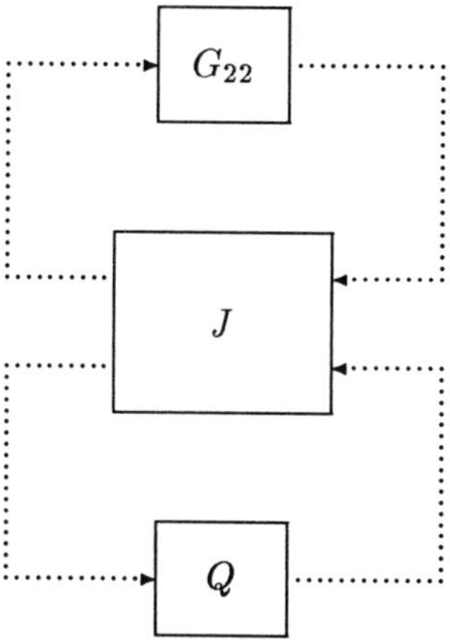

Figure 5.10: The LFT $\mathcal{F}(J,Q)$ controlling G_{22}.

$$\xi_G(k), \xi_J(k), \xi_Q(k) \to 0 \text{ as } k \to \infty \text{ for all initial conditions.}$$

The upper two blocks in Figure 5.10 can be combined as a new system, P, to give Figure 5.11. In this figure,

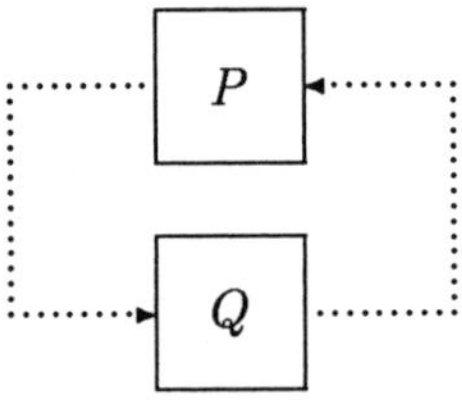

Figure 5.11: Q controlling P.

$$\xi_P(k), \xi_Q(k) \to 0 \text{ as } k \to \infty \text{ for all initial conditions.}$$

But in fact $P = 0$: Based on the realizations of G_{22} and J in (5.4) and (5.6), we have

$$\begin{aligned}
\hat{p}(\lambda) &= \left[\begin{array}{cc|c} A & B_2F & -B_2 \\ -LC_2 & A+B_2F+LC_2 & -B_2 \\ \hline C_2 & -C_2 & 0 \end{array}\right] \\
&= \left[\begin{array}{cc|c} A+LC_2 & 0 & 0 \\ -LC_2 & A+B_2F & -B_2 \\ \hline C_2 & 0 & 0 \end{array}\right] \\
&= 0;
\end{aligned}$$

the second realization is obtained using the similarity transformation

$$\begin{bmatrix} I & I \\ I & 0 \end{bmatrix}.$$

Thus, Q stands alone and

$$\xi_Q(k) \to 0 \text{ as } k \to \infty \text{ for all initial conditions.}$$

Hence Q is stable.

Proof of the converse is similar. ■

If a realization of Q is known, a realization of the controller K can be obtained by application of the formula in Section 5.1.

Closed-Loop Transfer Matrix

The preceding theorem gives every stabilizing K as an LFT of a parameter $\hat{q}$ in $\mathcal{RH}_\infty(\mathbb{D})$. Now we want to find the transfer matrix from ω to ζ in Figure 5.8 in terms of this parameter. It has the form

$$\hat{t}_1 + \hat{t}_2\hat{q}\hat{t}_3,$$

where $\hat{t}_i$ are fixed transfer matrices in $\mathcal{RH}_\infty(\mathbb{D})$ depending only on $\hat{g}$.

To get these $\hat{t}_i$s explicitly, we substitute the controller parametrization in the preceding theorem into Figure 5.8 to get that the controlled system is modeled as in Figure 5.12. If Q is memoryless, that is, $\hat{q}(\lambda) = Q_0$, a constant

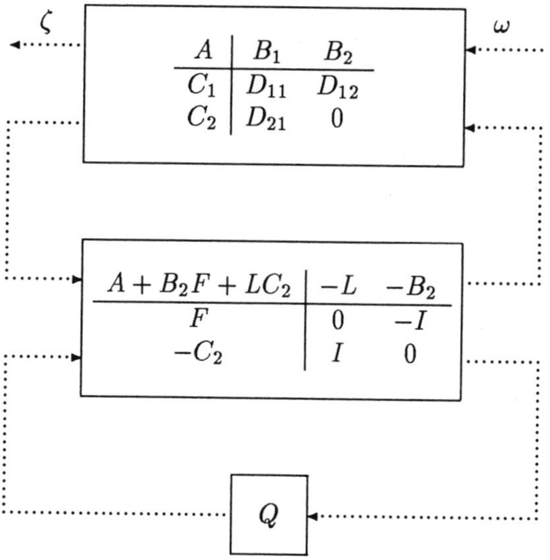

Figure 5.12: Generalized plant with parametrized controller.

matrix, then

$$\hat{k}(\lambda) = \left[\begin{array}{c|c} A + B_2F + LC_2 + B_2Q_0C_2 & -B_2Q_0 - L \\ \hline F + Q_0C_2 & -Q_0 \end{array}\right]. \tag{5.8}$$

In general, however, Figure 5.12 can be converted to Figure 5.13, where T is stable (since J_{11} internally stabilizes G_{22}) and its transfer matrix is given by

$$\hat{t}(\lambda) = \left[\begin{array}{cc|cc} A + B_2F & B_2F & B_1 & -B_2 \\ 0 & A + LC_2 & -B_1 - LD_{21} & 0 \\ \hline C_1 + D_{12}F & D_{12}F & D_{11} & -D_{12} \\ 0 & -C_2 & D_{21} & 0 \end{array}\right].$$

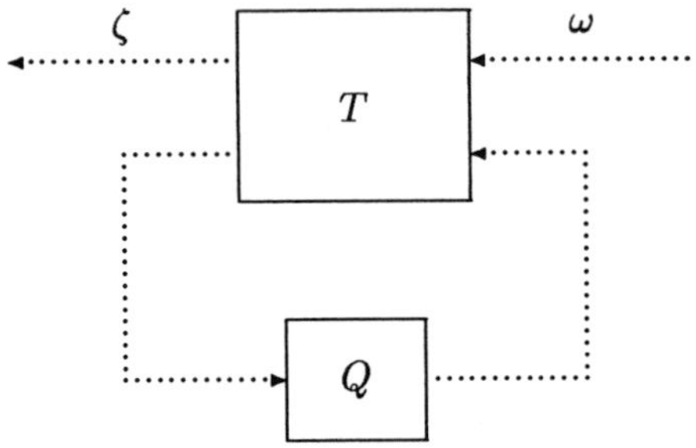

Figure 5.13: Closed-loop system with controller parameter Q.

Thus, the input-output system from ω to ζ is an LFT of Q. We read off that

$$\begin{aligned}
\hat{t}_{11}(\lambda) &= \left[\begin{array}{cc|c} A+B_2F & B_2F & B_1 \\ 0 & A+LC_2 & -B_1-LD_{21} \\ \hline C_1+D_{12}F & D_{12}F & D_{11} \end{array}\right] \\
\hat{t}_{12}(\lambda) &= \left[\begin{array}{cc|c} A+B_2F & B_2F & -B_2 \\ 0 & A+LC_2 & 0 \\ \hline C_1+D_{12}F & D_{12}F & -D_{12} \end{array}\right] \\
&= \left[\begin{array}{c|c} A+B_2F & -B_2 \\ \hline C_1+D_{12}F & -D_{12} \end{array}\right] \\
\hat{t}_{21}(\lambda) &= \left[\begin{array}{cc|c} A+B_2F & B_2F & B_1 \\ 0 & A+LC_2 & -B_1-LD_{21} \\ \hline 0 & -C_2 & D_{21} \end{array}\right] \\
&= \left[\begin{array}{c|c} A+LC_2 & B_1+LD_{21} \\ \hline C_2 & D_{21} \end{array}\right] \\
\hat{t}_{22}(\lambda) &= 0.
\end{aligned}$$

Defining

$$\hat{t}_1 = \hat{t}_{11}, \quad \hat{t}_2 = \hat{t}_{12}, \quad \hat{t}_3 = \hat{t}_{21},$$

we get that the transfer matrix from ω to ζ equals $\hat{t}_1 + \hat{t}_2\hat{q}\hat{t}_3$ with $\hat{t}_i$ all in $\mathcal{RH}_\infty(\mathbb{D})$.

Special Case: Stable Plants

As a special case of the preceding results, suppose G is already stable, that is, $\hat{g} \in \mathcal{RH}_\infty(\mathbb{D})$. Then in Theorem 5.4.1 we may take $F = 0$ and $L = 0$. This leads to the following simplifications:

$$\hat{\jmath} = \begin{bmatrix} 0 & -I \\ I & \hat{g}_{22} \end{bmatrix};$$

the controller parametrization is

$$\begin{aligned}\hat{k} &= -\hat{q}(I - \hat{g}_{22}\hat{q})^{-1} \\ &= -(I - \hat{q}\hat{g}_{22})^{-1}\hat{q}, \quad \hat{q} \in \mathcal{RH}_\infty(\mathbb{D});\end{aligned}$$

and the closed-loop transfer matrix from ω to ζ is

$$\hat{g}_{11} - \hat{g}_{12}\hat{q}\hat{g}_{21}.$$

5.5 Step Tracking

One learns in an undergraduate course how to design a controller so that the output of the plant will track a step command signal. One learns that the basic mechanism to do this is integral feedback control.

In general, a system is step-tracking if certain designated signals can track step command inputs. In the standard discrete-time system

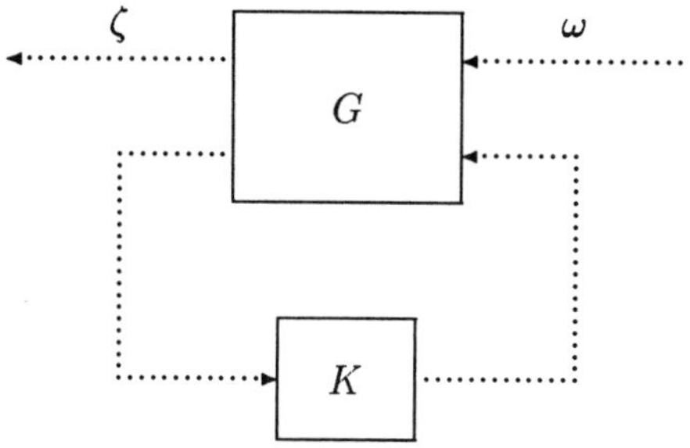

ζ is interpreted as an error signal, so the definition of *step-tracking* is that $\zeta(k) \to 0$ as $k \to \infty$ for every step input ω. A multivariable step input is a signal of the form $\omega(k) = \omega_0 1_d(k)$, where ω_0 is a constant vector and $1_d(k)$ is the 1-dimensional unit step function. In the definition of step-tracking we will also require internal stability.

Let us begin with an example.

Example 5.5.1 Consider the usual unity-feedback system in Figure 5.14. Assume K and P are SISO and P is strictly causal. The problem is to design a causal, internally stabilizing K so that the system is step-tracking, that is, $\varepsilon(k) \to 0$ as $k \to \infty$ when ρ is the unit step. Such a controller exists iff $\hat{p}$ does not have a zero at $\lambda = 1$, where $\hat{\rho}$ has a single pole.

If $\hat{p}$ already has a pole at $\lambda = 1$, then we just have to design K to internally stabilize P—step-tracking would be automatic. We could take K to be an observer-based controller. If $\hat{p}$ does not have such a pole, then $\hat{k}$ must introduce it, that is, K must have the form $K = K_1K_2$, where $\hat{k}_1(\lambda) = 1/(1-\lambda)$. Now imagine Figure 5.14 with P replaced by $P_{tmp} := PK_1$ and K replaced by K_2. It just remains to design K_2 to internally stabilize P_{tmp}; again, we could take K_2 to be an observer-based controller.

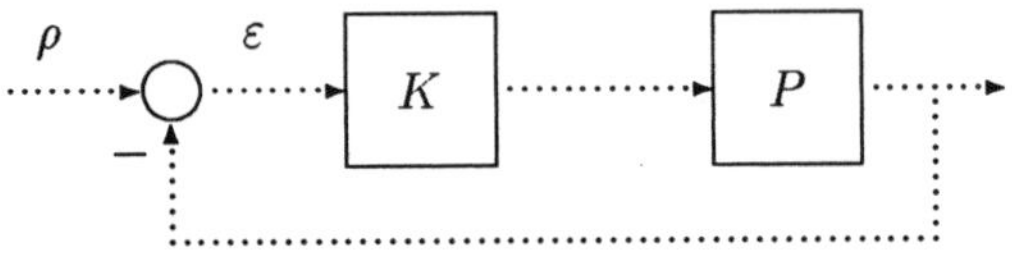

Figure 5.14: Unity-feedback system.

Another, more methodical approach is first to parametrize all stabilizing controllers and then to choose the parameter to get step-tracking. For simplicity, assume further that P is already stable. The parametrization now takes a simple form:

$$\hat{k} = \hat{q}(1 - \hat{p}\hat{q})^{-1}, \quad \hat{q} \in \mathcal{RH}_\infty(\mathbb{D}).$$

Then we have

$$\hat{\varepsilon} = (1 - \hat{p}\hat{q})\hat{\rho}.$$

To achieve step-tracking, it is sufficient (and necessary) to choose $\hat{q}$ in $\mathcal{RH}_\infty(\mathbb{D})$ to satisfy the equation

$$1 - \hat{p}(1)\hat{q}(1) = 0.$$

If $\hat{p}(1) = 0$ ($\hat{p}$ has a zero at $\lambda = 1$), no stable $\hat{q}$ exists to solve this equation; if $\hat{p}(1) \neq 0$, then $\hat{q}$ must be chosen to satisfy the interpolation condition

$$\hat{q}(1) = 1/\hat{p}(1).$$

We could take, for example, $\hat{q}(\lambda) \equiv 1/\hat{p}(1)$.

The second method just discussed works in general. For the standard discrete-time setup, parametrize K and then write the transfer matrix from ω to ζ in terms of the parameter Q:

$$\hat{\zeta} = (\hat{t}_1 + \hat{t}_2\hat{q}\hat{t}_3)\hat{\omega}.$$

Then step-tracking is achieved iff

$$\hat{t}_1(1) + \hat{t}_2(1)\hat{q}(1)\hat{t}_3(1) = 0. \tag{5.9}$$

This is a linear matrix equation in $\hat{q}(1)$.

Example 5.5.2 Figure 5.15 is a block diagram of a very simple master/slave system. There are two masses, a master (left-hand) and a slave (right-hand).

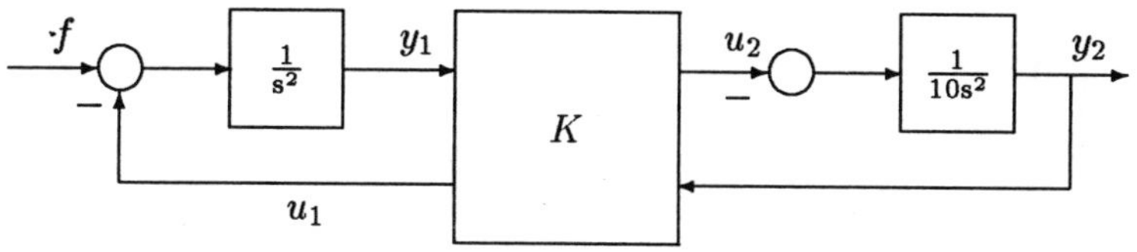

Figure 5.15: Master/slave system.

A human applies a force f to the master. The controller K applies an opposing force u_1 to the master and also a force u_2 to the slave. The controller inputs the two positions y_1 and y_2.

It is desired that the system function as a telerobot: When a human applies a force, the master should move appropriately in the direction of the applied force and the slave should follow the master. With this in mind, the performance goals are stated to be

1. internal stability,
2. for f a unit step, $y_1(\infty) = 1$ and $y_2(\infty) = y_1(\infty)$.

The system can be put in the standard form

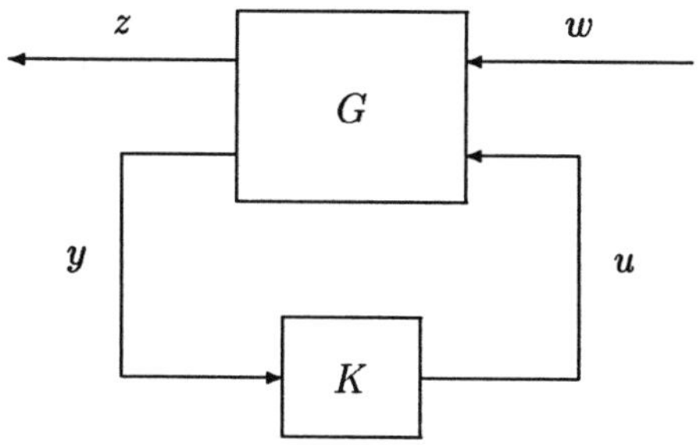

with the signals defined to be

$$z = \begin{bmatrix} y_1 - f \\ y_1 - y_2 \end{bmatrix}, \quad w = f, \quad y = \begin{bmatrix} y_1 \\ y_2 \end{bmatrix}, \quad u = \begin{bmatrix} u_1 \\ u_2 \end{bmatrix}.$$

Defining the state variables of G to be

$$x_1 = y_1, \quad x_2 = \dot{y}_1, \quad x_3 = y_2, \quad x_4 = \dot{y}_2,$$

we get that

$$\hat{g}(\lambda) = \left[\begin{array}{c|cc} A & B_1 & B_2 \\ \hline C_1 & D_{11} & 0 \\ C_2 & 0 & 0 \end{array}\right]$$

where

$$A = \begin{bmatrix} 0 & 1 & 0 & 0 \\ 0 & 0 & 0 & 0 \\ 0 & 0 & 0 & 1 \\ 0 & 0 & 0 & 0 \end{bmatrix}, \quad B_1 = \begin{bmatrix} 0 \\ 1 \\ 0 \\ 0 \end{bmatrix}, \quad B_2 = \begin{bmatrix} 0 & 0 \\ -1 & 0 \\ 0 & 0 \\ 0 & -0.1 \end{bmatrix}$$

$$C_1 = \begin{bmatrix} 1 & 0 & 0 & 0 \\ 1 & 0 & -1 & 0 \end{bmatrix}, \quad C_2 = \begin{bmatrix} 1 & 0 & 0 & 0 \\ 0 & 0 & 1 & 0 \end{bmatrix}, \quad D_{11} = \begin{bmatrix} -1 \\ 0 \end{bmatrix}.$$

Let us (arbitrarily) choose $h = 0.1$ and contemplate using a digital controller:

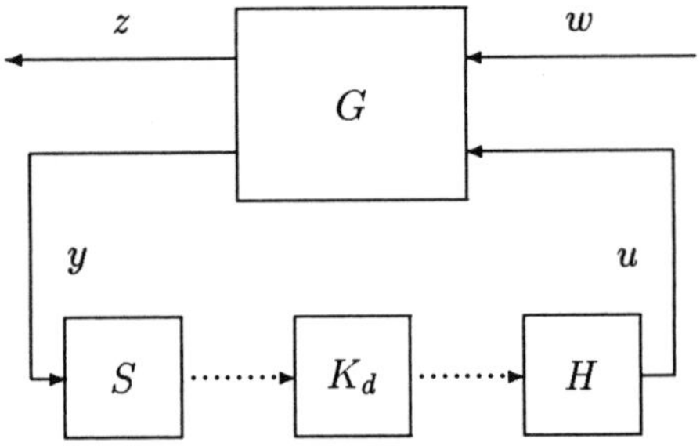

Discretizing G, we end up with the discrete-time setup

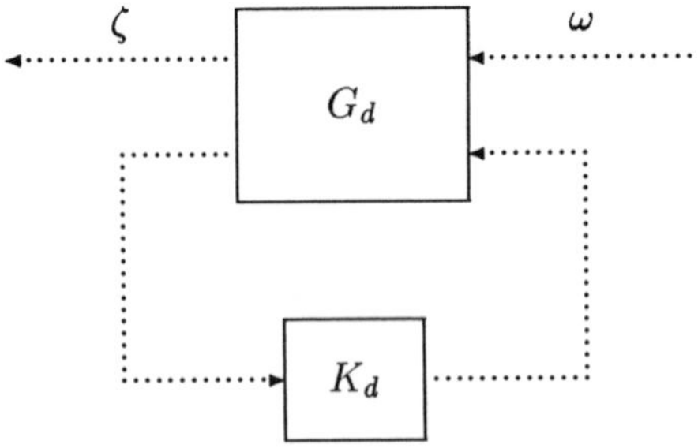

The state matrices for G_d are

$$\hat{g}_d(\lambda) = \left[\begin{array}{c|cc} A_d & B_{1d} & B_{2d} \\ \hline C_1 & D_{11} & 0 \\ C_2 & 0 & 0 \end{array}\right]$$

We are now set up to parametrize K_d as in the preceding section and then to choose a suitable parameter. The steps are as follows:

1. Choose F and L to stabilize $A_d + B_{2d}F$ and $A_d + LC_2$.

2. Get a realization of $\hat{t}_1$:

$$\hat{t}_1(\lambda) = \left[\begin{array}{c|c} A_{11} & B_{11} \\ \hline C_{11} & D_{11} \end{array}\right] := \left[\begin{array}{cc|c} A_d + B_{2d}F & B_{2d}F & B_{1d} \\ 0 & A_d + LC_2 & -B_{1d} \\ \hline C_1 & 0 & D_{11} \end{array}\right].$$

3. Get the DC gains of $\hat{t}_1$, $\hat{t}_2$, and $\hat{t}_3$:

$$\begin{aligned} \hat{t}_1(1) &= D_{11} + C_{11}(I - A_{11})^{-1}B_{11} \\ \hat{t}_2(1) &= -C_1(I - A_d - B_{2d}F)^{-1}B_{2d} \\ \hat{t}_3(1) &= C_2(I - A_d - LC_2)^{-1}B_{1d}. \end{aligned}$$

4. Solve

$$\hat{t}_1(1) + \hat{t}_2(1)\hat{q}(1)\hat{t}_3(1) = 0$$

for $\hat{q}(1)$ and set $Q_0 = \hat{q}(1)$.

5. Get the controller from (5.8):

$$\hat{k}_d(\lambda) = \left[\begin{array}{c|c} A_d + B_{2d}F + LC_2 + B_{2d}Q_0C_2 & -B_{2d}Q_0 - L \\ \hline F + Q_0C_2 & -Q_0 \end{array}\right].$$

Figure 5.16 shows step-response plots of the discretized signals for F and L obtained by the MATLAB commands

$$\begin{aligned} F &= -dlqr\ (A_d, B_{2d}, I, I) \\ L &= -dlqr\ (A_d', C_2', I, I)'. \end{aligned}$$

The master follows the commanded step and the slave follows the master, as required. Of course, the responses are not optimal in any sense.

We have implicitly assumed that if we achieve step-tracking of the discretized system, then it will also be achieved for the continuous-time SD system. This assumption will be justified in Section 11.3.

Exercises

5.1 Derive the other four formulas in Section 1.

5.2 Consider the discrete-time linear system with input v, output ψ, modeled by the difference equation

$$\psi(k) - \psi(k-1) + 3\psi(k-2) = 2v(k) - v(k-2).$$

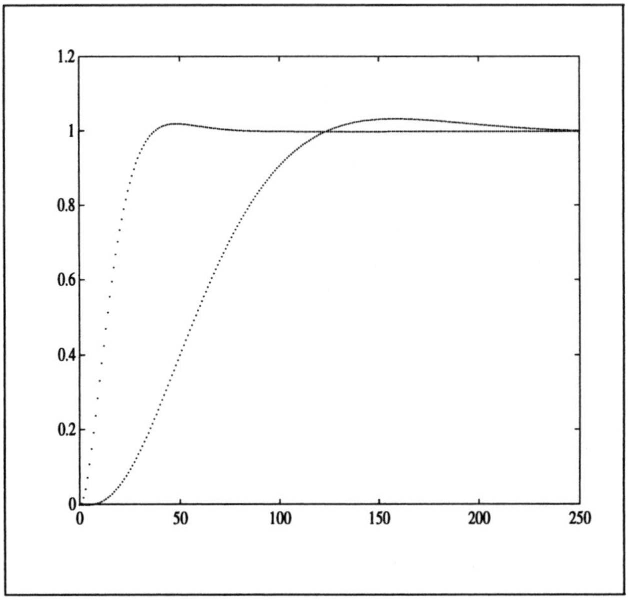

Figure 5.16: Step-response of master/slave example: $y_1(kh)$ (leftmost plot) and $y_2(kh)$ (rightmost plot) versus k.

Find a state model

$$\begin{aligned} \dot{\xi} &= A\xi + Bv \\ \psi &= C\xi + Dv. \end{aligned}$$

Choose F to put all the eigenvalues of $A + BF$ at the origin. For the control law

$$v = F\xi + \rho,$$

plot the response of ψ to a unit step in ρ.

5.3 Consider the discrete-time system

$$\begin{aligned} \dot{\xi} &= A\xi + Bv \\ \psi &= C\xi + Dv, \end{aligned}$$

with

$$A = \begin{bmatrix} -1 & -1.5 \\ 3 & 3.5 \end{bmatrix}, \quad B = \begin{bmatrix} -1 \\ 2 \end{bmatrix}, \quad C = \begin{bmatrix} 1 & 1 \end{bmatrix}, \quad D = 2.$$

Verify that an observer exists. Design an observer, writing it in the form

$$\dot{\xi}_o = A_0\xi_0 + B_{o1}v + B_{o2}\psi.$$

Can you achieve convergence of the error, that is, $\xi_o(k) - \xi(k) \to 0$, in finite time?

5.4 Consider the system

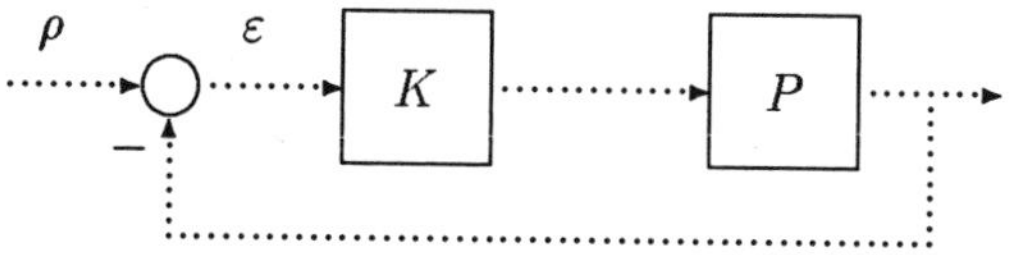

with

$$\hat{p}(\lambda) = \frac{\lambda}{\lambda - 2}.$$

Design a causal, internally stabilizing K so that $\varepsilon(k) \to 0$ as $k \to \infty$ when ρ is the unit ramp. First, use the two-step procedure: Parametrize all internally stabilizing controllers; select the parameter to achieve the desired tracking. Second, use an observer-based controller.

5.5 Same setup as in previous problem but with

$$\hat{p}(\lambda) = \frac{\lambda(\lambda + 1)}{2\lambda - 1}.$$

For ρ the unit step, design a causal, internally stabilizing K so that $\varepsilon(k) \to 0$ in finite time [i.e., $\varepsilon(k) = 0$ for all k greater than some finite value]. Such a response is called *deadbeat.*

5.6 What are necessary and sufficient conditions for the existence of a matrix $\hat{q}$ in $\mathcal{RH}_\infty(\mathbb{D})$ solving (5.9)?

5.7 Consider the following setup:

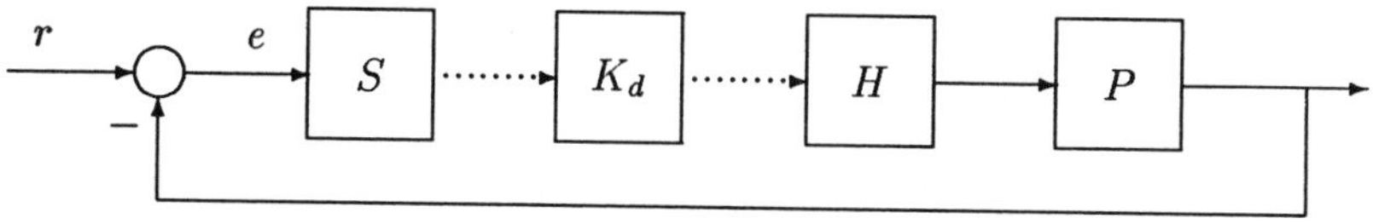

The plant and controller are SISO. The plant transfer function is

$$\hat{p}(s) = e^{-0.1s} \frac{1}{s - 1}.$$

Design a causal, internally stabilizing K_d to achieve

internal stability,

$e(t) \to 0$ as $t \to \infty$ when $r(t)$ is a step

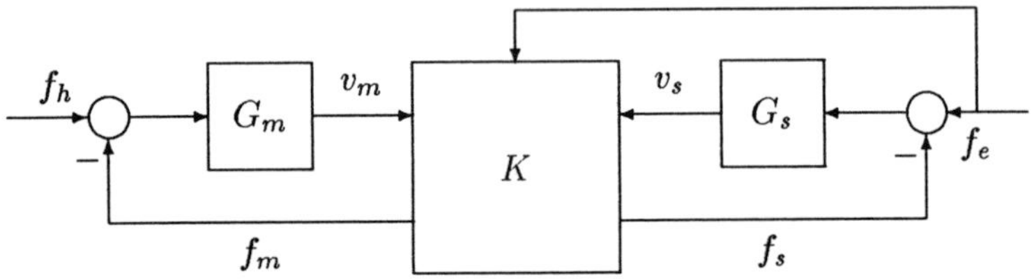

Figure 5.17: Bilateral hybrid telerobot.

using this procedure: Select h; compute $P_d = SPH$; in discrete time, design an internally stabilizing, step-tracking K_d for P_d.

5.8 Figure 5.17 shows the bilateral hybrid telerobot of Example 2.2.1. Take G_m and G_s to be SISO with transfer functions

$$\hat{g}_m(s) = \frac{1}{s}, \quad \hat{g}_s(s) = \frac{1}{10s}.$$

Put the system in the standard form by defining

$$z = \begin{bmatrix} v_s - v_m \\ f_m - f_e \end{bmatrix}, \quad w = \begin{bmatrix} f_h \\ f_e \end{bmatrix}.$$

Following Example 5.5.2, try to design a sampled-data controller so that the discretized system is internally stable and step-tracking. Conclude that step tracking is not achievable. Explain physically.

5.9 This exercise is to extend Theorem 5.4.1 to the case $D_{22} \neq 0$. In this case, for well-posedness of the feedback system in Figure 5.8 (i.e., for the closed-loop system to exist and to be causal), we require that $I - D_{22}\hat{k}(0)$ be invertible. Under the same assumptions as in Theorem 5.4.1, show that the set of all stabilizing Ks is parametrized by

$$\hat{k} = \mathcal{F}(\hat{\jmath}, \hat{q}), \quad \hat{q} \in \mathcal{RH}_\infty(\mathbb{D}), \quad I - D_{22}\hat{q}(0) \text{ invertible},$$

where

$$\hat{\jmath}(\lambda) = \left[\begin{array}{c|cc} A + B_2F + LC_2 + LD_{22}F & -L & -(B_2 + LD_{22}) \\ \hline F & 0 & -I \\ -(C_2 + D_{22}F) & I & D_{22} \end{array}\right].$$

Find the closed-loop transfer matrix $\omega \mapsto \zeta$ in terms of this parameter $\hat{q}$.

5.10 Consider the discrete-time system

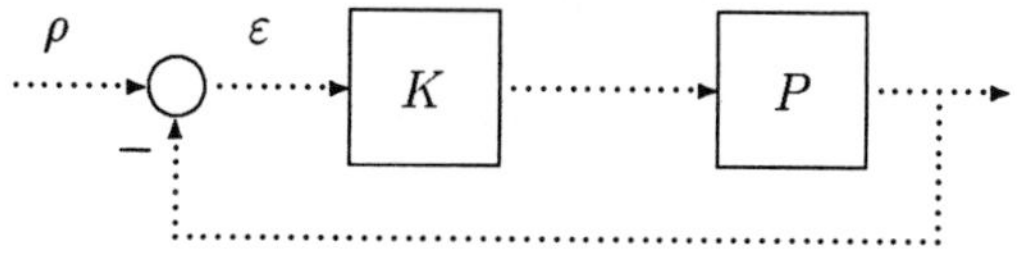

where the plant has the transfer function

$$\hat{p}(\lambda) = \frac{\lambda - 0.5}{\lambda + 2}$$

and the reference signal is

$$\rho = \left\{1, \frac{1}{a}, \frac{1}{a^2}, \frac{1}{a^3}, \cdots\right\}, \quad |a| \leq 1, \quad a \neq 0.$$

We are interested in the following control problem: Design an LTI and causal controller K to achieve internal stability and asymptotic tracking, namely, $\varepsilon(k) \to 0$ as $k \to \infty$.

1. For what value(s) of a, does this control problem have no solution?

2. Suppose a is chosen such that the problem has solutions; find a solution.

5.11 Consider the following sampled-data control system:

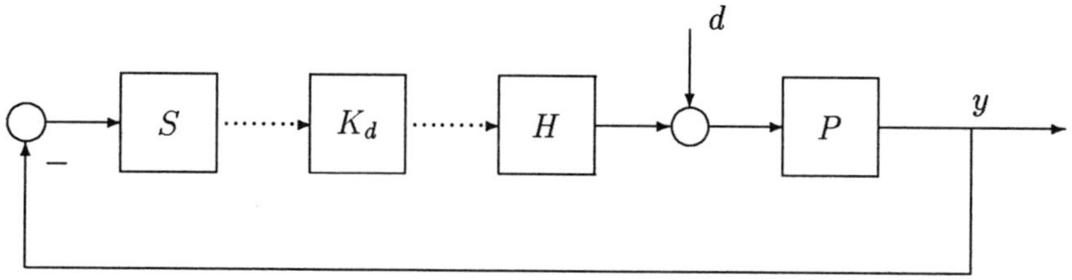

Let

$$\hat{p}(s) = \frac{1}{10s + 1}, \qquad h = 1.$$

1. Suppose $d(t) = 1(t)$, the unit step. Design a controller K_d so that the feedback system is internally stable and $y(t)$ converges to zero at the sampling instants, that is, $y(kh) \to 0$ as $k \to \infty$. Use the method of controller parametrization. (For internal stability, internally stabilize the discretized feedback system.)

2. Repeat but with $d(t) = \sin(10t)1(t)$.

Notes and References

The idea of parametrizing all internally stabilizing controllers is due to Youla, Jabr, and Bongiorno [155] and Kucera [95] via coprime factorizations, for which state-space formulas are given in [89] and [114]. Proofs of controller parametrization via coprime factorization for the continuous-time case can also be found in [51]; the discrete-time case is identical modulo the obvious changes. The parametrization in Theorem 5.4.1 is from [45]; the simple proof here is adapted from [21].

Chapter 6

Discrete-Time $\mathcal{H}_2$-Optimal Control

This chapter gives a state-space approach to a discrete-time $\mathcal{H}_2$-optimal control problem. This problem concerns the standard setup:

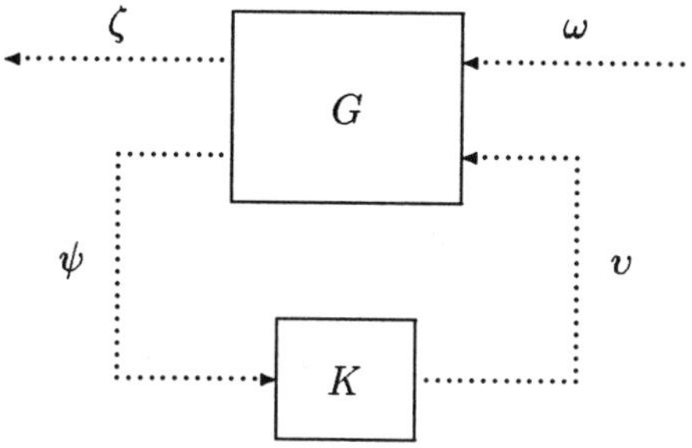

The input ω is standard white noise—zero mean, unit covariance matrix. The problem is to design a K that stabilizes G and minimizes the root-mean-square value of ζ; equivalently, the problem is to minimize the $\mathcal{H}_2(\mathbb{D})$-norm of the transfer matrix from ω to ζ.

6.1 The LQR Problem

Before attacking the general $\mathcal{H}_2$-optimal control problem, we shall devote some time to an easier problem, the *Linear Quadratic Regulator* (LQR) *Problem*, which can be stated as follows. We consider the usual state-space model in discrete time:

$$\begin{aligned} \xi(0) &= \xi_0 \\ \dot{\xi}(k) &= A\xi(k) + Bv(k), \quad k \geq 0. \end{aligned}$$

The initial state, ξ_0, is fixed. The control sequence $v(k)$ is to be chosen to minimize the weighted sum

$$J = \sum_{k=0}^{\infty} \left[\xi(k)'Q\xi(k) + v(k)'Rv(k)\right].$$

The matrices Q and R, called *weighting matrices*, are symmetric, with Q positive semidefinite and R positive definite.

As we will see, it turns out under some mild technical assumptions that the sequence that minimizes J has the form $v(k) = F\xi(k)$, that is, the optimal control law is state feedback. Furthermore, $A + BF$ is stable. So the solution of the LQR problem provides an alternative way to stabilize an unstable plant; in fact, this way is more sound numerically than eigenvalue assignment.

The matrix F is uniquely determined by the data (A, B, Q, R). The MATLAB command is

$$F = -dlqr(A, B, Q, R)$$

Typically, Q and R are used as design parameters: One proceeds as follows:

1. Choose any Q, R.

2. Compute F by solving the LQR problem.

3. Simulate the controlled system.

4. To improve the response, modify Q, R and return to step 2.

It is interesting to note that the LQR problem is a special 2-norm optimization problem. To see this, first note that the initial condition can be absorbed into the state equation by bringing in the unit impulse function, $\delta_d(k)$, as follows:

$$\begin{aligned} \xi(k) &= 0, \quad k < 0 \\ \dot{\xi} &= A\xi + \dot{\delta}_d\xi_0 + Bv. \end{aligned}$$

Second, note that

$$\xi'Q\xi + v'Rv = \begin{bmatrix} \xi \\ v \end{bmatrix}' \begin{bmatrix} Q & 0 \\ 0 & R \end{bmatrix} \begin{bmatrix} \xi \\ v \end{bmatrix}.$$

Being positive semidefinite, Q and R have positive semidefinite square roots. Defining

$$C = \begin{bmatrix} Q^{1/2} \\ 0 \end{bmatrix}, \quad D = \begin{bmatrix} 0 \\ R^{1/2} \end{bmatrix},$$

we get

$$\begin{bmatrix} Q & 0 \\ 0 & R \end{bmatrix} = \begin{bmatrix} C' \\ D' \end{bmatrix} \begin{bmatrix} C & D \end{bmatrix}.$$

The further definition

$$\zeta = C\xi + Dv,$$

gives

$$\xi' Q \xi + v' R v = \zeta' \zeta$$

and therefore

$$J = \sum_{0}^{\infty} \zeta(k)' \zeta(k) = \|\zeta\|_2^2.$$

In this way, the LQR problem can be restated as follows: Given the system depicted as

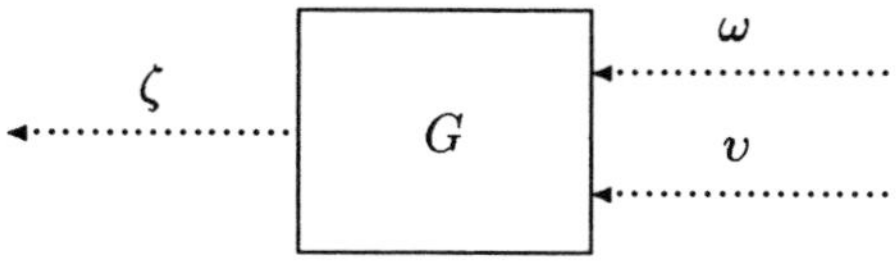

where the transfer matrix for G is

$$\hat{g}(\lambda) = \left[\begin{array}{c|cc} A & \xi_0 & B \\ \hline C & 0 & D \end{array}\right]$$

and where the disturbance input is $\omega = \dot{\delta}_d$, find the control sequence v to minimize $\|\zeta\|_2$.

We will pursue this latter direction in Section 6.4.

Heuristic Derivation of the Optimal Feedback

A rigorous derivation of the optimal feedback F will be given later in this chapter. In this section we will give a heuristic derivation (that is, with lots of hand-waving) to provide some motivation for the formulas.

First, let us modify the cost function slightly by halving it:

$$J = \frac{1}{2} \sum_{k=0}^{\infty} \left[\xi(k)' Q \xi(k) + v(k)' R v(k) \right].$$

Obviously, this doesn't change what the optimal control sequence is. We can view the problem as that of minimizing J subject to the equality constraint

$$-\dot{\xi} + A\xi + Bv = 0.$$

This suggests the method of Lagrange multipliers. So introduce a vector sequence $\lambda(k+1)$ (the time value is $k+1$ instead of k to get a nicer equation below) and then define the Lagrangian L to be

$$\sum_{k=0}^{\infty}\left\{\frac{1}{2}\xi(k)'Q\xi(k)+\frac{1}{2}v(k)'Rv(k)+\right.$$
$$\left.\lambda(k+1)'\left[-\xi(k+1)+A\xi(k)+Bv(k)\right]\right\}.$$

Regarding L as a function in turn of $\lambda(k+1)$, $v(k)$, and $\xi(k)$, we set the three partial derivatives to zero as necessary conditions for optimality:

$$\begin{aligned}\frac{\partial L}{\partial\lambda(k+1)} &= 0\\ \frac{\partial L}{\partial v(k)} &= 0\\ \frac{\partial L}{\partial\xi(k)} &= 0.\end{aligned}$$

These evaluate to

$$\begin{aligned}-\dot{\xi}+A\xi+Bv &= 0\\ v'R+\dot{\lambda}'B &= 0\\ \xi'Q-\lambda'+\dot{\lambda}'A &= 0.\end{aligned}$$

Substituting the second equation, that is,

$$v=-R^{-1}B'\dot{\lambda}, \tag{6.1}$$

into the first and cleaning up, we get

$$\begin{bmatrix} I & BR^{-1}B' \\ 0 & A' \end{bmatrix}\begin{bmatrix} \dot{\xi} \\ \dot{\lambda} \end{bmatrix}=\begin{bmatrix} A & 0 \\ -Q & I \end{bmatrix}\begin{bmatrix} \xi \\ \lambda \end{bmatrix}.$$

In order to proceed, we shall *assume* that A is invertible. Then in the preceding equation we can define[1]

$$S_2=\begin{bmatrix} I & BR^{-1}B' \\ 0 & A' \end{bmatrix}^{-1}\begin{bmatrix} A & 0 \\ -Q & I \end{bmatrix}$$

to get

$$\begin{bmatrix} \dot{\xi} \\ \dot{\lambda} \end{bmatrix}=S_2\begin{bmatrix} \xi \\ \lambda \end{bmatrix}. \tag{6.2}$$

Let us return to the definition of J:

[1]The subscript "2" on S is meant to link it to the $\mathcal{H}_2$ problem.

$$J = \frac{1}{2} \sum_{k=0}^{\infty} \left[\xi(k)' Q \xi(k) + v(k)' R v(k) \right].$$

Since R is positive definite, for J to be finite, it must necessarily be true that $v(k) \to 0$ as $k \to \infty$. From (6.1), a sufficient condition for this is that $\lambda(k) \to 0$. Also, another necessary condition for J to be finite is that $Q^{1/2}\xi(k) \to 0$. In light of these observations, we shall impose upon solutions $\xi(k), \lambda(k)$ of (6.2) that they converge to zero.

Define

$$\phi = \begin{bmatrix} \xi \\ \lambda \end{bmatrix}.$$

The solution of (6.2) is

$$\phi(k) = S_2^k \phi(0).$$

Suppose $\phi(0)$ is an eigenvector of S_2 corresponding to an eigenvalue μ with $|\mu| < 1$. Then

$$S_2 \phi(0) = \mu \phi(0)$$

and this implies that

$$S_2^k \phi(0) = \mu^k \phi(0).$$

Thus

$$\phi(k) = \mu^k \phi(0) \to 0.$$

More generally, $\phi(k) \to 0$ iff the initial state, $\phi(0)$, lies in the space spanned by the generalized eigenvectors of S_2 corresponding to eigenvalues strictly inside the unit disk.

Now we need two additional assumptions on S_2. First, *assume* it has no eigenvalues on the unit circle. It is possible to show that S_2 has the property that μ is an eigenvalue iff $1/\mu$ is an eigenvalue. So the first assumption implies that half the eigenvalues of S_2 lie strictly inside the unit disk—that is, they are *stable*—and the other half lie strictly outside—they are *unstable*. Let n denote the dimension of the vector ξ; then S_2 is $2n \times 2n$. So n eigenvalues of S_2 are stable and n are unstable. Define T to be the $2n \times n$ matrix whose columns are the generalized eigenvectors of S_2 corresponding to stable eigenvalues. Partition T as follows:

$$T = \begin{bmatrix} T_1 \\ T_2 \end{bmatrix}, \quad T_i : n \times n.$$

Secondly, *assume* that T_1 is invertible. Defining $X = T_2 T_1^{-1}$, we have

$$T = \begin{bmatrix} I \\ X \end{bmatrix} T_1.$$

So the matrices T and $\begin{bmatrix} I \\ X \end{bmatrix}$ have the same column span. We conclude that $\phi(k) \to 0$ iff $\phi(0)$ belongs to the column span of $\begin{bmatrix} I \\ X \end{bmatrix}$.

Claim *If $\phi(0)$ belongs to the column span of $\begin{bmatrix} I \\ X \end{bmatrix}$, then $\lambda(k) = X\xi(k)$ for all k.*

Proof Assume $\phi(0)$ belongs to the column span of $\begin{bmatrix} I \\ X \end{bmatrix}$, that is,

$$\phi(0) = \begin{bmatrix} I \\ X \end{bmatrix} v$$

for some vector v. Now the columns of $\begin{bmatrix} I \\ X \end{bmatrix}$ span the same space as the generalized eigenvectors of S_2 corresponding to stable eigenvalues. Suppose for simplicity that $\phi(0)$ is in fact one of these eigenvectors; then

$$S_2\phi(0) = \mu\phi(0)$$

for some $|\mu| < 1$. Thus

$$\begin{aligned} \phi(k) &= S_2^k\phi(0) \\ &= \mu^k\phi(0) \\ &= \mu^k \begin{bmatrix} I \\ X \end{bmatrix} v. \end{aligned}$$

This implies that

$$\begin{bmatrix} -X & I \end{bmatrix} \phi(k) = 0.$$

In this way we get

$$\begin{bmatrix} -X & I \end{bmatrix} \begin{bmatrix} \xi(k) \\ \lambda(k) \end{bmatrix} = 0,$$

that is, $\lambda(k) = X\xi(k)$. ∎

Finally, substitute $\lambda(k) = X\xi(k)$ into (6.1) to get the optimal control law as follows:

$$\begin{aligned} v &= -R^{-1}B'\dot{\lambda} \\ &= -R^{-1}B'X\dot{\xi} \\ &= -R^{-1}B'X[A\xi + Bv] \\ &= -R^{-1}B'XA\xi - R^{-1}B'XBv \\ \Longrightarrow v &= -(R + B'XB)^{-1}B'XA\xi \\ \Longrightarrow F &= -(R + B'XB)^{-1}B'XA. \end{aligned}$$

Summary

- Given data: (A, B, Q, R) with A invertible.
- Define

$$S_2 = \begin{bmatrix} I & BR^{-1}B' \\ 0 & A' \end{bmatrix}^{-1} \begin{bmatrix} A & 0 \\ -Q & I \end{bmatrix}.$$

- Assume S_2 has no eigenvalues on the unit circle.
- Compute a matrix T whose columns are the generalized eigenvectors of S_2 corresponding to stable eigenvalues. Partition T as $\begin{bmatrix} T_1 \\ T_2 \end{bmatrix}$.
- Assume T_1 is invertible.
- Compute $X = T_2 T_1^{-1}$.
- Conclusion: The optimal feedback is $F = -(R + B'XB)^{-1}B'XA$.

Example 6.1.1 A very simple example is

$$\begin{aligned} \dot{\xi} &= \xi + v \\ J &= \frac{1}{2}\sum \left[\xi(k)^2 + v(k)^2\right], \end{aligned}$$

for which

$$A = B = Q = R = 1.$$

Then

$$S_2 = \begin{bmatrix} 2 & -1 \\ -1 & 1 \end{bmatrix}.$$

The MATLAB command

$$[V, D] = eig(S_2)$$

yields eigenvalues along the diagonal of the matrix D and corresponding eigenvectors as the columns of V:

$$V = \begin{bmatrix} -0.8507 & -0.5257 \\ 0.5257 & -0.8507 \end{bmatrix}, \quad D = \begin{bmatrix} 2.6180 & 0 \\ 0 & 0.3820 \end{bmatrix}.$$

Thus

$$T = \begin{bmatrix} -0.5257 \\ -0.8507 \end{bmatrix}, \quad X = \frac{-0.8507}{-0.5257} = 1.6180.$$

Then $F = -0.6180$. This is exactly the same as that produced by the MATLAB command

$$F = -dlqr(1, 1, 1, 1).$$

6.2 Symplectic Pair and Generalized Eigenproblem

In the preceding section, we saw that the LQR problem, which is closely related to $\mathcal{H}_2$-optimal control, can be solved via computing the stable eigenspace of an associated matrix, called a *symplectic matrix*:

$$\begin{bmatrix} I & BR^{-1}B' \\ 0 & A' \end{bmatrix}^{-1} \begin{bmatrix} A & 0 \\ -Q & I \end{bmatrix}.$$

For this matrix to exist, A must be nonsingular. Such an assumption can be very restrictive in many applications; in many problems the A-matrix that appears in the symplectic matrix is *not* the plant A-matrix (which is normally nonsingular, being obtained from the step-invariant transformation)—it is an intermediate matrix. It is possible to remove this assumption by considering the generalized eigenproblem for the matrix pair

$$\begin{bmatrix} A & 0 \\ -Q & I \end{bmatrix}, \quad \begin{bmatrix} I & BR^{-1}B' \\ 0 & A' \end{bmatrix}. \tag{6.3}$$

Let M_l and M_r be two $n \times n$ matrices and M the ordered matrix pair (M_l, M_r). [2] The set of *generalized eigenvalues* of the pair M, denoted $\sigma(M)$, are those numbers λ satisfying

$$M_l x = \lambda M_r x$$

for some nonzero vector x, called a *generalized eigenvector*. It is easy to see that the generalized eigenvalues are the roots of the (generalized) characteristic equation

$$\det(M_l - \lambda M_r) = 0.$$

Note that if M_r is nonsingular, the generalized eigenproblem reduces to the eigenproblem for the matrix $M_r^{-1}M_l$. If M_r is singular, $\sigma(M)$ may be finite, empty, or infinite:

$$M_l = \begin{bmatrix} 1 & 0 \\ 2 & 3 \end{bmatrix}, \quad M_r = \begin{bmatrix} 1 & 0 \\ 0 & 0 \end{bmatrix}, \quad \sigma(M) = \{1\}$$

$$M_l = \begin{bmatrix} 1 & 0 \\ 2 & 3 \end{bmatrix}, \quad M_r = \begin{bmatrix} 0 & 0 \\ 1 & 0 \end{bmatrix}, \quad \sigma(M) = \emptyset$$

$$M_l = \begin{bmatrix} 1 & 0 \\ 2 & 0 \end{bmatrix}, \quad M_r = \begin{bmatrix} 1 & 0 \\ 0 & 0 \end{bmatrix}, \quad \sigma(M) = \mathbb{C}.$$

In general, $\det(M_l - \lambda M_r)$ has the form

$$\det(M_l - \lambda M_r) = \prod_{i=1}^{n} (\beta_i - \alpha_i \lambda). \tag{6.4}$$

[2] The subscripts l and r stand for "left" and "right."

For example, the second case above corresponds to $\alpha_1 = \alpha_2 = 0$ but $\beta_1, \beta_2 \neq 0$; the third case corresponds to $\alpha_i = \beta_i = 0$ for some i and thus $\det(M_l - \lambda M_r)$ is identically zero.

With the expression in (6.4), we can define the multiplicity of a generalized eigenvalue in the obvious way. Furthermore, if λ is a generalized eigenvalue with multiplicity $r > 1$, then the set of vectors $\{x_1, x_2, \cdots, x_l\}$ satisfying

$$\begin{aligned} M_l x_1 &= \lambda M_r x_1 \\ (M_l - \lambda M_r) x_k &= M_r x_{k-1}, \quad k = 2, 3, \cdots, l; \quad l \leq r, \end{aligned}$$

is a *chain of generalized principal vectors.*

What can be said about the generalized eigenvalues of the matrix pair in (6.3)? For a general discussion, let A, P, Q be real $n \times n$ matrices with P and Q symmetric. Define the ordered pair of $2n \times 2n$ matrices

$$S = (S_l, S_r) := \left(\begin{bmatrix} A & 0 \\ -Q & I \end{bmatrix}, \begin{bmatrix} I & P \\ 0 & A' \end{bmatrix} \right).$$

A pair of matrices of this form is called a *symplectic pair.* [3] As mentioned before, if A is nonsingular, the generalized eigenproblem of the symplectic pair S reduces to the eigenproblem of the symplectic matrix

$$\begin{bmatrix} I & P \\ 0 & A' \end{bmatrix}^{-1} \begin{bmatrix} A & 0 \\ -Q & I \end{bmatrix}.$$

However, we shall not make this simplification.

It is a nice property (to be shown later) that the generalized eigenvalues of a symplectic pair are symmetric about the unit circle. To state this fact precisely, we need to introduce generalized eigenvalues at infinity.

Note that if $\lambda \neq 0$, then

$$\det(M_l - \lambda M_r) = 0 \Leftrightarrow \det\left(M_r - \frac{1}{\lambda} M_l\right) = 0.$$

In other words, λ is a generalized eigenvalue of the pair (M_l, M_r) iff $1/\lambda$ is a generalized eigenvalue of the pair (M_r, M_l) with the same multiplicity. Based on this, let us define that $\lambda = \infty$ is a generalized eigenvalue with multiplicity r of (M_l, M_r) if $\lambda = 0$ is a generalized eigenvalue with multiplicity r of (M_r, M_l). It follows that (M_l, M_r) has generalized eigenvalue at 0 iff M_l is singular and has generalized eigenvalue at ∞ iff M_r is singular.

Theorem 6.2.1 *The generalized eigenvalues of the symplectic pair S are symmetric about the unit circle, that is, $\lambda \in \sigma(S)$ iff $1/\lambda \in \sigma(S)$ and both have the same multiplicity.*

Proof Define

[3] The symbol S is used both for a symplectic pair and for the sampling operator; context should prevent any confusion.

$$J = \begin{bmatrix} 0 & -I \\ I & 0 \end{bmatrix}.$$

Then it is easily checked that the symplectic pair (S_l, S_r) has the algebraic property

$$S_l J S_l' = S_r J S_r'. \tag{6.5}$$

For simplicity, consider $\lambda \in \sigma(S)$ with multiplicity 1. We may assume that $\lambda \neq 0$ and $\lambda \neq \infty$, for otherwise the theorem holds:

$$\begin{aligned} 0 \in \sigma(S) &\Leftrightarrow \det(S_l) = 0 \\ &\Leftrightarrow \det(A) = 0 \\ &\Leftrightarrow \det(S_r) = 0 \\ &\Leftrightarrow \infty \in \sigma(S). \end{aligned}$$

Thus it suffices to prove that for a finite, nonzero λ, $\lambda \in \sigma(S)$ implies $1/\lambda \in \sigma(S)$. To show this, note that

$$\begin{aligned} \lambda \in \sigma[(S_l, S_r)] &\Rightarrow \lambda \in \sigma[(S_l', S_r')] \\ &\Rightarrow \frac{1}{\lambda} \in \sigma[(S_r', S_l')]. \end{aligned}$$

The latter condition implies that there exists a nonzero vector x such that

$$S_r' x = \frac{1}{\lambda} S_l' x. \tag{6.6}$$

We claim that $S_l' x \neq 0$. To see this, write

$$x = \begin{bmatrix} x_1 \\ x_2 \end{bmatrix}$$

and suppose $S_l' x = 0$. Then from (6.6) $S_r' x = 0$. From the definition of S_l, $S_l' x = 0$ implies $x_2 = 0$; similarly, $S_r' x = 0$ implies $x_1 = 0$. Thus $x = 0$, a contradiction.

Now pre-multiply (6.6) by $S_r J$ and use (6.5) to get

$$S_l J S_l' x = \frac{1}{\lambda} S_r J S_l' x.$$

This implies $1/\lambda \in \sigma(S)$ since $J S_l' x$ is a nonzero vector. ∎

The MATLAB command

$$[V, D] = eig(S_l, S_r),$$

computes the generalized eigenvalues of the pair (S_l, S_r) along the diagonal of the matrix D and corresponding eigenvectors as the columns of V.

6.3 Symplectic Pair and Riccati Equation

The solution to the general $\mathcal{H}_2$ problem requires some basics about Riccati equations whose solutions can be obtained via the generalized eigenproblem for symplectic pairs. From now on, we shall drop the adjective "generalized" whenever no confusion will arise. We start with a special case of the Riccati equation, namely, a Lyapunov equation.

Lyapunov Equation

The equation

$$A'XA - X + Q = 0$$

is called a (discrete-time) *Lyapunov equation*. Here A, Q, X are all square matrices, say $n \times n$, with Q symmetric. (The MATLAB function to solve this equation is *dlyap*.)

One situation is where A and Q are given and the equation is to be solved for X. Existence and uniqueness are easy to establish in principle. Define the linear map

$$\Phi : \mathbb{R}^{n\times n} \to \mathbb{R}^{n\times n}, \quad \Phi(X) = A'XA - X.$$

Then the Lyapunov equation has a solution X iff Q belongs to $\operatorname{Im}\Phi$, the image space (range) of Φ; if this condition holds, the solution is unique iff Φ is injective, hence bijective. Let σ denote the spectrum of a linear transformation, that is, the set of eigenvalues. It can be shown that

$$\sigma(\Phi) = \{\lambda_1\lambda_2 - 1 : \lambda_1, \lambda_2 \in \sigma(A)\}.$$

(If $\lambda_1, \lambda_2 \in \sigma(A)$, then $\lambda_1, \bar{\lambda}_2 \in \sigma(A')$. Thus there exist non-zero vectors x_1 and x_2 such that $A'x_1 = \lambda_1 x_1$ and $A'x_2 = \bar{\lambda}_2 x_2$. Letting $X := x_1 x_2^* \neq 0$, we have

$$\begin{aligned} \Phi(X) &= A'x_1x_2^*A - x_1x_2^* \\ &= (\lambda_1\lambda_2 - 1)X, \end{aligned}$$

so $\lambda_1\lambda_2 - 1 \in \sigma(\Phi)$.) Thus the Lyapunov equation has a unique solution iff A has the property that no two of its eigenvalues are reciprocals. For example, if A is stable, the unique solution is

$$X = \sum_0^\infty A'^k Q A^k.$$

We will be more interested in another situation, where we want to infer stability of A.

Theorem 6.3.1 *Suppose A, Q, X satisfy the Lyapunov equation, (Q, A) is detectable, and Q and X are positive semidefinite. Then A is stable.*

Proof For a proof by contradiction, suppose A has some eigenvalue λ with $|\lambda| \geq 1$. Let x be a corresponding eigenvector. Pre-multiply the Lyapunov equation by x^* and post-multiply by x to get

$$(|\lambda|^2 - 1)x^*Xx + x^*Qx = 0.$$

Both terms on the left are ≥ 0. Hence $x^*Qx = 0$, which implies that $Qx = 0$ since $Q \geq 0$. Thus

$$\begin{bmatrix} A - \lambda \\ Q \end{bmatrix} x = 0.$$

By detectability we must have $x = 0$, a contradiction. ■

Riccati Equation

Let A, P, Q be real $n \times n$ matrices with P and Q symmetric. The equation

$$A'X(I + PX)^{-1}A - X + Q = 0$$

is called the (discrete-time algebraic) *Riccati equation*. We will be interested in solutions X satisfying two properties, namely, X is symmetric and $(I + PX)^{-1}A$ is stable. Such solutions can be found by looking at the eigenproblem for the associated symplectic pair

$$S = (S_l, S_r) = \left(\begin{bmatrix} A & 0 \\ -Q & I \end{bmatrix}, \begin{bmatrix} I & P \\ 0 & A' \end{bmatrix} \right),$$

which was studied in Section 6.2.

Counting multiplicities and the eigenvalue at infinity, the pair S always has $2n$ eigenvalues. Now we *assume* S has no eigenvalues on $\partial\mathbb{D}$, the unit circle. Then by Theorem 6.2.1 it must have n inside and n outside. Let us focus on the n stable eigenvalues and denote by $\mathcal{X}_i(S)$ the stable generalized eigenspace of S, namely, the subspace in $\mathbb{R}^{2n}$ spanned by all the eigenvectors and principal vectors of S corresponding to the stable eigenvalues. This space $\mathcal{X}_i(S)$ has dimension n. Finding a basis for $\mathcal{X}_i(S)$, stacking the basis vectors up to form a matrix, and partitioning the matrix, we get

$$\mathcal{X}_i(S) = \operatorname{Im} \begin{bmatrix} X_1 \\ X_2 \end{bmatrix}, \tag{6.7}$$

where $X_1, X_2 \in \mathbb{R}^{n \times n}$. It follows that there exists a stable $n \times n$ matrix S_i, whose eigenvalues correspond to the stable eigenvalues of S, such that

$$S_l \begin{bmatrix} X_1 \\ X_2 \end{bmatrix} = S_r \begin{bmatrix} X_1 \\ X_2 \end{bmatrix} S_i. \tag{6.8}$$

Some properties of the matrix $X_1'X_2$ are useful.

Lemma 6.3.1 *Suppose S has no eigenvalues on $\partial\mathbb{D}$. Then* *(i)* $X_1'X_2$ *is symmetric;*

(ii) $X_1'X_2 \geq 0$ *if* $P \geq 0$ *and* $Q \geq 0$.

Proof (i) Write (6.8) as two equations:

$$AX_1 = X_1S_i + PX_2S_i, \tag{6.9}$$
$$X_2 = QX_1 + A'X_2S_i. \tag{6.10}$$

We need to show that $X_1'X_2 - X_2'X_1 = 0$. To do this, substitute X_2 from (6.10) to get

$$X_1'X_2 - X_2'X_1 = (AX_1)'X_2S_i - S_i'X_2'(AX_1)$$

and then replace AX_1 on the right-hand side by (6.9):

$$X_1'X_2 - X_2'X_1 = S_i'(X_1'X_2 - X_2'X_1)S_i.$$

This is a Lyapunov equation; since S_i is stable, the unique solution is

$$X_1'X_2 - X_2'X_1 = 0.$$

(ii) Define $M := X_1'X_2 = X_2'X_1$ and pre-multiply (6.9) by $S_i'X_2'$ to get

$$S_i'X_2'AX_1 = S_i'MS_i + S_i'X_2'PX_2S_i. \tag{6.11}$$

Take transpose of equation (6.10) and then post-multiply by X_1 to get

$$M = X_1'QX_1 + S_i'X_2'AX_1. \tag{6.12}$$

Thus equations (6.11) and (6.12) give

$$S_i'MS_i - M + S_i'X_2'PX_2S_i + X_1'QX_1 = 0.$$

This is a Lyapunov equation. Since S_i is stable, the unique solution is

$$M = \sum_0^\infty S_i'^k(S_i'X_2'PX_2S_i + X_1'QX_1)S_i^k,$$

which is ≥ 0 since P and Q are ≥ 0. ■

Now *assume* further that X_1 is nonsingular, that is, the two subspaces

$$\mathcal{X}_i(S), \quad \operatorname{Im}\begin{bmatrix} 0 \\ I \end{bmatrix}$$

are complementary. Set $X := X_2X_1^{-1}$. Since

$$\begin{bmatrix} X_1 \\ X_2 \end{bmatrix} = \begin{bmatrix} I \\ X \end{bmatrix} X_1,$$

we get

$$\mathcal{X}_i(S) = \text{Im}\begin{bmatrix} I \\ X \end{bmatrix} \tag{6.13}$$

The $n \times n$ matrix X is uniquely determined by S (though X_1 and X_2 are not), that is, $S \mapsto X$ is a function. We shall denote this function by *Ric* and write $X = Ric(S)$.

To recap, *Ric* is a function mapping a symplectic pair S to a matrix X, where X is defined by equation (6.13). The domain of *Ric*, denoted *dom Ric*, consists of all symplectic pairs S with two properties, namely, S has no eigenvalues on $\partial\mathbb{D}$ and the two subspaces

$$\mathcal{X}_i(S), \quad \text{Im}\begin{bmatrix} 0 \\ I \end{bmatrix}$$

are complementary.

The function *Ric* stands for "Riccati"; the reason is apparent from the lemma below.

Lemma 6.3.2 *Suppose $S \in$ dom Ric and $X = Ric(S)$. Then*

(i) X is symmetric;

(ii) X satisfies the Riccati equation

$$A'X(I+PX)^{-1}A - X + Q = 0;$$

(iii) $(I+PX)^{-1}A$ is stable.

Proof With $X_1 = I$ and $X_2 = X$, (i) follows from Lemma 6.3.1; moreover, (6.9) and (6.10) simplify to the following two equations

$$\begin{aligned} A &= (I+PX)S_i, & (6.14)\\ X &= Q + A'XS_i. & (6.15) \end{aligned}$$

Now we need to show that the matrix $I + PX$ is nonsingular. It is easier to see this with the additional assumption that $P \geq 0$ and $Q \geq 0$. (This is the case throughout the chapter.) Then by Lemma 6.3.1 X is also ≥ 0. Thus the matrix PX has no negative eigenvalues, which implies that $I+PX$ is nonsingular. The general proof for the nonsingularity of $I + PX$ is much harder; see [149]. Therefore from (6.14)

$$S_i = (I+PX)^{-1}A. \tag{6.16}$$

This proves (iii) since S_i is stable. Substitute (6.16) into (6.15) to get the Riccati equation. ■

The following result gives verifiable conditions under which S belongs to *dom Ric*.

Theorem 6.3.2 *Suppose S has the form*

$$S = \left(\begin{bmatrix} A & 0 \\ -C'C & I \end{bmatrix}, \begin{bmatrix} I & BB' \\ 0 & A' \end{bmatrix} \right)$$

with (A, B) stabilizable and (C, A) having no unobservable modes on $\partial\mathbb{D}$. Then $S \in$ dom Ric and $Ric(S) \geq 0$. If (C, A) is observable, then $Ric(S) > 0$.

Before proving the theorem, let us note the relationship between Riccati and Lyapunov equations. Suppose $S \in$ *dom Ric* and $X := Ric(S)$. The associated Riccati equation is

$$A'X(I + BB'X)^{-1}A - X + C'C = 0. \tag{6.17}$$

Define

$$\begin{aligned} F &:= -(I + B'XB)^{-1}B'XA \\ &= -B'X(I + BB'X)^{-1}A. \end{aligned}$$

It is easily verified that

$$\begin{aligned} A_F &:= A + BF \\ &= (I + BB'X)^{-1}A. \end{aligned}$$

Then the first term in (6.17) can be written as follows:

$$\begin{aligned} A'X(I + BB'X)^{-1}A &= A'XA_F \\ &= A_F'(I + XBB')XA_F \\ &= A_F'XA_F + A_F'XBB'XA_F \\ &= A_F'XA_F + F'F. \end{aligned}$$

Thus the Riccati equation can be rewritten as the Lyapunov equation

$$A_F'XA_F - X + C'C + F'F = 0.$$

Noting that A_F is stable (Lemma 6.3.2), we have

$$X = \sum_0^{\infty} A_F'^k (C'C + F'F) A_F^k. \tag{6.18}$$

The MATLAB commands to compute X and F are these:

$$\begin{aligned} [F_{tmp}, X] &= dlqr\,(A, B, C'C, I), \\ F &= -F_{tmp}. \end{aligned}$$

This computation is based on the eigenproblem of symplectic matrices instead of the generalized eigenproblem of symplectic pairs and hence it implicitly assumes that the matrix A is nonsingular.

Proof of Theorem 6.3.2 We first show that S has no eigenvalues on the unit circle. Suppose, on the contrary, that $e^{j\theta}$ is an eigenvalue and $\begin{bmatrix} x \\ z \end{bmatrix}$ a corresponding eigenvector; that is,

$$\begin{bmatrix} A & 0 \\ -C'C & I \end{bmatrix} \begin{bmatrix} x \\ z \end{bmatrix} = e^{j\theta} \begin{bmatrix} I & BB' \\ 0 & A' \end{bmatrix} \begin{bmatrix} x \\ z \end{bmatrix}.$$

Write as two equations and re-arrange:

$$(A - e^{j\theta})x = e^{j\theta} BB'z, \tag{6.19}$$

$$e^{j\theta}(A' - e^{-j\theta})z = -C'Cx. \tag{6.20}$$

Pre-multiply (6.19) and (6.20) by $e^{-j\theta}z^*$ and x^* respectively to get

$$\begin{aligned} e^{-j\theta}z^*(A - e^{j\theta})x &= \|B'z\|^2 \\ e^{j\theta}x^*(A' - e^{-j\theta})z &= -\|Cx\|^2. \end{aligned}$$

Take complex-conjugate of the latter equation to get

$$\begin{aligned} -\|Cx\|^2 &= e^{-j\theta}z^*(A - e^{j\theta})x \\ &= \|B'z\|^2. \end{aligned}$$

Therefore $B'z = 0$ and $Cx = 0$. So from (6.19) and (6.20)

$$\begin{aligned} (A - e^{j\theta})x &= 0, \\ (A - e^{j\theta})^*z &= 0. \end{aligned}$$

We arrive at the equations

$$\begin{aligned} z^*[A - e^{j\theta} \quad B] &= 0 \\ \begin{bmatrix} A - e^{j\theta} \\ C \end{bmatrix} x &= 0. \end{aligned}$$

By controllability and observability of modes on $\partial\mathbb{D}$ it follows that $x = z = 0$, a contradiction.

Next, we will show that the two subspaces

$$\mathcal{X}_i(S), \quad \operatorname{Im}\begin{bmatrix} 0 \\ I \end{bmatrix}$$

are complementary. As in the proof of Lemma 6.3.1 bring in matrices X_1, X_2, S_i to get equations (6.9) and (6.10), re-written as below ($P = BB', Q = C'C$):

$$AX_1 = X_1S_i + BB'X_2S_i, \tag{6.21}$$

$$-C'CX_1 + X_2 = A'X_2S_i. \tag{6.22}$$

We want to show that X_1 is nonsingular, that is, $\operatorname{Ker} X_1 = 0$. First, it is claimed that $\operatorname{Ker} X_1$ is S_i-invariant. To prove this, let $x \in \operatorname{Ker} X_1$. Pre-multiply (6.21) by $x'S_i'X_2'$ and post-multiply by x to get

$$x'S_i'X_2'X_1S_ix + x'S_i'X_2'BB'X_2S_ix = 0.$$

Note that since $X_2'X_1 \geq 0$ (Lemma 6.3.1), both terms on the left are ≥ 0. Thus $B'X_2S_ix = 0$. Now post-multiply (6.21) by x to get $X_1S_ix = 0$, that is, $S_ix \in \operatorname{Ker} X_1$. This proves the claim.

Now to prove that X_1 is nonsingular, suppose on the contrary that

$$\operatorname{Ker} X_1 \neq 0.$$

Then S_i has an eigenvalue, μ, whose corresponding eigenvector, x, is in $\operatorname{Ker} X_1$:

$$S_ix = \mu x, \tag{6.23}$$

$$0 \neq |\mu| < 1, \quad 0 \neq x \in \operatorname{Ker} X_1.$$

Post-multiply (6.22) by x and use (6.23):

$$(\mu A' - 1)X_2x = 0.$$

Since $B'X_2x = 0$ too from $B'X_2S_ix = 0$ and (6.23), we have

$$x^*X_2' \left[\begin{array}{cc} A - \frac{1}{\mu} & B \end{array} \right] = 0.$$

Then stabilizability implies $X_2x = 0$. But if $X_1x = 0$ and $X_2x = 0$, then $x = 0$, a contradiction. This concludes the proof of complementarity.

Now set $X := Ric(S)$. By Lemma 6.3.1 ($P = BB', Q = C'C, X_1 = I, X_2 = X$), $X \geq 0$.

Finally, suppose (C, A) is observable. We will show using (6.18) that if $x'Xx = 0$, then $x = 0$; thus $X > 0$. Pre-multiply (6.18) by x' and post-multiply by x:

$$x'Xx = \sum_0^\infty ||CA_F^kx||^2 + \sum_0^\infty ||FA_F^kx||^2.$$

If $x'Xx = 0$, then

$$CA_F^kx = 0, \quad \forall k \geq 0;$$

$$FA_F^kx = 0, \quad \forall k \geq 0.$$

Based on the two equations, we can readily infer by induction that

$$CA^kx = 0, \quad \forall k \geq 0,$$

which implies that x belongs to the unobservable subspace of (C, A) and so $x = 0$. ∎

6.4 State Feedback and Disturbance Feedforward

First we allow the controller to have full information. In this case, as we will see, the optimal controller is a constant state feedback with a disturbance feedforward. With the exogenous input being some impulse function, say, $\omega = \delta_d \omega_0$ (ω_0 is a constant vector and δ_d the discrete unit impulse), we can even think of v as unconstrained. The precise problem is as follows:

- Given the system equations

$$G: \quad \begin{aligned} \dot{\xi} &= A\xi + B_1\omega + B_2 v, \quad \omega = \delta_d\omega_0, \ \xi(0) = 0 \\ \zeta &= C_1\xi + D_{11}\omega + D_{12}v \end{aligned}$$

  with the assumptions

  (i) (A, B_2) is stabilizable;

  (ii) $M := D_{12}' D_{12}$ is nonsingular;

  (iii) the matrix

$$\begin{bmatrix} A - \lambda & B_2 \\ C_1 & D_{12} \end{bmatrix}$$

  has full column rank $\forall \lambda \in \partial\mathbb{D}$;

- Solve the optimization problem

$$\min_{v \in \ell(\mathbb{Z}_+)} \|\zeta\|_2^2.$$

Note that for ease of presentation we initially allow v to be in $\ell(\mathbb{Z}_+)$, the *extended* space for $\ell_2(\mathbb{Z}_+)$ consisting of all sequences; however, the optimal v, to be seen later, will actually lie in $\ell_2(\mathbb{Z}_+)$. Assumptions (i) and (iii) are mild and standard restrictions. Assumption (ii) means that the system must have at least as many outputs to be controlled as control inputs and the control weighting is nonsingular.

The setup can be depicted as

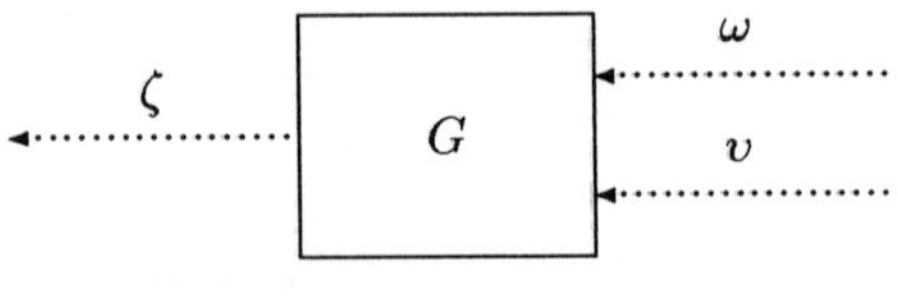

where the transfer matrix for G is

$$\hat{g}(\lambda) = \left[\begin{array}{c|cc} A & B_1 & B_2 \\ \hline C_1 & D_{11} & D_{12} \end{array}\right].$$

Define the symplectic pair

$$S_2 = \left(\begin{bmatrix} A - B_2 M^{-1} D_{12}' C_1 & 0 \\ -C_1'(I - D_{12} M^{-1} D_{12}') C_1 & I \end{bmatrix}, \right.$$

$$\left. \begin{bmatrix} I & B_2 M^{-1} B_2' \\ 0 & (A - B_2 M^{-1} D_{12}' C_1)' \end{bmatrix} \right).$$

If $D_{12}'C_1 = 0$, S_2 reduces to the symplectic pair associated with the corresponding LQR problem with $Q = C_1'C_1$ and $R = M$. In the general case, it is easy to show that the only possible eigenvalues of $D_{12}M^{-1}D_{12}'$ are 0 and 1; thus the matrix $I - D_{12}M^{-1}D_{12}'$ is positive semidefinite. If $D_{12}'C_1 = 0$, the symplectic pair S_2 reduces to the one we saw for the LQR problem in the preceding sections with $Q = C_1'C_1$ and $R = M$.

Lemma 6.4.1 *S_2 belongs to dom Ric.*

Proof First, it is easy to show that $(A - B_2M^{-1}D_{12}'C_1, B_2M^{-1/2})$ is stabilizable. Next, observe the following identity

$$\begin{bmatrix} I & -B_2 M^{-1} D_{12}' \\ 0 & I \end{bmatrix} \begin{bmatrix} A - \lambda & B_2 \\ C_1 & D_{12} \end{bmatrix} \begin{bmatrix} I & 0 \\ -M^{-1} D_{12}' C_1 & I \end{bmatrix}$$

$$= \begin{bmatrix} A - B_2 M^{-1} D_{12}' C_1 - \lambda & 0 \\ (I - D_{12} M^{-1} D_{12}') C_1 & D_{12} \end{bmatrix}.$$

Since the first and third matrices on the left are nonsingular, it follows from assumption (iii) that the resultant matrix also has full column rank $\forall \lambda \in \partial \mathbb{D}$. Thus

$$\operatorname{rank} \begin{bmatrix} A - B_2 M^{-1} D_{12}' C_1 - \lambda \\ (I - D_{12} M^{-1} D_{12}') C_1 \end{bmatrix} = n, \quad \forall \lambda \in \partial \mathbb{D},$$

where n is the dimension of the matrix A. It is not hard to show then that

$$\operatorname{rank} \begin{bmatrix} A - B_2 M^{-1} D_{12}' C_1 - \lambda \\ (I - D_{12} M^{-1} D_{12}')^{1/2} C_1 \end{bmatrix} = n, \quad \forall \lambda \in \partial \mathbb{D}.$$

In view of Theorem 6.3.2, this means that the required matrix pair has no unobservable modes on $\partial \mathbb{D}$; by Theorem 6.3.2, $S_2 \in$ *dom Ric*. ■

By this lemma, $X := Ric(S_2)$ is well-defined and is ≥ 0 (Theorem 6.3.2). Hence the matrix $M + B_2'XB_2$ is nonsingular. Define the matrices

$$\begin{aligned} F &= -(M + B_2' X B_2)^{-1} (B_2' X A + D_{12}' C_1) \\ F_0 &= -(M + B_2' X B_2)^{-1} (B_2' X B_1 + D_{12}' D_{11}) \\ A_F &= A + B_2 F \\ C_{1F} &= C_1 + D_{12} F \end{aligned}$$

and the transfer matrix

$$\hat{g}_c(\lambda) = \left[\begin{array}{c|c} A_F & B_1 + B_2F_0 \\ \hline C_{1F} & D_{11} + D_{12}F_0 \end{array}\right].$$

Some algebra shows that

$$A_F = (I + B_2M^{-1}B_2'X)^{-1}(A - B_2M^{-1}D_{12}'C_1).$$

By Lemma 6.3.2, A_F is stable and so $\hat{g}_c \in \mathcal{RH}_2(\mathbb{D})$.

If the matrix $A - B_2M^{-1}D_{12}'C_1$ is nonsingular, the MATLAB command for computing X is

$$[F_{tmp}, X] = dlqr\ (A - B_2M^{-1}D_{12}'C_1, B_2, C_1'(I - D_{12}M^{-1}D_{12}')C_1, M).$$

Theorem 6.4.1 *The unique optimal control is $v_{opt} = F\xi + F_0\omega$. Moreover,*

$$\min_v \|\zeta\|_2 = \|\hat{g}_c\omega_0\|_2.$$

In contrast with the full-information case in continuous time where the optimal control is a constant state feedback, the discrete-time optimal control law involves a disturbance feedforward term, and this is true even when $D_{11} = 0$. The optimal control is therefore

$$v_{opt}(k) = \begin{cases} F_0\omega_0, & k = 0 \\ F\xi(k), & k \geq 1. \end{cases}$$

It becomes a constant state feedback when $k \geq 1$.

Note that in the LQR problem in Section 6.1, the optimal control is a state feedback with no feedforward. This can be explained as follows. If we convert the LQR problem into a 2-norm optimization problem as in Section 6.1, the initial state ξ_0 is replaced by an impulse input $\delta_d\xi_0$. The impulse is applied at $k = -1$ whereas we are interested in $v(k)$ for $k \geq 0$; thus the possible feedforward, an impulse at $k = -1$, is out of consideration. (See also Exercise 6.6.)

A useful trick is to change variable. Start with the system equations

$$\begin{aligned} \dot{\xi} &= A\xi + B_1\omega + B_2v \\ \zeta &= C_1\xi + D_{11}\omega + D_{12}v, \end{aligned}$$

and define a new control variable

$$\rho := v - F\xi - F_0\omega.$$

The equations become

$$\begin{aligned} \dot{\xi} &= A_F\xi + (B_1 + B_2F_0)\omega + B_2\rho \\ \zeta &= C_{1F}\xi + (D_{11} + D_{12}F_0)\omega + D_{12}\rho. \end{aligned}$$

So in the frequency domain

$$\hat{\zeta} = \hat{g}_c \omega_0 + \hat{g}_i \hat{\rho},$$

where $\hat{g}_c$ is as above and $\hat{g}_i$ is seen to be

$$\hat{g}_i(\lambda) = \left[\begin{array}{c|c} A_F & B_2 \\ \hline C_{1F} & D_{12} \end{array}\right].$$

Two properties of the matrices $\hat{g}_c$ and $\hat{g}_i$ are crucial for the development to follow. For this we need a useful notation: For any transfer matrix $\hat{g}(\lambda)$, we write $\hat{g}^{\sim}(\lambda)$ as simply $\hat{g}(\lambda^{-1})'$. [Formulas of $\hat{g}^{\sim}$ in terms of state-space data can be easily obtained (Exercise 6.3).]

Lemma 6.4.2 *The matrix* $\hat{g}_i^{\sim}\hat{g}_c$ *belongs to* $\mathcal{RH}_2(\mathbb{D})^{\perp}$ *and*

$$\hat{g}_i^{\sim}\hat{g}_i = M + B_2'XB_2.$$

Proof We show $\hat{g}_i^{\sim}\hat{g}_c \in \mathcal{RH}_2(\mathbb{D})^{\perp}$. To simplify notation, define

$$\begin{aligned} B_{1F} &= B_1 + B_2F_0 \\ D_{11F} &= D_{11} + D_{12}F_0. \end{aligned}$$

Then

$$\begin{aligned} \hat{g}_i(\lambda) &= D_{12} + \lambda C_{1F}B_2 + \lambda^2 C_{1F}A_F B2 + \cdots \\ \hat{g}_i^{\sim}(\lambda) &= D_{12}' + \lambda^{-1}B_2'C_{1F}' + \lambda^{-2}B_2'A_F'C_{1F}' + \cdots && (6.24) \\ \hat{g}_c(\lambda) &= D_{11F} + \lambda C_{1F}B_{1F} + \lambda^2 C_{1F}A_F B_{1F} + \cdots. && (6.25) \end{aligned}$$

Multiplying the two series in (6.24) and (6.25) gives

$$\hat{g}_i^{\sim}(\lambda)\hat{g}_c(\lambda) = \cdots + \lambda^{-2}E_{-2} + \lambda^{-1}E_{-1} + E_0 + \lambda E_1 + \lambda^2 E_2 + \cdots.$$

Now we need to show that $E_0 = E_1 = \cdots = 0$. Note that

$$\begin{aligned} E_0 &= D_{12}'D_{11F} + B_2'C_{1F}'C_{1F}B_{1F} + B_2'A_F'C_{1F}'C_{1F}A_F B_{1F} + \cdots \\ &= D_{12}'D_{11F} + B_2'(C_{1F}'C_{1F} + A_F'C_{1F}'C_{1F}A_F + \cdots)B_{1F}. && (6.26) \end{aligned}$$

Using the equivalent Lyapunov equation (Exercise 6.4)

$$A_F'XA_F - X + C_{1F}'C_{1F} = 0,$$

we get that since A_F is stable, the unique X is the quantity in the parentheses of (6.26). Hence

$$E_0 = D_{12}'D_{11F} + B_2'XB_{1F}.$$

Now bring in the definitions of D_{11F}, B_{1F}, and then F_0 to get

$$E_0 = D_{12}'D_{11} + B_2'XB_1 + (M + B_2'XB_2)F_0 = 0.$$

Similarly,

$$\begin{aligned} E_1 &= D_{12}'C_{1F}B_{1F} + B_2'(C_{1F}'C_{1F} + A_F'C_{1F}'C_{1F}A_F + \cdots)A_FB_{1F} \\ &= D_{12}'C_{1F}B_{1F} + B_2'XA_FB_{1F} \\ &= (D_{12}'C_{1F} + B_2'XA_F)B_{1F} \\ &= 0 \end{aligned}$$

by definitions of C_{1F}, A_F, and F; and so on for $E_2 = E_3 = \cdots = 0$.

The proof of the second statement is similar. ∎

There is a useful time-domain interpretation of this lemma. For this, we need to regard G_c and G_i, the systems with transfer matrices $\hat{g}_c$ and $\hat{g}_i$, as systems defined on the whole time set $\mathbb{Z}$, that is, as linear transformations on $\ell_2(\mathbb{Z})$. In view of the decomposition

$$\ell_2(\mathbb{Z}) = \ell_2(\mathbb{Z}_-) \oplus \ell_2(\mathbb{Z}_+),$$

the matrix representation of, for example, G_c is

$$[G_c] = \left[\begin{array}{cc|ccc} & \vdots & \vdots & \vdots & \\ \cdots & g_c(0) & 0 & 0 & \cdots \\ \hline \cdots & g_c(1) & g_c(0) & 0 & \cdots \\ \cdots & g_c(2) & g_c(1) & g_c(0) & \cdots \\ & \vdots & \vdots & \vdots & \end{array}\right].$$

Here the vertical and horizontal lines separate the time intervals $\{k < 0\}$, $\{k \geq 0\}$, and $g_c(k)$ is the impulse-response function. Notice that the matrix is Toeplitz because G_c is time-invariant. Also, the impulse response is observed to be the first column of $[G_c]$ to the right of the vertical lines; the λ-transform of this column is therefore the transfer matrix $\hat{g}_c(\lambda)$.

Consider the linear system G_i^* whose matrix is the transpose $[G_i]'$. This is the *adjoint system*, a topic covered in some depth in Section 10.3. Now consider the system $G_i^*G_c$. The first column of $[G_i^*G_c]$ to the right of the vertical lines is the impulse response function, and the λ-transform of this is the transfer matrix $\hat{g}_i^{\sim}\hat{g}_c$. Because this transfer matrix belongs to $\mathcal{RH}_2(\mathbb{D})^\perp$, by Lemma 6.4.2, so the impulse response function of $G_i^*G_c$ belongs to $\ell_2(\mathbb{Z}_-)$, that is, the matrix $[G_i^*G_c]$ is strictly upper triangular.

Proof of Theorem 6.4.1 Since v is free in $\ell(\mathbb{Z}_+)$, so is ρ. Thus we can write (formally) in the time domain

$$\langle G_c\delta_d\omega_0, G_i\rho\rangle = \langle G_i^*G_c\delta_d\omega_0, \rho\rangle.$$

Since $G_i^*G_c\delta_d\omega_0 \in \ell_2(\mathbb{Z}_-)$, the right-hand side of the preceding equation equals 0 for all ρ in $\ell(\mathbb{Z}_+)$. This allows us to write in the frequency domain

$$\|\hat{\zeta}\|_2^2 = \|\hat{g}_c\omega_0\|_2^2 + \|\hat{g}_i\hat{\rho}\|_2^2.$$

Since $\hat{g}_i^{\sim}\hat{g}_i = M + B_2'XB_2$ by Lemma 6.4.2, we have

$$\|\hat{\zeta}\|_2^2 = \|\hat{g}_c\omega_0\|_2^2 + \|(M + B_2'XB_2)^{1/2}\hat{\rho}\|_2^2.$$

This equation gives the desired conclusion: The optimal $\hat{\rho}$ is $\hat{\rho} = 0$ (i.e., $v = F\xi + F_0\omega$) and the minimum norm of ζ equals $\|\hat{g}_c\omega_0\|_2$. ■

With v_{opt} applied, the resultant system is stable since A_F is stable; thus v_{opt} indeed lies in $\ell_2(\mathbb{Z}_+)$, as commented before.

There is a different interpretation of Theorem 6.4.1. Consider the setup

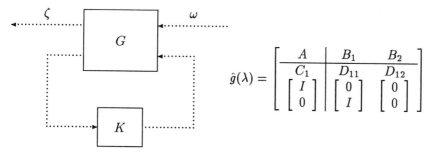

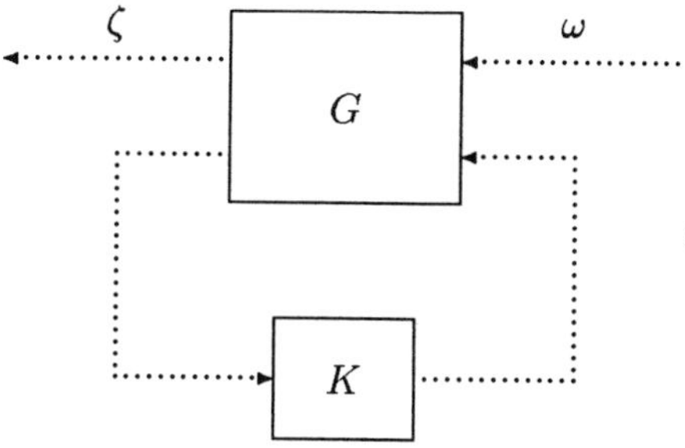

$$\hat{g}(\lambda) = \left[\begin{array}{c|cc} A & B_1 & B_2 \\ \hline C_1 & D_{11} & D_{12} \\ \begin{bmatrix} I \\ 0 \end{bmatrix} & \begin{bmatrix} 0 \\ I \end{bmatrix} & \begin{bmatrix} 0 \\ 0 \end{bmatrix} \end{array}\right]$$

Thus the controller input is $\begin{bmatrix} \xi \\ \omega \end{bmatrix}$, the state together with the disturbance. Theorem 6.4.1 implies that (under the assumptions stated at the beginning of this section) the unique internally stabilizing controller that minimizes the $\mathcal{H}_2$-norm from ω to ζ is the constant controller

$$\hat{k}(\lambda) = \begin{bmatrix} F & F_0 \end{bmatrix}.$$

Finally, what if the controller is allowed to process only the state, that is,

$$\hat{g}(\lambda) = \left[\begin{array}{c|cc} A & B_1 & B_2 \\ \hline C_1 & D_{11} & D_{12} \\ I & 0 & 0 \end{array}\right]?$$

It turns out that the unique internally stabilizing controller that minimizes the $\mathcal{H}_2$-norm from ω to ζ is the constant controller $\hat{k}(\lambda) = F$.

6.5 Output Feedback

This section studies the $\mathcal{H}_2$-optimal control problem posed at the start of this chapter, where the measured output ψ does not have full information and therefore dynamic feedback is necessary. All discussion pertains to the standard setup

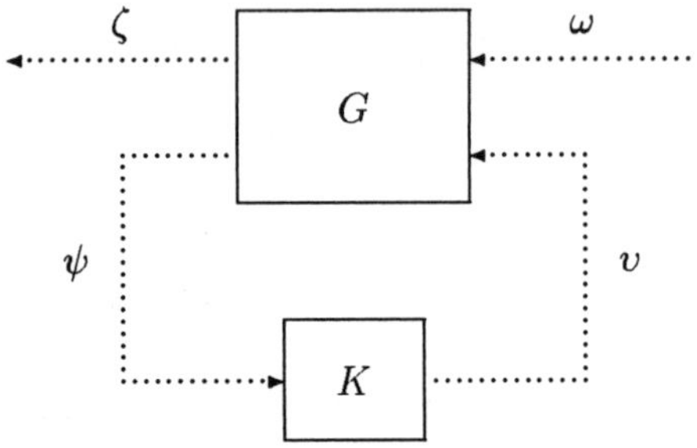

Let $T_{\zeta\omega}$ denote the closed-loop system from ω to ζ. We say a causal, FDLTI controller K is ***admissible*** if it achieves internal stability. Our goal is to find an admissible K to minimize $\|\hat{t}_{\zeta\omega}\|_2$.

Two Special Problems

For later benefit, we begin with two special $\mathcal{H}_2$-optimal control problems.

The ***first special problem*** has a plant of the form

$$\hat{g}(\lambda) = \left[\begin{array}{c|cc} A & B_1 & B_2 \\ \hline C_1 & D_{11} & D_{12} \\ C_2 & I & 0 \end{array}\right]$$

with the assumptions

(i) (A, B_2) is stabilizable;

(ii) $M := D_{12}'D_{12}$ is nonsingular;

(iii) the matrix

$$\begin{bmatrix} A-\lambda & B_2 \\ C_1 & D_{12} \end{bmatrix}$$

has full column rank $\forall \lambda \in \partial\mathbb{D}$;

(iv) $A - B_1C_2$ is stable.

Since $D_{21} = I$, the disturbance, ω, enters the measurement directly. Define

$$\begin{aligned} S_2 &= \left(\begin{bmatrix} A - B_2M^{-1}D_{12}'C_1 & 0 \\ -C_1'(I - D_{12}M^{-1}D_{12}')C_1 & I \end{bmatrix}, \right. \\ &\qquad \left. \begin{bmatrix} I & B_2M^{-1}B_2' \\ 0 & (A - B_2M^{-1}D_{12}'C_1)' \end{bmatrix} \right) \\ X &= Ric(S_2) \\ F &= -(M + B_2'XB_2)^{-1}(B_2'XA + D_{12}'C_1) \\ F_0 &= -(M + B_2'XB_2)^{-1}(B_2'XB_1 + D_{12}'D_{11}) \\ \hat{g}_c(\lambda) &= \left[\begin{array}{c|c} A + B_2F & B_1 + B_2F_0 \\ \hline C_1 + D_{12}F & D_{11} + D_{12}F_0 \end{array}\right]. \end{aligned}$$

The next result says that the optimal controller achieves the same performance as does the optimal state feedback and disturbance feedforward controller when the state and the disturbance are directly measured.

Theorem 6.5.1 *The unique optimal controller is*

$$\hat{k}_{opt}(\lambda) := \left[\begin{array}{c|c} A + B_2F - B_2F_0C_2 - B_1C_2 & B_1 + B_2F_0 \\ \hline F - F_0C_2 & F_0 \end{array}\right].$$

Moreover,

$$\min_K \|\hat{t}_{\zeta\omega}\|_2 = \|\hat{g}_c\|_2.$$

Proof Apply the controller K_{opt} and let η denote its state. Then the system equations are

$$\begin{aligned} \dot{\xi} &= A\xi + B_1\omega + B_2v \\ \zeta &= C_1\xi + D_{11}\omega + D_{12}v \\ \psi &= C_2\xi + \omega \\ \dot{\eta} &= (A + B_2F - B_2F_0C_2 - B_1C_2)\eta + (B_1 + B_2F_0)\psi \\ v &= (F - F_0C_2)\eta + F_0\psi, \end{aligned}$$

so

$$\dot{\eta} = A\eta + B_1\omega + B_2v + B_1(\psi - C_2\eta - \omega).$$

Defining $\varepsilon := \xi - \eta$, we get

$$\dot{\varepsilon} = (A - B_1C_2)\varepsilon.$$

It is now easy to infer internal stability from the stability of $A + B_2F$ and $A - B_1C_2$. For zero initial conditions on ξ and η, we have $\varepsilon(k) \equiv 0$, that is, $\eta(k) \equiv \xi(k)$. Hence

$$\begin{aligned} v &= (F - F_0C_2)\eta + F_0\psi \\ &= F\xi + F_0(\psi - C_2\xi) \\ &= F\xi + F_0\omega. \end{aligned}$$

This means that K_{opt} has the same action as the optimal state feedback and disturbance feedforward. Thus by Theorem 6.4.1 K_{opt} is optimal and in this case

$$\|\hat{t}_{\zeta\omega}\|_2 = \|\hat{g}_c\|_2.$$

The proof that K_{opt} is unique is left as Exercise 6.9. ■

The *second special problem* is the dual of the first; so G has the form

$$\hat{g}(\lambda) = \left[\begin{array}{c|cc} A & B_1 & B_2 \\ \hline C_1 & D_{11} & I \\ C_2 & D_{21} & 0 \end{array}\right]$$

with the assumptions

(i) (C_2, A) is detectable;

(ii) $N := D_{21}D_{21}'$ is nonsingular;

(iii) the matrix

$$\begin{bmatrix} A-\lambda & B_1 \\ C_2 & D_{21} \end{bmatrix}$$

has full row rank $\forall \lambda \in \partial \mathbb{D}$;

(iv) $A - B_2C_1$ is stable.

Define

$$\begin{aligned}
T_2 &= \left(\begin{bmatrix} (A - B_1D_{21}'N^{-1}C_2)' & 0 \\ -B_1(I - D_{21}'N^{-1}D_{21})B_1' & I \end{bmatrix}, \right. \\
&\qquad \left. \begin{bmatrix} I & C_2'N^{-1}C_2 \\ 0 & A - B_1D_{21}'N^{-1}C_2 \end{bmatrix} \right) \\
Y &= Ric(T_2) \\
L &= -(AYC_2' + B_1D_{21}')(N + C_2YC_2')^{-1} \\
L_0 &= -(C_1YC_2' + D_{11}D_{21}')(N + C_2YC_2')^{-1} \\
\hat{g}_f(\lambda) &= \left[\begin{array}{c|c} A + LC_2 & B_1 + LD_{21} \\ \hline C_1 + L_0C_2 & D_{11} + L_0D_{21} \end{array}\right].
\end{aligned}$$

Theorem 6.5.2 *The unique optimal controller is*

$$\hat{k}_{opt}(\lambda) := \left[\begin{array}{c|c} A + LC_2 - B_2L_0C_2 - B_2C_1 & L - B_2L_0 \\ \hline C_1 + L_0C_2 & L_0 \end{array}\right].$$

Moreover,

$$\min_K \|\hat{t}_{\zeta\omega}\|_2 = \|\hat{g}_f\|_2.$$

Proof Notice the duality: $\hat{k}_{opt}$ is the unique optimal controller for $\hat{g}$ iff $\hat{k}_{opt}'$ is the unique optimal controller for $\hat{g}'$. Then the results follow from Theorem 6.5.1. ■

The General Problem

Now consider the general output feedback case with

$$\hat{g} = \left[\begin{array}{c|cc} A & B_1 & B_2 \\ \hline C_1 & D_{11} & D_{12} \\ C_2 & D_{21} & 0 \end{array}\right]$$

and with the following assumptions:

(i) (A, B_2) is stabilizable and (C_2, A) is detectable;

(ii) $M := D_{12}'D_{12}$ and $N := D_{21}D_{21}'$ are nonsingular;

(iii) the matrices

$$\begin{bmatrix} A-\lambda & B_2 \\ C_1 & D_{12} \end{bmatrix}, \begin{bmatrix} A-\lambda & B_1 \\ C_2 & D_{21} \end{bmatrix}$$

have full column and row rank, respectively, $\forall \lambda \in \partial\mathbb{D}$;

The first parts of the three assumptions have been seen in Section 6.4; the second parts are dual to their first parts: Together they guarantee that the symplectic pair T_2 introduced above belongs to *dom Ric*. The second part of assumption (ii) means that the sensor noise weighting is nonsingular. Finally, assumption (iii) is related to the condition in the model-matching problem that $\hat{t}_2(\lambda)$ and $\hat{t}_3(\lambda)$ have no zeros on $\partial\mathbb{D}$.

Define

$$\begin{aligned}
S_2 &= \left(\begin{bmatrix} A - B_2M^{-1}D_{12}'C_1 & 0 \\ -C_1'(I - D_{12}M^{-1}D_{12}')C_1 & I \end{bmatrix}, \right. \\
&\qquad \left. \begin{bmatrix} I & B_2M^{-1}B_2' \\ 0 & (A - B_2M^{-1}D_{12}'C_1)' \end{bmatrix} \right) \\
X &= Ric(S_2) \\
F &= -(M + B_2'XB_2)^{-1}(B_2'XA + D_{12}'C_1) \\
F_0 &= -(M + B_2'XB_2)^{-1}(B_2'XB_1 + D_{12}'D_{11}) \\
\hat{g}_c(\lambda) &= \left[\begin{array}{c|c} A + B_2F & B_1 + B_2F_0 \\ \hline C_1 + D_{12}F & D_{11} + D_{12}F_0 \end{array}\right] \\
T_2 &= \left(\begin{bmatrix} (A - B_1D_{21}'N^{-1}C_2)' & 0 \\ -B_1(I - D_{21}'N^{-1}D_{21})B_1' & I \end{bmatrix}, \right. \\
&\qquad \left. \begin{bmatrix} I & C_2'N^{-1}C_2 \\ 0 & A - B_1D_{21}'N^{-1}C_2 \end{bmatrix} \right) \\
Y &= Ric(T_2) \\
L &= -(AYC_2' + B_1D_{21}')(N + C_2YC_2')^{-1} \\
L_0 &= (FYC_2' + F_0D_{21}')(N + C_2YC_2')^{-1}
\end{aligned}$$

$$\begin{aligned} R &= (M + B_2'XB_2)^{1/2} \\ \hat{g}_f(\lambda) &= \left[\begin{array}{c|c} A + LC_2 & B_1 + LD_{21} \\ \hline R(L_0C_2 - F) & R(L_0D_{21} - F_0) \end{array}\right]. \end{aligned}$$

Theorem 6.5.3 *The unique optimal controller is*

$$\hat{k}_{opt}(\lambda) := \left[\begin{array}{c|c} A + B_2F + LC_2 - B_2L_0C_2 & L - B_2L_0 \\ \hline L_0C_2 - F & L_0 \end{array}\right].$$

Moreover,

$$\min_K \|\hat{t}_{\zeta\omega}\|_2^2 = \|\hat{g}_c\|_2^2 + \|\hat{g}_f\|_2^2.$$

The first term in the minimum cost, $\|\hat{g}_c\|_2^2$, is associated with optimal control with state feedback and disturbance feedforward and the second, $\|\hat{g}_f\|_2^2$, with optimal filtering. These two norms can easily be computed as follows:

$$\begin{aligned} \|\hat{g}_c\|_2^2 &= \text{trace}\ [(D_{11} + D_{12}F_0)'(D_{11} \\ &\qquad +D_{12}F_0) + (B_1 + B_2F_0)'X(B_1 + B_2F_0)] \\ \|\hat{g}_f\|_2^2 &= \text{trace}\ \{R[(L_0D_{21} - F_0)(L_0D_{21} - F_0)' \\ &\qquad +(L_0C_2 - F)Y(L_0C_2 - F)']R'\}. \end{aligned}$$

Here X and Y also satisfy respectively the two Lyapunov equations

$$\begin{aligned} (A + B_2F)'X(A + B_2F) - X + (C_1 + D_{12}F)'(C_1 + D_{12}F) &= 0 \\ (A + LC_2)Y(A + LC_2)' - Y + (B_1 + LD_{21})(B_1 + LD_{21})' &= 0. \end{aligned}$$

Proof of Theorem 6.5.3 Let K be any admissible controller. Start with the system equations

$$\begin{aligned} \dot{\xi} &= A\xi + B_1\omega + B_2v \\ \zeta &= C_1\xi + D_{11}\omega + D_{12}v, \end{aligned}$$

and define a new control variable, $\rho := v - F\xi - F_0\omega$. The equations become

$$\begin{aligned} \dot{\xi} &= (A + B_2F)\xi + (B_1 + B_2F_0)\omega + B_2\rho \\ \zeta &= (C_1 + D_{12}F)\xi + (D_{11} + D_{12}F_0)\omega + D_{12}\rho, \end{aligned}$$

or in the frequency domain

$$\hat{\zeta} = \hat{g}_c\hat{\omega} + \hat{g}_i\hat{\rho},$$

where

$$\hat{g}_i(\lambda) = \left[\begin{array}{c|c} A + B_2F & B_2 \\ \hline C_1 + D_{12}F & D_{12} \end{array}\right].$$

This implies that

$$\hat{t}_{\zeta\omega} = \hat{g}_c + \hat{g}_i \hat{t}_{\rho\omega},$$

where $\hat{t}_{\rho\omega}$ is the transfer matrix from ω to ρ. So it follows from Lemma 6.4.2 that

$$\|\hat{t}_{\zeta\omega}\|_2^2 = \|\hat{g}_c\|_2^2 + \|R\hat{t}_{\rho\omega}\|_2^2.$$

Now we look at how ρ is generated:

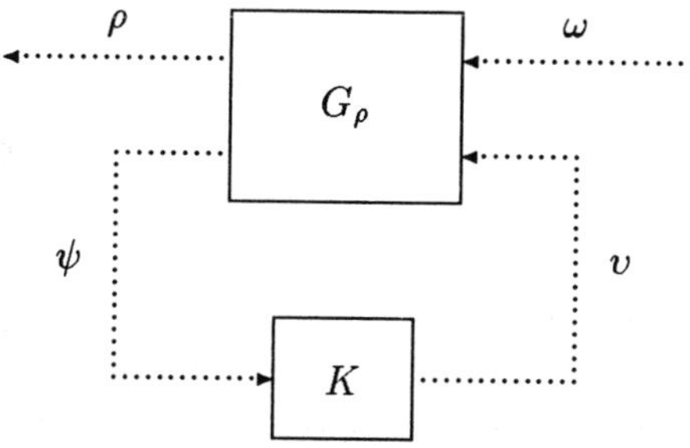

$$\hat{g}_\rho(\lambda) = \left[\begin{array}{c|cc} A & B_1 & B_2 \\ \hline -F & -F_0 & I \\ C_2 & D_{21} & 0 \end{array}\right].$$

Note that K stabilizes G iff K stabilizes G_ρ (the two closed-loop systems have identical A-matrices). So

$$\min_K \|\hat{t}_{\zeta\omega}\|_2^2 = \|\hat{g}_c\|_2^2 + \min_K \|R\hat{t}_{\rho\omega}\|_2^2.$$

Define

$$\begin{aligned} \rho_{new} &= R\rho \\ K_{new} &= RK \\ G_{\rho new} &= \begin{bmatrix} R & 0 \\ 0 & I \end{bmatrix} G_\rho \begin{bmatrix} I & 0 \\ 0 & R^{-1} \end{bmatrix}. \end{aligned}$$

Then minimizing $\|R\hat{t}_{\rho\omega}\|_2$ is exactly minimizing the norm on $\mathcal{H}_2(\mathbb{D})$ of the transfer matrix $\omega \mapsto \rho_{new}$:

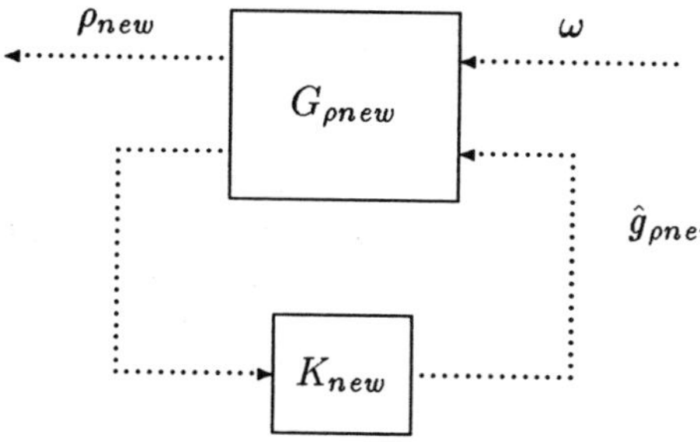

$$\hat{g}_{\rho new}(\lambda) = \left[\begin{array}{c|cc} A & B_1 & B_2 R^{-1} \\ \hline -RF & -RF_0 & I \\ C_2 & D_{21} & 0 \end{array}\right]$$

Now this is the second special problem. So by Theorem 6.5.2 the unique optimal controller is

$$\hat{k}_{new,opt}(\lambda) := \left[\begin{array}{c|c} A + LC_2 - B_2 L_0 C_2 + B_2 F & L - B_2 L_0 \\ \hline R(L_0 C_2 - F) & RL_0 \end{array}\right]$$

and the minimum cost is

$$\min \|R\hat{t}_{\rho w}\|_2 = \|\hat{g}_f\|_2.$$

Therefore for the original problem we have

$$\begin{aligned}\hat{k}_{opt}(\lambda) &= R^{-1}\hat{k}_{new,opt}(\lambda) \\ &= \left[\begin{array}{c|c} A + LC_2 - B_2L_0C_2 + B_2F & L - B_2L_0 \\ \hline L_0C_2 - F & L_0 \end{array}\right] \\ \min \|\hat{t}_{\zeta w}\|_2^2 &= \|\hat{g}_c\|_2^2 + \|\hat{g}_f\|_2^2.\end{aligned}$$

■

Example 6.5.1 Bilateral hybrid telerobot (cont'd). This example continues the telerobot problem begun in Example 2.2.1. There an analog controller was designed using $\mathcal{H}_2$-optimization in continuous time. Here we discretize the problem. The sampled output is $\zeta := Sz = ST_{zw}w$ and the corresponding SD system is shown in Figure 6.1. Instead of minimizing

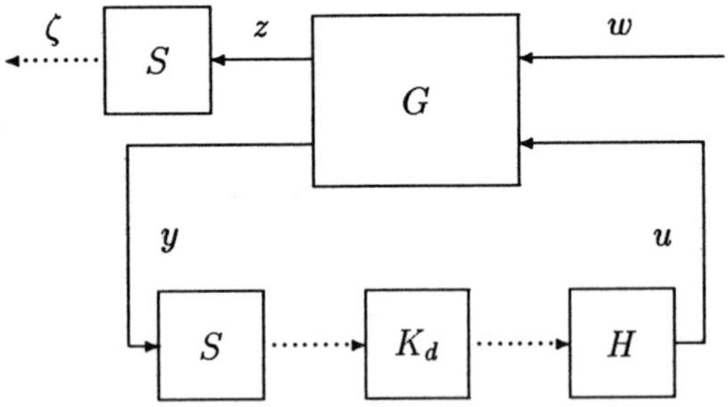

Figure 6.1: SD system with sampled output.

$\sum_i \|T_{zw}\delta e_i\|_2^2$ as before, we shall now minimize $\sum_i \|ST_{zw}\delta e_i\|_2^2$, namely, the average energy of ζ. Fix $w = \delta e_i$. Bringing the two samplers and the hold into G in Figure 6.1 leads to Figure 6.2. This is not quite a discrete-time system because of the continuous-time input w. Notice that $SG_{12}H$ and $SG_{22}H$ are the usual step-invariant transformations G_{12d} and G_{22d}. So the system from w to ζ is as follows:

$$\zeta = SG_{11}\delta e_i + G_{12d}K_d(I - G_{22d}K_d)^{-1}SG_{21}\delta e_i. \tag{6.27}$$

To convert this to a purely discrete-time system, we need to find discrete-time systems having the same outputs as $SG_{11}\delta e_i$ and $SG_{21}\delta e_i$. The next lemma provides this.

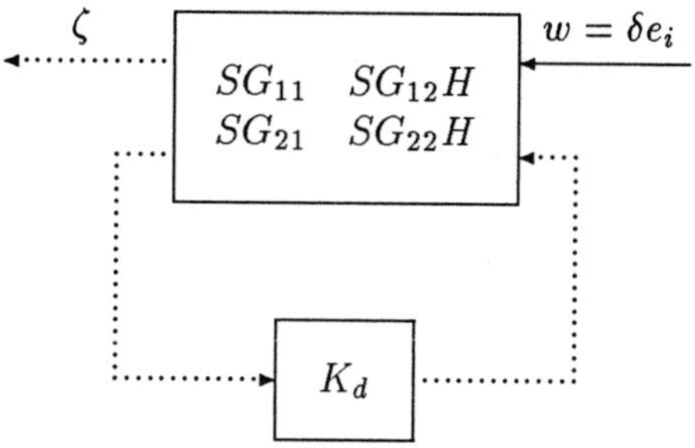

Figure 6.2: Discretized system with continuous-time input w.

Lemma 6.5.1 *Suppose G is a continuous-time system with transfer matrix*

$$\left[\begin{array}{c|c} A & B \\ \hline C & 0 \end{array}\right].$$

Let F_d denote the discrete-time system with transfer matrix

$$\left[\begin{array}{c|c} e^{hA} & e^{hA}B \\ \hline C & CB \end{array}\right].$$

Then $SG\delta e_i = F_d\delta_d e_i$.

The proof is left as Exercise 6.15.

Using this lemma in (6.27) and also using the realization

$$\left[\begin{array}{c|cc} A & B_1 & B_2 \\ \hline C_1 & 0 & D_{12} \\ C_2 & 0 & 0 \end{array}\right]$$

of G [see (2.3)], we arrive at the equivalent discrete-time system, Figure 6.3. It is emphasized that the discrete-time system $G_{eq,d}$ is *not* the discretization G_d of G. Let $T_{\zeta\omega}$ denote the system from ω to ζ in Figure 6.3. The optimization problem has been reduced to the minimization of $\sum_i \|T_{\zeta\omega}\delta_d e_i\|_2^2$. But this equals precisely $\|\hat{t}_{\zeta\omega}\|_2^2$, the square of the $\mathcal{H}_2$-norm of $\hat{t}_{\zeta\omega}(\lambda)$.

Let us summarize the derivation:

$$\begin{aligned} \sum_i \|ST_{zw}\delta e_i\|_2^2 \text{ in Figure 6.1} &= \sum_i \|T_{\zeta\omega}\delta_d e_i\|_2^2 \text{ in Figure 6.3} \\ &= \|\hat{t}_{\zeta\omega}\|_2^2 \text{ in Figure 6.3.} \end{aligned}$$

In this way we arrive at a discrete-time $\mathcal{H}_2$ problem, namely, the minimization of $\|\hat{t}_{\zeta\omega}\|_2$ in Figure 6.3.

For the computation below, we take $h = 0.2$ and the weights α_v, α_c, α_f, and α_s are as before. Some regularization is required. First, the poles of $\hat{g}_m(s)$ and $\hat{g}_s(s)$ at $s = 0$ are perturbed to $s = 10^{-3}$. Second, for the problem at hand, the matrix in the D_{21}-location of $G_{eq,d}$ equals

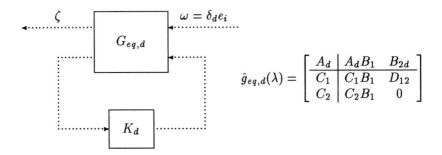

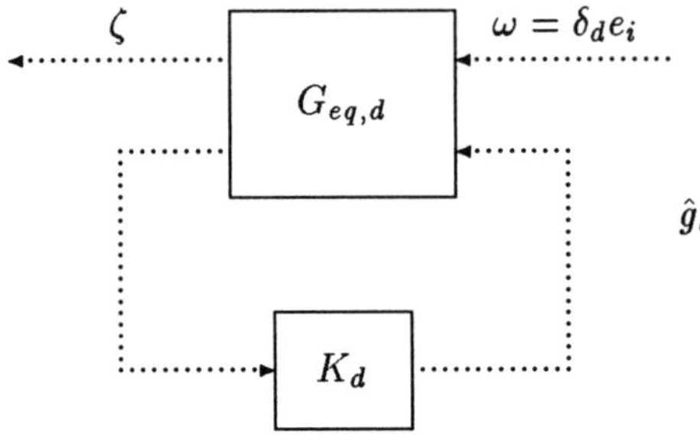

$$\hat{g}_{eq,d}(\lambda) = \left[\begin{array}{c|cc} A_d & A_d B_1 & B_{2d} \\ \hline C_1 & C_1 B_1 & D_{12} \\ C_2 & C_2 B_1 & 0 \end{array}\right]$$

Figure 6.3: Equivalent discrete-time system.

$$C_2 B_1 = \begin{bmatrix} 0 & 20 \\ 0 & 0 \\ 0 & 0 \end{bmatrix}.$$

To get the rank up to 3, the number of rows, two additional columns are added to give

$$\begin{bmatrix} 0 & 0 & 0 & 20 \\ \epsilon & 0 & 0 & 0 \\ 0 & \epsilon & 0 & 0 \end{bmatrix}.$$

Then $C_1 B_1$ and $A_d B_1$ in the D_{11}- and B_1-locations are padded with two zero columns. Again, ϵ was set to 0.1.

The resulting responses are shown in Figures 6.4 and 6.5. For comparison, Figures 6.6 and 6.7 show the responses for K_d equal to the discretization of the optimal analog controller computed before. Since $h = 0.2$ is large, the latter controller is quite inferior to the optimal discrete-time one.

6.6 $\mathcal{H}_2$-Optimal Step Tracking

In Section 5.5 we saw how to design a step-tracking controller, that is, a controller K so that the system

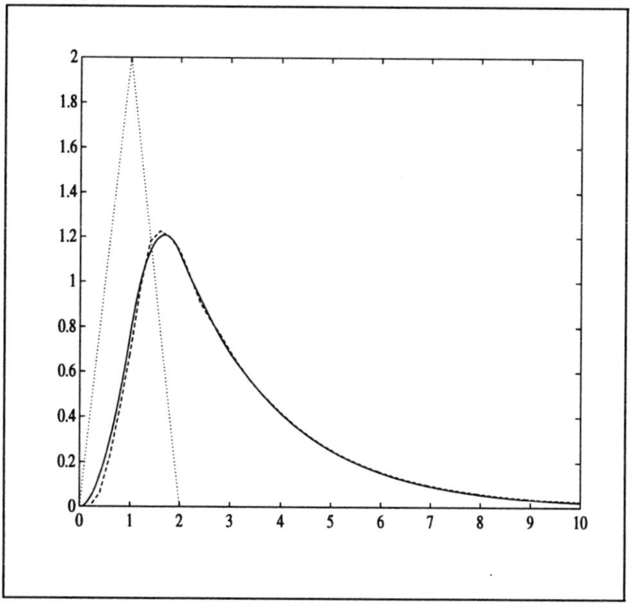

Figure 6.4: Design for discretized problem: v_s (solid), v_m (dash), and f_h (dot).

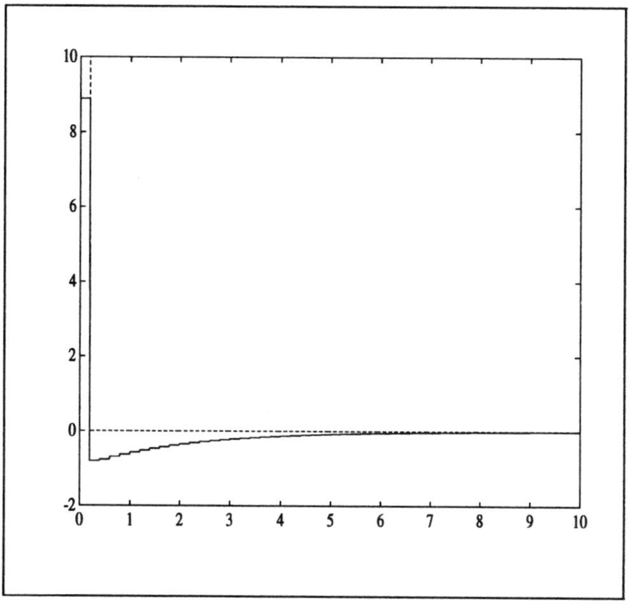

Figure 6.5: Design for discretized problem: f_m (solid) and f_e (dash).

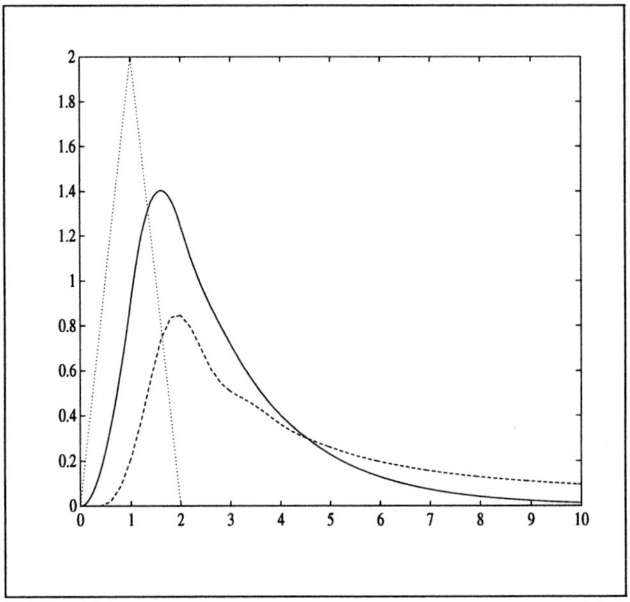

Figure 6.6: Discretization of optimal analog controller: v_s (solid), v_m (dash), and f_h (dot).

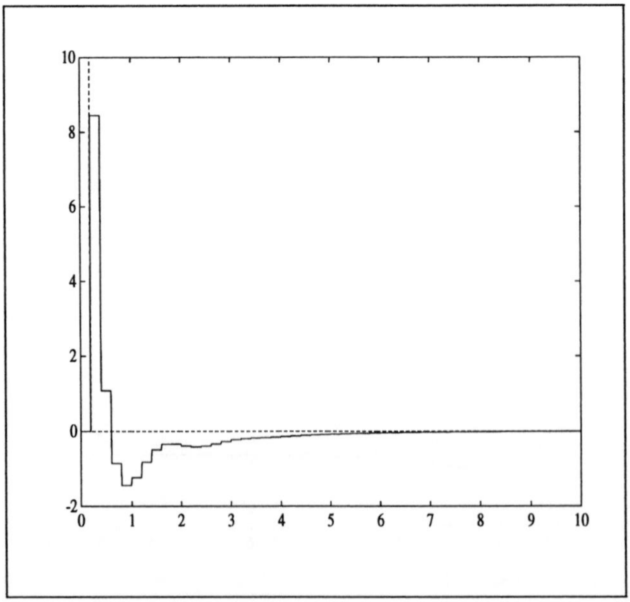

Figure 6.7: Discretization of optimal analog controller: f_m (solid) and f_e (dash).

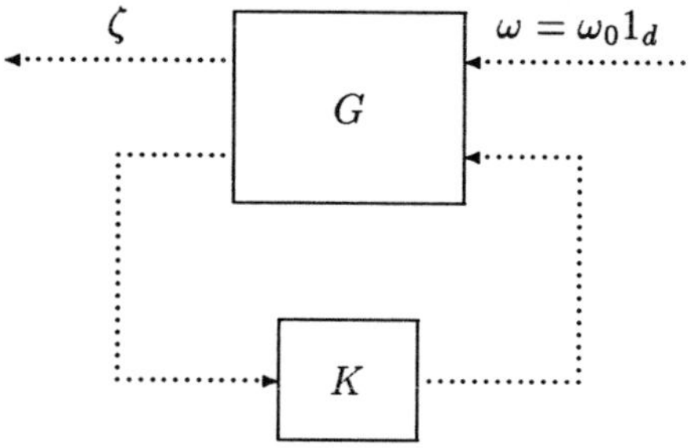

has the property that $\zeta(k) \to 0$ for every constant vector ω_0; 1_d is the scalar-valued unit-step function. In this section we fix ω and strengthen the convergence criterion by requiring that K minimize $\|\zeta\|_2$. By absorbing ω_0 into G, we can assume from the start that ω is the scalar-valued step, 1_d.

We shall outline a solution that reduces the problem to the standard $\mathcal{H}_2$-optimal control problem. Begin with a state model for G:

$$\hat{g}(\lambda) = \left[\begin{array}{c|cc} A & B_1 & B_2 \\ \hline C_1 & D_{11} & D_{12} \\ C_2 & D_{21} & 0 \end{array}\right].$$

It is assumed, as usual, that (A, B_2) is stabilizable and (C_2, A) is detectable. Also, since ω is 1-dimensional, the three matrices B_1, D_{11}, D_{21} have only one column each. As in Section 5.4, parametrize K and then write the transfer matrix from ω to ζ in terms of the parameter Q:

$$\hat{\zeta} = (\hat{t}_1 + \hat{t}_2\hat{q}\hat{t}_3)\hat{\omega}.$$

Then $\zeta(k) \to 0$ iff [see also (5.9)]

$$\hat{t}_1(1) + \hat{t}_2(1)\hat{q}(1)\hat{t}_3(1) = 0.$$

This is a linear matrix equation in $\hat{q}(1)$. Assuming a solution exists, let Q_0 be constant matrix solution. Then the following formula parametrizes all solutions in $\mathcal{RH}_\infty(\mathbb{D})$:

$$\hat{q}(\lambda) = Q_0 + (1-\lambda)\hat{q}_1(\lambda), \quad \hat{q}_1 \in \mathcal{RH}_\infty(\mathbb{D}).$$

Letting N denote the SISO system with transfer function $1-\lambda$, we get the following formula for Q: $Q = Q_0 + NQ_1$. With this formula and the representation $1_d = N^{-1}\delta_d$, we get the following time-domain relationship from ω to ζ:

$$\begin{aligned} \zeta &= (T_1 + T_2QT_3)\omega \\ &= [T_1 + T_2(Q_0 + NQ_1)T_3]N^{-1}\delta_d \\ &= (T_1 + T_2Q_0T_3)N^{-1}\delta_d + T_2Q_1T_3\delta_d. \end{aligned}$$

Defining $R := (T_1 + T_2Q_0T_3)N^{-1}$, we get

$\zeta = (R + T_2 Q_1 T_3)\delta_d.$

The latter equation corresponds to the block diagram

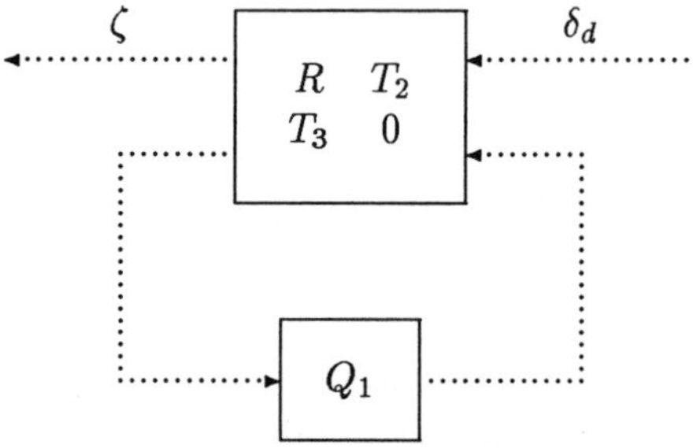

which is a standard $\mathcal{H}_2$-optimal control problem.

Let us summarize the steps as follows:

Step 1 Starting with a realization of G, obtain a parametrization of all stabilizing controllers; get the three matrices $\hat{t}_i$ $(i = 1, 2, 3)$ so that

$$\zeta = (T_1 + T_2 Q T_3)\omega.$$

Step 2 Solve the following constant matrix equation for Q_0:

$$\hat{t}_1(1) + \hat{t}_2(1) Q_0 \hat{t}_3(1) = 0.$$

(If no solution exists, then step tracking is not possible.)

Step 3 Define $\hat{n}(\lambda) = 1 - \lambda$ and get a state model for $R := (T_1 + T_2 Q_0 T_3)N^{-1}$. (Exercise 6.13 is useful here.)

Step 4 Get a state model for

$$G_{tmp} = \begin{bmatrix} R & T_2 \\ T_3 & 0 \end{bmatrix}$$

and solve the corresponding $\mathcal{H}_2$ problem. Denote the optimal controller by Q_1.

Step 5 Back-solve for Q from $Q = Q_0 + NQ_1$; then get K from Q.

Example 6.6.1 Consider the sampled-data setup of Figure 6.8. The plant P is a stable, SISO, second-order system with transfer function

$$\hat{p}(s) = \frac{1}{(10s + 1)(25s + 1)}$$

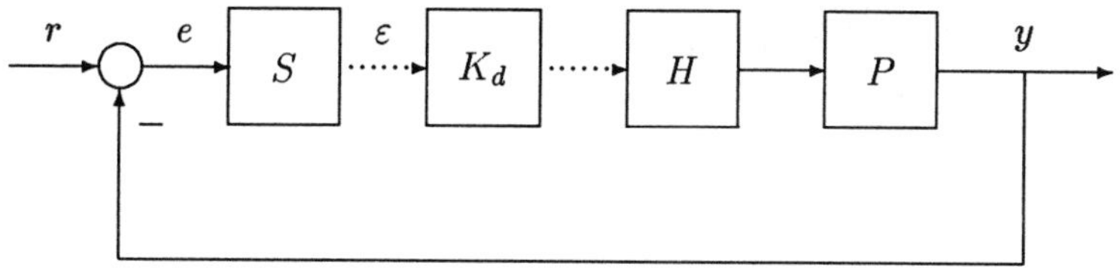

Figure 6.8: A sampled-data tracking system.

and the reference input r is the unit step. The goal is that the plant output y should track r "optimally."

We shall design the controller by discretizing the plant at the sampling period $h = 1$ s, which is much smaller than the time constants of the plant (10 and 25 s). The discretized plant $P_d = SPH$ has the transfer function

$$\begin{aligned} \hat{p}_d(\lambda) &= \frac{2.0960 \times 10^{-3}\lambda(\lambda + 1.0478)}{(\lambda - 1.0408)(\lambda - 1.1052)} \\ &= \left[\begin{array}{cc|c} 0.8675 & -0.0037 & 0.9325 \\ 0.9325 & 0.9981 & 0.4773 \\ \hline 0 & 0.0040 & 0 \end{array}\right] \end{aligned}$$

and the discretized system is shown in Figure 6.9. Here $\rho := Sr$, $\varepsilon := Se$, and $\psi := Sy$. Note that since r is the continuous-time unit step, ρ is the discrete-time unit step, 1_d. The optimal tracking problem we pose is to design an LTI K_d to achieve internal stability and minimize $\|\varepsilon\|_2$. This performance criterion ignores intersample behaviour: $\|\varepsilon\|_2$ could be small and yet $\|e\|_2$ could be large. This is an important point and we shall return to it later.

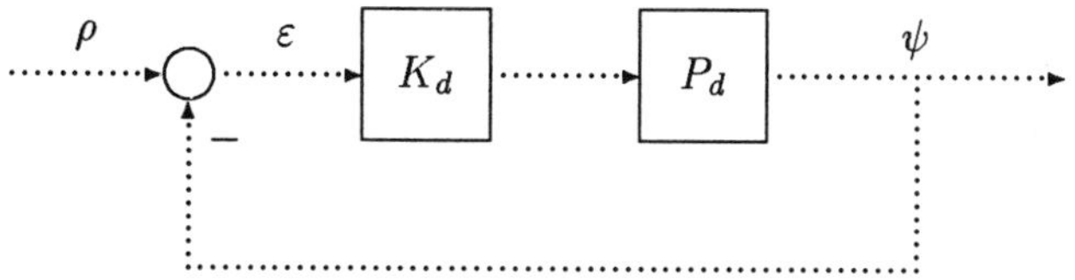

Figure 6.9: The discretized tracking system.

This discrete-time problem can be solved by the general procedure given above, but here we give a slight variation. We begin by parametrizing the family of controllers that achieve internal stability. The formula for $\hat{k}_d$ is

$$\hat{k}_d = \frac{\hat{q}}{1 - \hat{p}_d\hat{q}}, \qquad \hat{q} \in \mathcal{RH}_\infty(\mathbb{D}).$$

Then the transfer function from ρ to ε is $1 - \hat{p}_d\hat{q}$. Thus

$$\hat{\varepsilon}(\lambda) = [1 - \hat{p}_d(\lambda)\hat{q}(\lambda)]\frac{1}{1-\lambda},$$

so $\hat{q}(\lambda)$ must be of the form

$$\hat{q}(\lambda) = 1 + (1-\lambda)\hat{q}_1(\lambda), \qquad \hat{q}_1 \in \mathcal{RH}_\infty(\mathbb{D})$$

and then

$$\hat{\varepsilon} = \hat{t}_1 - \hat{t}_2\hat{q}_1,$$

where

$$\hat{t}_1(\lambda) = \frac{1-\hat{p}_d(\lambda)}{1-\lambda}, \quad \hat{t}_2(\lambda) = \hat{p}_d(\lambda).$$

The time-domain equation is

$$\varepsilon = (T_1 - T_2 Q_1)\omega,$$

where ω is the unit impulse. This problem *is* in the standard form, namely,

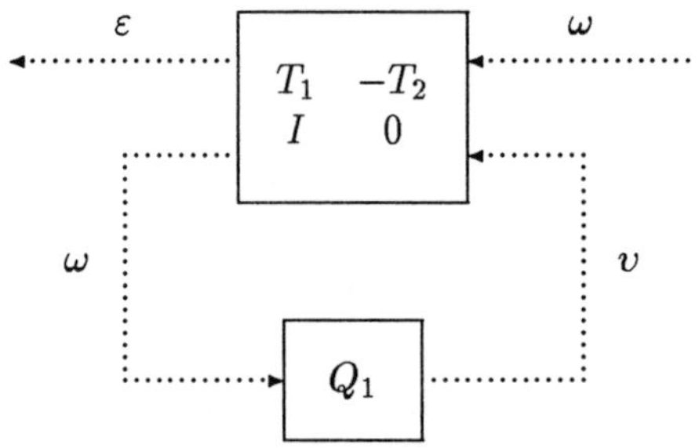

Now bring in realizations of T_1 and T_2:

$$\hat{t}_1(\lambda) = \left[\begin{array}{c|c} A_{t1} & B_{t1} \\ \hline C_{t1} & D_{t1} \end{array}\right], \quad \hat{t}_2(\lambda) = \left[\begin{array}{c|c} A_{t2} & B_{t2} \\ \hline C_{t2} & D_{t2} \end{array}\right].$$

The induced realization for

$$G_{tmp} := \begin{bmatrix} T_1 & -T_2 \\ I & 0 \end{bmatrix}$$

is

$$\left[\begin{array}{c|cc} A & B_1 & B_2 \\ \hline C_1 & D_{11} & D_{12} \\ 0 & I & 0 \end{array}\right] = \left[\begin{array}{cc|cc} A_{t1} & 0 & B_{t1} & 0 \\ 0 & A_{t2} & 0 & B_{t2} \\ \hline C_{t1} & -C_{t2} & D_{t1} & -D_{t2} \\ 0 & 0 & I & 0 \end{array}\right].$$

For the data at hand, the numbers are

$$A = \begin{bmatrix} 0.8675 & -0.0037 & 0 & 0 \\ 0.9325 & 0.9981 & 0 & 0 \\ 0 & 0 & 0.8675 & -0.0037 \\ 0 & 0 & 0.9325 & 0.9981 \end{bmatrix},$$

$$B_1 = \begin{bmatrix} 0.9325 \\ -249.5227 \\ 0 \\ 0 \end{bmatrix}, \quad B_2 = \begin{bmatrix} 0 \\ 0 \\ 0.9325 \\ 0.4773 \end{bmatrix},$$

$$C_1 = \begin{bmatrix} 0 & -0.0040 & 0 & -0.0040 \end{bmatrix}, \quad D_{11} = 1, \quad D_{12} = 0.$$

Since $D_{12} = 0$, the problem does not satisfy the assumptions of Section 6.5. But

$$\|\varepsilon\|_2^2 = \|\varepsilon(0)\|^2 + \|\dot{\varepsilon}\|_2^2.$$

If ξ denotes the state of G_{tmp}, then

$$\varepsilon = C_1\xi + D_{11}\omega,$$

so

$$\varepsilon(0) = D_{11}\omega(0).$$

Hence

$$\|\varepsilon\|_2^2 = \|D_{11}\omega(0)\|^2 + \|\dot{\varepsilon}\|_2^2$$

and an equivalent optimization problem is to minimize $\|\dot{\varepsilon}\|_2$. For $k \geq 0$ we have

$$\begin{aligned} \dot{\varepsilon} &= C_1\dot{\xi} \\ &= C_1A\xi + C_1B_1\omega + C_1B_2v. \end{aligned}$$

Thus the equivalent problem pertains to the generalized plant

$$\left[\begin{array}{c|cc} A & B_1 & B_2 \\ \hline C_1A & C_1B_1 & C_1B_2 \\ 0 & I & 0 \end{array}\right].$$

In this case, the values are

$$\begin{aligned} C_1A &= \begin{bmatrix} -0.0037 & -0.0040 & -0.0037 & -0.0040 \end{bmatrix} \\ C_1B_1 &= 0.9981 \\ C_1B_2 &= -0.0019. \end{aligned}$$

This model satisfies all four assumptions required for Theorem 6.5.1. We compute

$$\begin{aligned} M &= (C_1B_2)'(C_1B_2) \\ &= 3.6451 \times 10^{-6} \end{aligned}$$

and

$$A - B_2 M^{-1} B_2' C_1' C_1 A = \begin{bmatrix} 0.8675 & -0.0037 & 0 & 0 \\ 0.9325 & 0.9981 & 0 & 0 \\ -1.8219 & -1.9500 & -0.9544 & -1.9538 \\ -0.9325 & -0.9981 & 0 & 0 \end{bmatrix}.$$

Note that the latter matrix is singular; so we have to work with the generalized eigenproblem for the symplectic pair. The computations give

$$\begin{aligned} X &= 0_{4\times 4} \\ F &= -(M + B_2' X B_2)^{-1}(B_2' X A + B_2' C_1' C_1 A) \\ &= \begin{bmatrix} -1.9538 & -2.0911 & -1.9538 & -2.0911 \end{bmatrix} \\ F_0 &= -(M + B_2' X B_2)^{-1}(B_2' X B_1 + B_2' C_1' C_1 B_1) \\ &= 522.7759. \end{aligned}$$

Then from Theorem 6.5.1

$$\hat{q}_1(\lambda) = \left[\begin{array}{c|c} A + B_2 F & B_1 + B_2 F_0 \\ \hline F & F_0 \end{array}\right].$$

The computation of $\hat{q}$ and then $\hat{k}_d$ proceeds by back-substitution. The optimal controller is

$$\hat{k}_d(\lambda) = \frac{-477.1019(\lambda - 1.1052)(\lambda - 1.0408)}{(\lambda + 1.0478)(\lambda - 1)}.$$

Notice that $\hat{k}_d$ contains the pole at $\lambda = 1$ required for step tracking. In addition, it cancels all the stable poles and zeros of $\hat{p}_d$. For this controller, the sampled error, ε, is the unit impulse, $\varepsilon = \delta_d$. The discretized system has a deadbeat response, the plant output ψ requiring only one discrete-time step (1 s in real time) to reach its final value.

The sampled-data system of Figure 6.8 was simulated for a step input (simulation of sampled-data systems is treated in Chapter 8). The continuous-time output $y(t)$ is plotted in Figure 6.10. Notice that $y(t) = 1$ at the sampling instants, but considerable intersample ripple is present, and the settling time is quite long too. This example indicates that optimal design based on discrete performance specs alone may be ill-posed because intersample behaviour is completely ignored and the behaviour at sampling instants is over-emphasized.

6.7 Transfer Function Approach

So far we have focused on a state-space approach to $\mathcal{H}_2$ optimization; this approach has the advantage that all computations involve only matrices. Before we conclude this chapter, it is enlightening to look at briefly an alternative

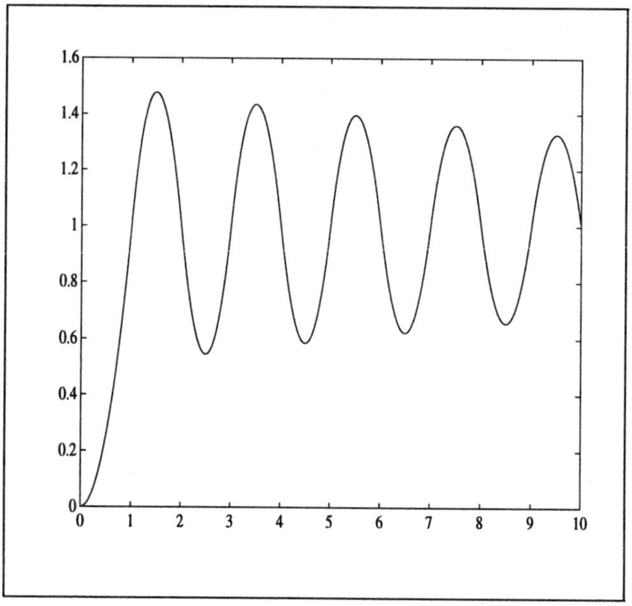

Figure 6.10: Step-response of example.

approach, based on transfer functions, which continues with the method developed in Chapter 5, namely, to parametrize all stabilizing controllers and then to seek a suitable parameter in the frequency domain.

In general, one can use the controller parametrization of Section 5.4 to reduce an $\mathcal{H}_2$-optimal control problem to one of the form

$$\min_{\hat{q}\in\mathcal{RH}_\infty(\mathbb{D})} \|\hat{t}_1 + \hat{t}_2\hat{q}\hat{t}_3\|_2.$$

Let us now address this latter problem; for simplicity, we restrict to the SISO case where $\hat{t}_i$ and $\hat{q}$ are 1×1. Then we may as well assume that $\hat{t}_3 = 1$. Also, the problem is more suggestive as one of approximation when it has a minus sign:

$$\min_{\hat{q}\in\mathcal{RH}_\infty(\mathbb{D})} \|\hat{t}_1 - \hat{t}_2\hat{q}\|_2.$$

We shall use some function spaces from Section 4.5. Fix $\hat{t}_1$ and $\hat{t}_2$ in $\mathcal{RH}_2(\mathbb{D}) = \mathcal{RH}_\infty(\mathbb{D})$. Think of $\hat{t}_1$ as a model to be matched by the product $\hat{t}_2\hat{q}$ by designing $\hat{q}$ in $\mathcal{RH}_2(\mathbb{D})$. The model-matching error is precisely the $\mathcal{RH}_2(\mathbb{D})$-norm $\|\hat{t}_1 - \hat{t}_2\hat{q}\|_2$. The trivial case is when $\hat{t}_2^{-1} \in \mathcal{RH}_2(\mathbb{D})$, for then the unique optimal $\hat{q}$ is obviously $\hat{t}_2^{-1}\hat{t}_1$. Thus the more interesting case is when $\hat{t}_2^{-1} \notin \mathcal{RH}_2(\mathbb{D})$. For simplicity (and also for the existence of the optimal $\hat{q}$, see Exercise 6.20) we shall *assume* that $\hat{t}_2$ has no zeros on the unit circle.

It is now important to introduce the notions of inner and outer functions. A function in $\mathcal{RH}_2(\mathbb{D})$ is an *inner function* if its magnitude equals 1 everywhere on the unit circle. It is not difficult to show that such a function has pole-zero symmetry in the sense that a point λ_0 is a zero iff its conjugate reciprocal, $1/\overline{\lambda}_0$, is a pole. Consequently, all its zeros lie inside the unit disk $\mathbb{D}$, hence the adjective "inner." An inner function is, up to sign, the product of factors of the form

$$\frac{\lambda - a}{1 - \overline{a}\lambda}, \qquad |a| < 1.$$

Examples of inner functions are

$$1, \quad \lambda, \quad \frac{\lambda - 0.5}{1 - 0.5\lambda}.$$

A function in $\mathcal{RH}_2(\mathbb{D})$ is an *outer function* if its zeros are all outside $\mathbb{D}$. Examples of outer functions are

$$1, \quad \lambda - 1, \quad \frac{\lambda - 2}{\lambda + 3}.$$

It is a useful fact that every function in $\mathcal{RH}_2(\mathbb{D})$ can be written as the product of two such factors: For example

$$\frac{\lambda(\lambda-1)(2\lambda+3)(2\lambda+1)}{\lambda-4} = \left[\frac{\lambda(\lambda+0.5)}{1+0.5\lambda}\right]\left[\frac{2(\lambda-1)(2\lambda+3)(1+0.5\lambda)}{\lambda-4}\right].$$

Using this factorization, we can solve the $\mathcal{H}_2$ model-matching problem as follows. Fix $\hat{q}$ in $\mathcal{RH}_2(\mathbb{D})$. Let $\hat{t}_2 = \hat{t}_{2i}\hat{t}_{2o}$ be a factorization into inner and outer factors. Then

$$\begin{aligned} \|\hat{t}_1 - \hat{t}_2\hat{q}\|_2^2 &= \|\hat{t}_1 - \hat{t}_{2i}\hat{t}_{2o}\hat{q}\|_2^2 \\ &= \|\hat{t}_{2i}(\hat{t}_1\hat{t}_{2i}^{-1} - \hat{t}_{2o}\hat{q})\|_2^2 \\ &= \|\hat{t}_1\hat{t}_{2i}^{-1} - \hat{t}_{2o}\hat{q}\|_2^2. \end{aligned}$$

The last equality follows from the fact that $|\hat{t}_{2i}(\lambda)| = 1$ whenever $|\lambda| = 1$. The function $\hat{t}_1\hat{t}_{2i}^{-1}$ lives in $\mathcal{RL}_2(\partial\mathbb{D})$. Project it into $\mathcal{RH}_2(\mathbb{D})$ and $\mathcal{RH}_2(\mathbb{D})^\perp$:

$$\hat{t}_1\hat{t}_{2i}^{-1} = (\hat{t}_1\hat{t}_{2i}^{-1})_{\mathcal{RH}_2(\mathbb{D})^\perp} + (\hat{t}_1\hat{t}_{2i}^{-1})_{\mathcal{RH}_2(\mathbb{D})}.$$

Now use the fact that $\|\hat{f} + \hat{g}\|_2^2 = \|\hat{f}\|_2^2 + \|\hat{g}\|_2^2$ for $\hat{f}$ in $\mathcal{RH}_2(\mathbb{D})^\perp$ and $\hat{g}$ in $\mathcal{RH}_2(\mathbb{D})$:

$$\begin{aligned} \|\hat{t}_1 - \hat{t}_2\hat{q}\|_2^2 &= \|(\hat{t}_1\hat{t}_{2i}^{-1})_{\mathcal{RH}_2(\mathbb{D})^\perp} + (\hat{t}_1\hat{t}_{2i}^{-1})_{\mathcal{RH}_2(\mathbb{D})} - \hat{t}_{2o}\hat{q}\|_2^2 \\ &= \|(\hat{t}_1\hat{t}_{2i}^{-1})_{\mathcal{RH}_2(\mathbb{D})^\perp}\|_2^2 + \|(\hat{t}_1\hat{t}_{2i}^{-1})_{\mathcal{RH}_2(\mathbb{D})} - \hat{t}_{2o}\hat{q}\|_2^2. \end{aligned}$$

Observing that $\hat{t}_{2o}^{-1} \in \mathcal{RH}_2(\mathbb{D})$, we conclude that the unique optimal $\hat{q}$ is given by

$$\hat{q} = \hat{t}_{2o}^{-1}(\hat{t}_1\hat{t}_{2i}^{-1})_{\mathcal{RH}_2(\mathbb{D})}.$$

Example 6.7.1 In Example 6.6.1, an $\mathcal{H}_2$-optimal step tracking problem was reduced to an $\mathcal{H}_2$ model matching problem:

$$\min_{\hat{q}_1 \in \mathcal{RH}_\infty(\mathbb{D})} \|\hat{t}_1 - \hat{t}_2\hat{q}_1\|_2$$

with

$$\hat{t}_1(\lambda) := \frac{1-\hat{p}_d(\lambda)}{1-\lambda} = \frac{-0.9979(\lambda - 1.1527)}{(\lambda - 1.1052)(\lambda - 1.0408)}$$

and

$$\hat{t}_2(\lambda) := \hat{p}_d(\lambda) = \frac{2.0960 \times 10^{-3}\lambda(\lambda + 1.0478)}{(\lambda - 1.0408)(\lambda - 1.1052)}.$$

The inner factor of $\hat{t}_2$ is simply $\hat{t}_{2i}(\lambda) = \lambda$, and then $\hat{t}_{2o} = \hat{t}_2/\hat{t}_{2i}$. The projections of $\hat{t}_1\hat{t}_{2i}^{-1}$ are

$$(\hat{t}_1\hat{t}_{2i}^{-1})_{\mathcal{RH}_2(\mathbb{D})^\perp}(\lambda) = \frac{1}{\lambda}, \quad (\hat{t}_1\hat{t}_{2i}^{-1})_{\mathcal{RH}_2(\mathbb{D})}(\lambda) = \frac{-\lambda + 1.1481}{(\lambda - 1.1052)(\lambda - 1.0408)}.$$

Thus the optimal $\hat{q}_1$ is

$$\hat{t}_{2o}^{-1}(\hat{t}_1\hat{t}_{2i}^{-1})_{\mathcal{RH}_2(\mathbb{D})}(\lambda) = \frac{-477.1019(\lambda - 1.1481)}{\lambda + 1.0478}.$$

Back-substitute to get the same optimal controller as in Example 6.6.1.

Exercises

6.1 Consider

$$\dot{\xi} = A\xi + Bv, \quad \xi(0) = \xi_0,$$

with A stable. Prove true or false: For every v in $\ell_2(\mathbb{Z}_+)$, $\xi(k)$ tends to 0 as k tends to ∞.

6.2 Consider a discrete-time system with control input v, disturbance input ω, and output ζ_1, modeled in the frequency domain by the equation

$$\hat{\zeta}_1(\lambda) = \frac{\lambda(\lambda+1)}{-\lambda^2 - \lambda + 6}\hat{\omega}(\lambda) + \frac{\lambda^2}{-2\lambda^2 - 3\lambda + 2}\hat{v}(\lambda).$$

1. Set

$$\zeta_2 = v, \quad \zeta = \begin{bmatrix} \zeta_1 \\ \zeta_2 \end{bmatrix}.$$

Derive a state model for the system from $\begin{bmatrix} \omega \\ v \end{bmatrix}$ to ζ. Let ξ denote the state.

2. Now consider state-feedback control, that is, $\psi = \xi$:

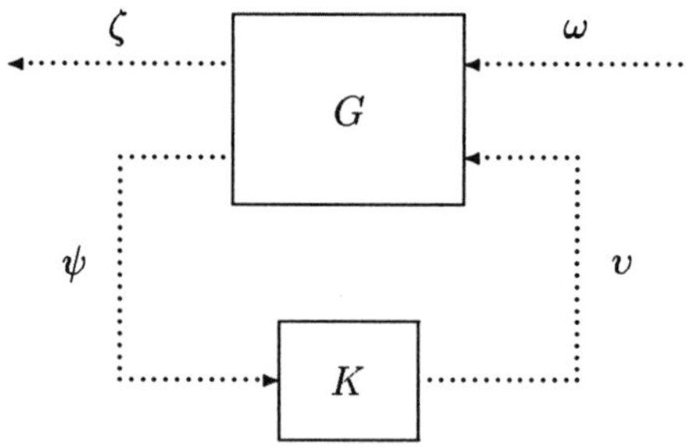

Compute the optimal internally stabilizing K to minimize $\|\zeta\|_2$ when ω is the unit impulse. Also, compute the minimum value of $\|\zeta\|_2$.

6.3 Given

$$\hat{g}(\lambda) = \left[\begin{array}{c|c} A & B \\ \hline C & D \end{array}\right]$$

with A nonsingular, show that

$$\hat{g}^{\sim}(\lambda) = \left[\begin{array}{c|c} A'^{-1} & -A'^{-1}C' \\ \hline B'A'^{-1} & D' - B'A'^{-1}C' \end{array}\right].$$

6.4 Define S_2, X, F, A_F, C_{1F} as in Section 6.4. Write down the Riccati equation associated with X. Show that this Riccati equation can be rewritten as a Lyapunov equation:

$$A_F'XA_F - X + C_{1F}'C_{1F} = 0.$$

You might want to establish the following two identities first:

$$A_F = (I + B_2M^{-1}B_2'X)^{-1}(A - B_2M^{-1}D_{12}'C_1),$$

$$-B_2'XA_F = MF + D_{12}'C_1.$$

6.5 Prove the second statement in Lemma 6.4.2.

6.6 Recall in the LQR problem in Section 6.1, the optimal control is a constant state feedback alone. If one converts the LQR problem into an $\mathcal{H}_2$-optimization problem as in Section 6.1, one gets an impulse input of the form $\delta_d(k+1)\omega_0$. This exercise is to study full-information control in this situation.

The system equations are

$$\begin{aligned} \dot{\xi} &= A\xi + B_1\omega + B_2 v, \quad \omega = \dot{\delta}_d\omega_0, \; \xi(-1) = 0 \\ \zeta &= C_1\xi + D_{11}\omega + D_{12}v \end{aligned}$$

with the same assumptions as in Section 6.4. The optimization problem is again

$$\min_{v \in \ell(\mathbb{Z}_+)} ||\zeta||_2.$$

Define $S_2, X, F, \hat{g}_i$ as in Theorem 6.4.1 and Lemma 6.4.2, but define

$$\hat{g}_c(\lambda) = \left[\begin{array}{c|c} A + B_2F & B_1 \\ \hline C_1 + D_{12}F & D_{11} \end{array}\right].$$

Prove

1. $\hat{g}_c^{\sim}\hat{g}_i \in \mathcal{RH}_2(\mathbb{D})$. Thus $\hat{g}_i^{\sim}\hat{g}_c\hat{\omega} \in \mathcal{RH}_2(\mathbb{D})^{\perp}$.
2. The unique optimal control is $v_{opt} = F\xi$. Moreover,

$$\min_v ||\zeta||_2 = ||\hat{g}_c\omega_0||_2.$$

6.7 Suppose v and ψ are scalar-valued signals and the transfer function from v to ψ is $\lambda^2/(\lambda-1)^2$. For the standard canonical realization (A, B, C) consider the optimization problem

$$\min_{v=F\xi} \sum_{k=0}^{\infty} \rho\psi(k)^2 + v(k)^2,$$

where ρ is positive. Plot the eigenvalues of $A + BF$ as ρ varies from 0 to ∞.

6.8 Pertaining to the standard setup, prove the following two statements:

1. Given the plant

$$\hat{g}(\lambda) = \left[\begin{array}{c|cc} A & B_1 & B_2 \\ \hline C_1 & D_{11} & D_{12} \\ \begin{bmatrix} C_2 \\ 0 \end{bmatrix} & \begin{bmatrix} 0 \\ I \end{bmatrix} & \begin{bmatrix} 0 \\ 0 \end{bmatrix} \end{array}\right]$$

with exactly the same assumptions on the matrices as in Theorem 6.5.1, the $\mathcal{H}_2$-optimal controller is

$$\hat{k}_{opt}(\lambda) = \left[\begin{array}{c|cc} A + B_2F - B_1C_2 & B_1 & B_1 + B_2F_0 \\ \hline F & 0 & F_0 \end{array}\right],$$

where the matrices F and F_0 are as defined in Theorem 6.5.1.

2. Given the plant

$$\hat{g}(\lambda) = \left[\begin{array}{c|cc} A & B_1 & \begin{bmatrix} B_2 & 0 \end{bmatrix} \\ \hline C_1 & D_{11} & \begin{bmatrix} 0 & I \end{bmatrix} \\ C_2 & D_{21} & \begin{bmatrix} 0 & 0 \end{bmatrix} \end{array}\right]$$

with the same assumptions as in Theorem 6.5.2, the $\mathcal{H}_2$-optimal controller is

$$\hat{k}_{opt}(\lambda) = \left[\begin{array}{c|c} A + LC_2 - B_2C_1 & L \\ \hline C_1 & 0 \\ C_1 + L_0C_2 & L_0 \end{array}\right],$$

where L and L_0 are as defined in Theorem 6.5.2.

6.9 Prove uniqueness in Theorem 6.5.1. (Hint: for every admissible controller the equation

$$\|\hat{t}_{\zeta\omega}\|_2^2 = \|\hat{g}_c\|_2^2 + \|(M + B_2'XB_2)^{1/2}\hat{t}_{\rho\omega}\|_2^2$$

is still valid; see the proof of Theorem 6.5.3. Now show that the unique solution of $\hat{t}_{\rho\omega} = 0$ is the controller given.)

6.10 In addition to the four assumptions made in Theorem 6.5.3, let us assume further the orthogonality and normality conditions:

(a) $D_{12}' \begin{bmatrix} C_1 & D_{12} \end{bmatrix} = \begin{bmatrix} 0 & I \end{bmatrix}$,

(b) $\begin{bmatrix} B_1 \\ D_{21} \end{bmatrix} D_{21}' = \begin{bmatrix} 0 \\ I \end{bmatrix}$.

For example, (b) concerns how the exogenous signal ω enter G: The plant disturbance and the sensor noise are orthogonal and the sensor noise weighting is normalized. Simplify the formulas in Theorem 6.5.3. What simplification can you make for Theorem 6.4.1 under condition (a)?

6.11 Consider the following setup:

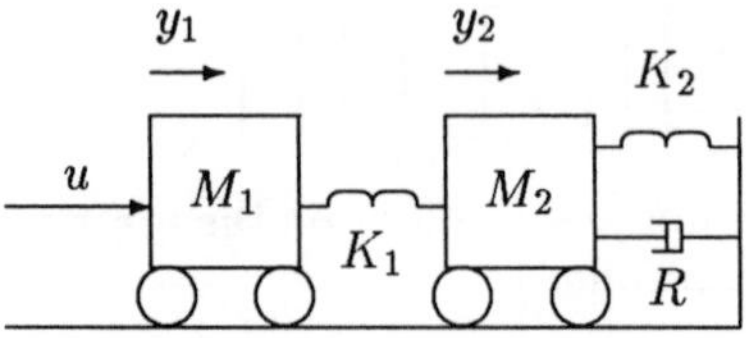

Shown are two carts, of masses M_1 and M_2, two springs, with spring constants K_1 and K_2, and a damper, with constant R. A force u is applied and y_1, y_2 denote displacements.

Consider a controller with input y_2 and output u. The control task is to have the righthand cart follow a given step reference signal r; the tracking error is therefore $r - y_2$.

1. Taking the controller to be of the form SK_dH, put the system into the standard SD form and find a state model for G with $x = (y_1, \dot{y}_1, y_2, \dot{y}_2)$.

2. Outline a procedure to design a stabilizing controller K_d to minimize $\|S(r - y_2)\|_2$.

6.12 Consider $\mathcal{H}_2$-optimal control pertaining to the standard setup, where the system G has the following data:

$$\hat{g}(\lambda) = \left[\begin{array}{c|cc} A & B_1 & B_2 \\ \hline C_1 & 0 & 1 \\ C_2 & 0 & 0 \end{array}\right],$$

$$A = \begin{bmatrix} 0 & 1 \\ -0.5 & -1 \end{bmatrix}, \quad B_1 = \begin{bmatrix} 2 \\ 1 \end{bmatrix}, \quad B_2 = \begin{bmatrix} 0 \\ 1 \end{bmatrix},$$

$$C_1 = \begin{bmatrix} 2 & -1 \end{bmatrix}, \quad C_2 = \begin{bmatrix} 1 & 0 \end{bmatrix}.$$

Note that $D_{21} = 0$; so the problem has a singular sensor noise weighting. Often this can be fixed by introducing a time advance in the ω channel. Define

$$\hat{g}_{new}(\lambda) := \hat{g}(\lambda) \begin{bmatrix} \lambda^{-1} & 0 \\ 0 & 1 \end{bmatrix}.$$

1. Show that the $\mathcal{H}_2(\mathbb{D})$-norm of the closed-loop system is unchanged upon the introduction of λ^{-1} in the ω channel.

2. Obtain a state-space realization for $\hat{g}_{new}$.

3. Compute the optimal controller for $\hat{g}_{new}$. (This is also the optimal controller for $\hat{g}$.)

6.13 Show that if

$$\hat{g}(\lambda) = \left[\begin{array}{c|c} A & B \\ \hline C & D \end{array}\right]$$

and $\hat{g}(1) = 0$, then

$$\frac{1}{1-\lambda}\hat{g}(\lambda) = \left[\begin{array}{c|c} A & A(A-I)^{-1}B \\ \hline C & D \end{array}\right].$$

6.14 This is a discrete-time optimal observer problem. All signals are defined on the time set $\mathbb{Z}_+$.

The system is modelled as

$$\dot{\xi} = A\xi, \quad \psi = C\xi.$$

The state estimate, η, is to be the output of a FDLTI stable system K with input ψ. Since $\xi(0)$ would be unknown, we average the estimation error over the cases $\xi(0) = e_i$, where e_i is the ith basis vector. Thus the observer K is to minimize

$$\sum_i \{\|\eta - \xi\|_2^2 : \xi(0) = e_i\}.$$

Set this problem up as an $\mathcal{H}_2$ optimal control problem. Give conditions for a solution to exist. Give a procedure to compute a solution.

6.15 Prove Lemma 6.5.1.

6.16 In the telerobot example, $\hat{f}_e(s)$ and $\hat{f}_h(s)$ are not rational, containing time-delay terms. They were approximated by rational functions, $\hat{g}_e(s)$ and $\hat{g}_h(s)$. This simplifies the analog design because it makes the problem finite-dimensional. But actually it is not necessary for the discretized system because time delays discretize to finite-dimensional systems. Do the telerobot example by discretizing the problem but without approximating $\hat{f}_e(s)$ and $\hat{f}_h(s)$.

6.17 Solve the $\mathcal{H}_2$ model-matching problem for

$$\hat{t}_1(\lambda) = \frac{\lambda + 0.5}{\lambda - 2}, \quad \hat{t}_2(\lambda) = \frac{\lambda(3\lambda - 1)}{\lambda + 4}.$$

6.18 For the system in Figure 6.9, take

$$\hat{p}_d(\lambda) = \frac{\lambda}{(\lambda - 2)(3\lambda + 2)}.$$

Compute the LTI controller $\hat{k}_d$ to achieve internal stability and minimize $\|\varepsilon\|_2$ when ρ is the unit impulse. Use this method: Parametrize all stabilizing controllers; reduce to an $\mathcal{H}_2$ model-matching problem.

6.19 Repeat, but with

$$\hat{p}_d(\lambda) = \frac{\lambda}{\lambda - 2}, \quad \hat{\rho}(\lambda) = \frac{1}{\lambda + a}, \quad a > 1.$$

6.20 This problem looks at what happens in $\mathcal{H}_2$ model matching when $\hat{t}_2$ has a zero on the unit circle. Take $\hat{t}_1(\lambda) = 1$ and $\hat{t}_2(\lambda) = 1 - \lambda$. Find a sequence of $\hat{q}$s in $\mathcal{RH}_\infty(\mathbb{D})$ such that the associated model matching errors converge to zero. From this conclude that an optimal $\hat{q}$ does not exist.

Notes and References

The $\mathcal{H}_2$-optimal control problem posed in this chapter is deterministic, but it is mathematically equivalent to the minimum variance control problem and the LQG problem. The material in Section 6.1 is standard and is treated in many books, for example [54]. For more on $\mathcal{H}_2$-optimal control, and the role of the Riccati equation, see for example [10], [38], [97], [4]. The idea to use the generalized eigenproblem for symplectic pairs to study Riccati equations is due to Pappas, Laub, and Sandell, Jr. [118]; Theorem 6.2.1 is from [118]. Lemma 6.3.2 is standard; see for example [118], [149], and [75]. Theorem 6.3.2 is due to Kucera [97]. An indirect proof is given in [118] that X_1 is invertible; the proof given here follows that in [51]. The derivations in Sections 6.4 and 6.5 are new; they follow the continuous-time approach taken in [40]. The solution technique in Section 6.7 is due to Youla, Jabr, and Bongiorno [155] and Kucera [96].

Chapter 7

Introduction to Discrete-Time $\mathcal{H}_\infty$-Optimal Control

The problem studied in this chapter is the minimization of the $\mathcal{H}_\infty(\mathbb{D})$-norm from ω to ζ in the standard setup:

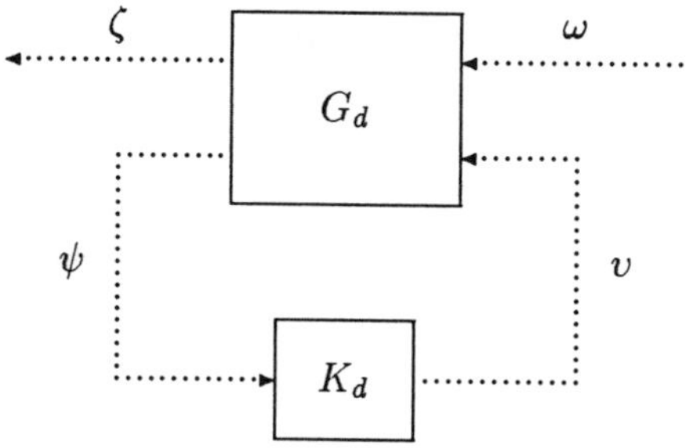

Recall (Theorem 4.4.2) that the $\mathcal{H}_\infty(\mathbb{D})$-norm equals the least upper bound of $\|\zeta\|_2$ over all ω with $\|\omega\|_2 \leq 1$; thus, the $\mathcal{H}_\infty(\mathbb{D})$-norm measures the system gain in the sense of energy. It is a useful alternative to the $\mathcal{H}_2(\mathbb{D})$-norm when the exogenous inputs are not fixed.

A complete development of the theory of discrete-time $\mathcal{H}_\infty$-optimal control would take many pages. Instead, this chapter presents only an introduction.

7.1 Computing the $\mathcal{H}_\infty$-Norm

Recall from Section 4.5 that the $\mathcal{H}_\infty$-norm of a stable transfer matrix $\hat{g}_d(\lambda)$ is defined to be the peak magnitude on the unit circle:

$$\|\hat{g}_d\|_\infty := \sup_{\theta} \ \sigma_{\max}\left[\hat{g}_d\left(e^{j\theta}\right)\right].$$

One way to compute this is to take a finite grid of frequencies, $0 \leq \theta_1 < \cdots < \theta_n \leq \pi$, and compute

$$\max_{i} \ \sigma_{\max}\left[\hat{g}_d\left(e^{j\theta_i}\right)\right].$$

There is another way, one that uses state data.

Begin with a state model

$$\hat{g}_d(\lambda) = \left[\begin{array}{c|c} A & B \\ \hline C & D \end{array}\right],$$

with $\rho(A) < 1$. The formula below for $\|\hat{g}_d\|_\infty$ involves the symplectic pair

$$\begin{aligned} S &= (S_l, S_r) \\ &= \left(\begin{bmatrix} A + BD'(\gamma^2 - DD')^{-1}C & 0 \\ -\gamma C'(\gamma^2 - DD')^{-1}C & I \end{bmatrix}, \right. \\ &\qquad \left. \begin{bmatrix} I & -\gamma B(\gamma^2 - D'D)^{-1}B' \\ 0 & [A + BD'(\gamma^2 - DD')^{-1}C]' \end{bmatrix} \right), \end{aligned}$$

where γ is a positive number. For $D = 0$ and $\gamma = 1$, this is just like the one in $\mathcal{H}_2$ theory except for a sign change in the $(1,2)$-entry of S_r (see Theorem 6.3.2). The matrices $\gamma^2 - DD'$, $\gamma^2 - D'D$ are invertible provided they are positive definite, equivalently, γ^2 is greater than the largest eigenvalue of DD' (or $D'D$), equivalently, $\gamma > \sigma_{max}(D)$.

Theorem 7.1.1 *Assume $\gamma > \sigma_{max}(D)$ and $0 \leq \theta < 2\pi$. Then γ is a singular value of $\hat{g}_d\left(e^{j\theta}\right)$ iff $e^{-j\theta}$ is an eigenvalue of S.*

Proof ($\Longrightarrow$) Assume γ is a singular value of $\hat{g}_d\left(e^{j\theta}\right)$. Then γ^2 is an eigenvalue of $\hat{g}_d\left(e^{j\theta}\right)^* \hat{g}_d\left(e^{j\theta}\right)$. So there exists a nonzero vector u such that

$$\hat{g}_d\left(e^{j\theta}\right)^* \hat{g}_d\left(e^{j\theta}\right) u = \gamma^2 u.$$

Defining $v = \gamma^{-1}\hat{g}_d\left(e^{j\theta}\right) u$, we have the pair of equations

$$\hat{g}_d\left(e^{j\theta}\right) u = \gamma v, \quad \hat{g}_d\left(e^{j\theta}\right)^* v = \gamma u.$$

Thus in terms of state matrices

$$\begin{aligned} Du + e^{j\theta} C\left(I - e^{j\theta}A\right)^{-1} Bu &= \gamma v \\ D'v + e^{-j}\ B'\left(I - e^{-j\theta}A'\right)^{-1} C'v &= {}^{\theta}\gamma u. \end{aligned}$$

Define

$$r := e^{j\theta}\left(I - e^{j\theta}A\right)^{-1} Bu, \quad s := \left(I - e^{-j\theta}A'\right)^{-1} C'v \tag{7.1}$$

so that

$$Du + Cr = \gamma v, \quad D'v + e^{-j\theta} B's = \gamma u.$$

Using the fact that

$$\begin{bmatrix} \gamma & -D' \\ -D & \gamma \end{bmatrix}^{-1} = \begin{bmatrix} \gamma(\gamma^2 - D'D)^{-1} & D'(\gamma^2 - DD')^{-1} \\ D(\gamma^2 - D'D)^{-1} & \gamma(\gamma^2 - DD')^{-1} \end{bmatrix},$$

we can solve the latter equations for u and v:

$$u = D'(\gamma^2 - DD')^{-1}Cr + e^{-j\theta}\gamma(\gamma^2 - D'D)^{-1}B's \quad (7.2)$$
$$v = \gamma(\gamma^2 - DD')^{-1}Cr + e^{-j\theta}D(\gamma^2 - D'D)^{-1}B's. \quad (7.3)$$

From (7.1) we have

$$e^{-j\theta}r = Ar + Bu, \quad s = e^{-j\theta}A's + C'v.$$

Substituting from (7.2) and (7.3) for u and v gives

$$\begin{aligned} e^{-j\theta}r &= Ar + BD'(\gamma^2 - DD')^{-1}Cr + e^{-j\theta}\gamma B(\gamma^2 - D'D)^{-1}B's \\ s &= e^{-j\theta}A's + \gamma C'(\gamma^2 - DD')^{-1}Cr + e^{-j\theta}C'(\gamma^2 - DD')^{-1}DB's, \end{aligned}$$

that is,

$$S_l \begin{bmatrix} r \\ s \end{bmatrix} = e^{-j\theta} S_r \begin{bmatrix} r \\ s \end{bmatrix}.$$

Since $\begin{bmatrix} r \\ s \end{bmatrix} \neq 0$ (since $u \neq 0$), $e^{-j\theta}$ is an eigenvalue of S.

($\Longleftarrow$) Reverse the argument—start with r, s and get u. ■

Corollary 7.1.1 *Let γ_{max} denote the maximum γ such that S has an eigenvalue on the unit circle. Then $\|\hat{g}_d\|_\infty = \max\{\sigma_{max}(D), \gamma_{max}\}$.*

Proof By the maximum modulus theorem,

$$\|\hat{g}_d\|_\infty \geq \sigma_{max}[\hat{g}_d(0)] = \sigma_{max}(D).$$

If $\|\hat{g}_d\|_\infty > \sigma_{max}(D)$, then from the theorem

$$\begin{aligned} \|\hat{g}_d\|_\infty &= \gamma_{max} \\ &= \max\{\sigma_{max}(D), \gamma_{max}\}. \end{aligned}$$

If $\|\hat{g}_d\|_\infty = \sigma_{max}(D)$, then for all $\gamma > \sigma_{max}(D)$, S does not have an eigenvalue on the unit circle, so

$$\begin{aligned} \|\hat{g}_d\|_\infty &= \sigma_{max}(D) \\ &= \max\{\sigma_{max}(D), \gamma_{max}\}. \end{aligned}$$

■

The computation of γ_{max} can be performed effectively by bisection search.

Example 7.1.1 As a very simple example, consider

$$\hat{g}_d(\lambda) = \left[\begin{array}{c|c} 0.5 & 1 \\ \hline 1 & 1 \end{array}\right] = \frac{1+0.5\lambda}{1-0.5\lambda}.$$

The Bode magnitude plot of $\hat{g}_d$ is maximum at DC ($\lambda = 1$) and that maximum value equals 3. Thus $\|\hat{g}_d\|_\infty = 3$.

Figure 7.1 is a plot, versus γ^{-1}, of the distance from the unit circle to the closest eigenvalue of S. Thinking in terms of root loci, one can see that

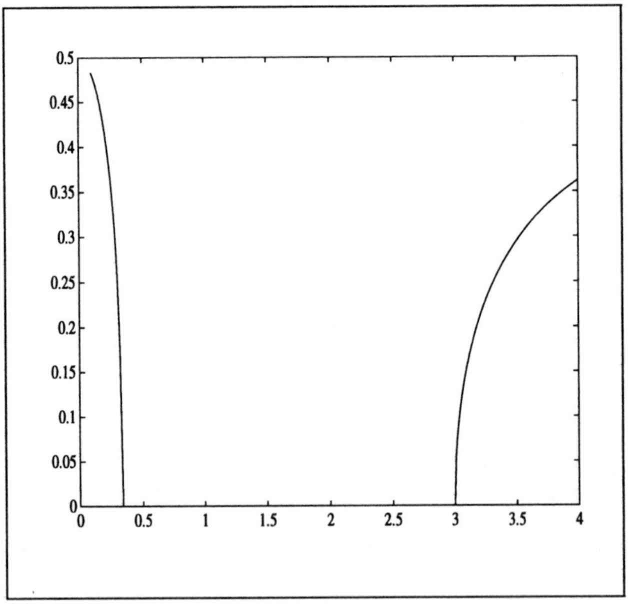

Figure 7.1: The distance from the unit circle to the closest eigenvalue of S plotted versus γ^{-1}.

as γ^{-1} increases from 0, some branch approaches the unit circle, remains on the unit circle, and then moves away from the unit circle again. There is an eigenvalue on the unit circle for γ^{-1} in the range $[1/3, 3]$. The minimum value equals $1/3$, so γ_{max} equals the reciprocal, 3. Since $D = 1$ we can apply the corollary to get

$$\|\hat{g}_d\|_\infty = \max\{1, 3\} = 3.$$

Example 7.1.2 This is an MIMO example:

$$\hat{g}_d(\lambda) = \begin{bmatrix} \frac{\lambda}{3+\lambda} & \frac{\lambda}{2-\lambda} \\ \frac{\lambda}{1+0.1\lambda} & \frac{5\lambda}{(4+\lambda)(4-\lambda)} \end{bmatrix}.$$

A realization is

$$A = \begin{bmatrix} 0.4 & 0.05 & 0 & 0 & 0 \\ 1 & 0 & 0 & 0 & 0 \\ 0 & 0 & -0.3333 & 0.0625 & 0.0208 \\ 0 & 0 & 1 & 0 & 0 \\ 0 & 0 & 0 & 1 & 0 \end{bmatrix}, \quad B = \begin{bmatrix} 1 & 0 \\ 0 & 0 \\ 0 & 1 \\ 0 & 0 \\ 0 & 0 \end{bmatrix}$$

$$C = \begin{bmatrix} 0.5 & 0.05 & 0.3333 & 0 & -0.0208 \\ 1 & -0.5 & 0.3125 & 0.1042 & 0 \end{bmatrix}, \quad D = \begin{bmatrix} 0 & 0 \\ 0 & 0 \end{bmatrix}.$$

Figure 7.2 is a plot of $\sigma_{\max}\left[\hat{g}_d\left(e^{j\theta}\right)\right]$ versus θ. The peak magnitude equals

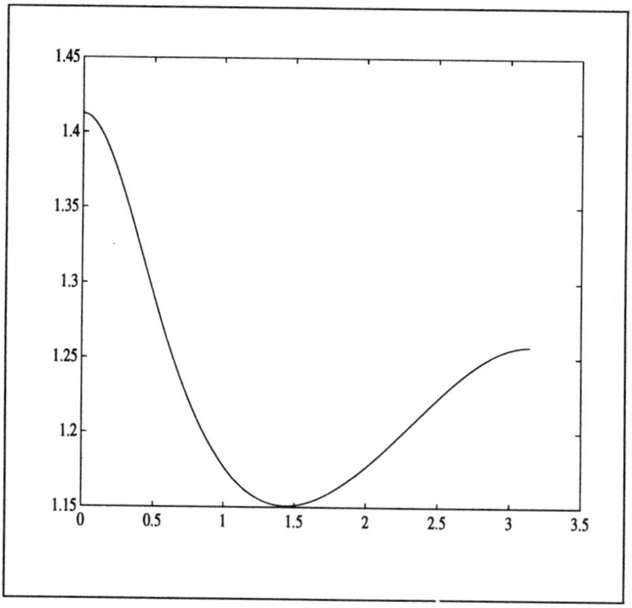

Figure 7.2: $\sigma_{\max}\left[\hat{g}_d\left(e^{j\theta}\right)\right]$ versus θ.

1.4122, so this equals the $\mathcal{H}_\infty(\mathbb{D})$-norm. The value of the function ranges over the interval $[1.1509, 1.4122]$. According to Theorem 7.1.1, S has an eigenvalue on the unit circle for all values of γ in the range $[1.1509, 1.4122]$, that is, for all values of γ^{-1} in the range $[0.7081, 0.8689]$.

Figure 7.3 is a plot, versus γ^{-1}, of the distance from the unit circle to the closest eigenvalue of S. There is an eigenvalue on the unit circle for γ^{-1} in the range $[0.7081, 0.8689]$. The minimum value equals 0.7081; by the corollary the reciprocal of this equals $\|\hat{g}_d\|_\infty$.

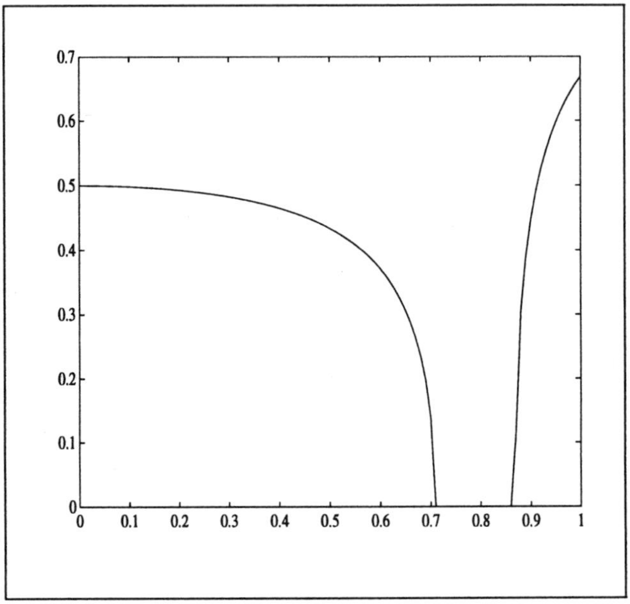

Figure 7.3: The distance from the unit circle to the closest eigenvalue of S plotted versus γ^{-1}.

7.2 Discrete-Time $\mathcal{H}_\infty$-Optimization by Bilinear Transformation

Since MATLAB has the function *hinfsyn* for continuous-time $\mathcal{H}_\infty$-optimal controller design, perhaps the simplest way to do discrete-time $\mathcal{H}_\infty$-optimal controller design is by converting to a continuous-time problem via bilinear transformation. The reason this works is that the bilinear transformation preserves $\mathcal{H}_\infty$-norms (whereas it does not preserve $\mathcal{H}_2$-norms, for example). More precisely, let $\hat{g}_d(\lambda)$ be a transfer matrix in $\mathcal{H}_\infty(\mathbb{D})$. Define $\hat{g}_c(s)$ via

$$\hat{g}_c(s) = \hat{g}_d\left(\frac{1-s}{1+s}\right).$$

Then $\hat{g}_c \in \mathcal{H}_\infty(\mathbb{C}_+)$ and $\|\hat{g}_c\|_\infty = \|\hat{g}_d\|_\infty$. The mapping between $\mathbb{C}_+$ and $\mathbb{D}$ used here is

$$\lambda = \frac{1-s}{1+s}, \quad s = \frac{1-\lambda}{1+\lambda}.$$

Notice that any other conformal mapping could be used with equal benefit. Notice also that the continuous-time system G_c so constructed is entirely artificial: It does not represent any physical system naturally associated with G_d.

Design of an $\mathcal{H}_\infty$-optimal discrete-time controller K_d for the system in Figure 7.4 via bilinear transformation goes as follows:

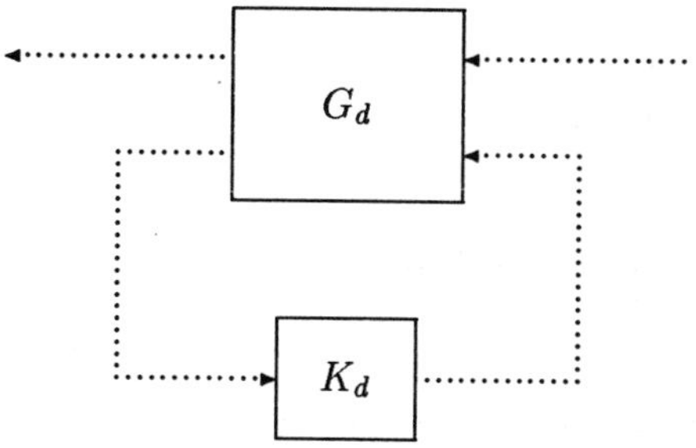

Figure 7.4: Discrete-time system; K_d is to be designed.

Step 1 Let the given state model for G_d be

$$\hat{g}_d(\lambda) = \left[\begin{array}{c|c} A_d & B_d \\ \hline C_d & D_d \end{array}\right] = \left[\begin{array}{c|cc} A_d & B_{d1} & B_{d2} \\ \hline C_{d1} & D_{d11} & D_{d12} \\ C_{d2} & D_{d21} & D_{d22} \end{array}\right].$$

Step 2 Define the artificial continuous-time system G_c by

$$\hat{g}_c(s) = \left[\begin{array}{c|c} A_c & B_c \\ \hline C_c & D_c \end{array}\right] = \left[\begin{array}{c|cc} A_c & B_{c1} & B_{c2} \\ \hline C_{c1} & D_{c11} & D_{c12} \\ C_{c2} & D_{c21} & D_{c22} \end{array}\right],$$

where

$$\begin{aligned} A_c &= (A_d - I)(A_d + I)^{-1} \\ B_c &= (I - A_c)B_d \\ C_c &= C_d(A_d + I)^{-1} \\ D_c &= D_d - C_c B_d. \end{aligned}$$

Step 3 Design an $\mathcal{H}_\infty(\mathbb{C}_+)$-optimal controller K_c for G_c. Let the state model for K_c be

$$\hat{k}_c(s) = \left[\begin{array}{c|c} A_{Kc} & B_{Kc} \\ \hline C_{Kc} & D_{Kc} \end{array}\right].$$

Step 4 Define the discrete-time system K_d by

$$\hat{k}_d(\lambda) = \left[\begin{array}{c|c} A_{Kd} & B_{Kd} \\ \hline C_{Kd} & D_{Kd} \end{array}\right],$$

where

$$\begin{aligned} A_{Kd} &= (I - A_{Kc})^{-1}(I + A_{Kc}) \\ B_{Kd} &= (I - A_{Kc})^{-1}B_{Kc} \\ C_{Kd} &= C_{Kc}(I + A_{Kd}) \\ D_{Kd} &= D_{Kc} + C_{Kc}B_{Kd}. \end{aligned}$$

Then K_d is $\mathcal{H}_\infty(\mathbb{D})$-optimal for G_d.

Clearly a necessary assumption for validity of the procedure is that $A_d + I$ and $I - A_{Kc}$ be invertible, that is, that -1 not be an eigenvalue of A_d and that 1 not be an eigenvalue of A_{Kc}. Also, the continuous-time $\mathcal{H}_\infty$ problem in Step 3 must be regular.

Example 7.2.1 Let us continue Examples 2.3.1 and 3.3.2. There we designed an analog controller K by $\mathcal{H}_\infty$-optimization in continuous time and discretized it. Here we shall design a discrete-time controller by discretizing the problem and designing in discrete time.

Figure 7.5 shows the feedback system with the discrete-time controller K_d to be designed. Figure 7.5 can be converted to the standard SD setup

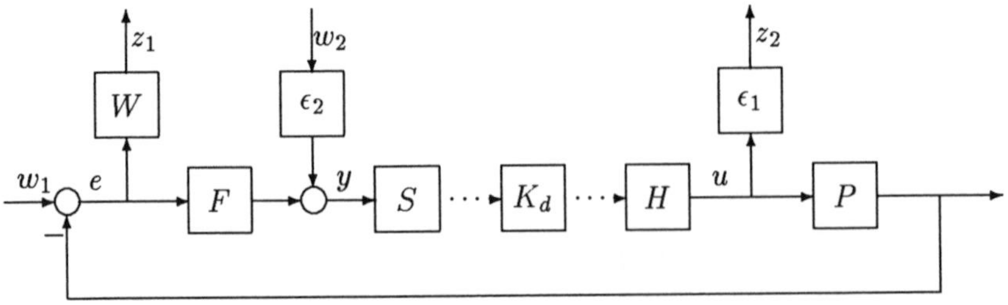

Figure 7.5: SD feedback system.

of Figure 7.6. Discretize the input and output as in Figure 7.7. Defining $G_d := SGH$, we arrive at Figure 7.4. Then K_d can be designed by the procedure above.

Figure 7.8 shows the results for the sampling period $h = 0.5$. The solid line is the Bode magnitude plot of $1/(1 + \hat{p}\hat{k}\hat{f})$, where $\hat{k}$ is the continuous-time controller obtained by $\mathcal{H}_\infty$-optimization in continuous time; the dashed line is the Bode magnitude plot of $1/(1 + \hat{p}\hat{r}\hat{k}_d\hat{f})$, where $\hat{k}_d$ is the discrete-time controller obtained by $\mathcal{H}_\infty$-optimization in discrete time. There is a

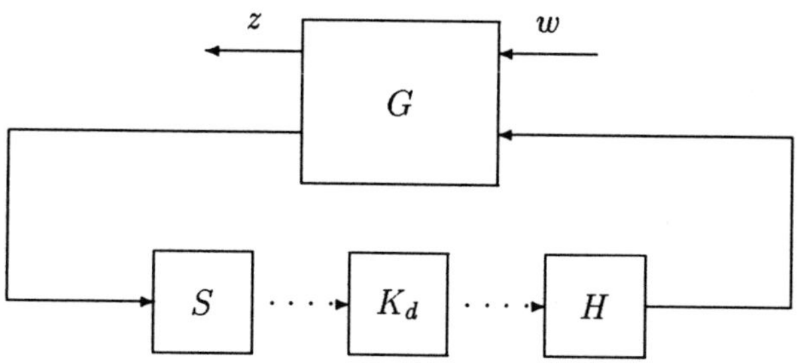

Figure 7.6: Standard SD system.

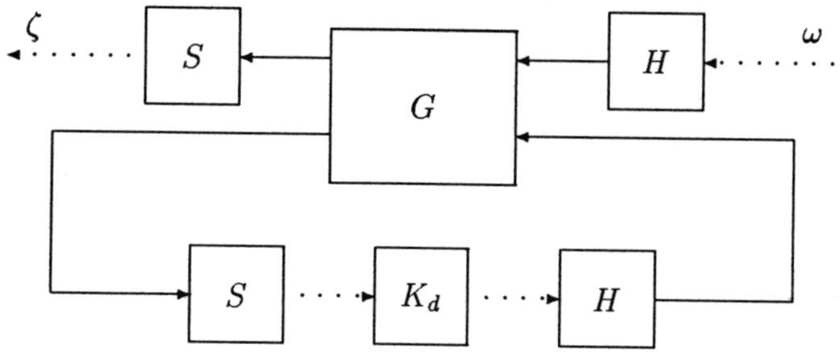

Figure 7.7: Discretization.

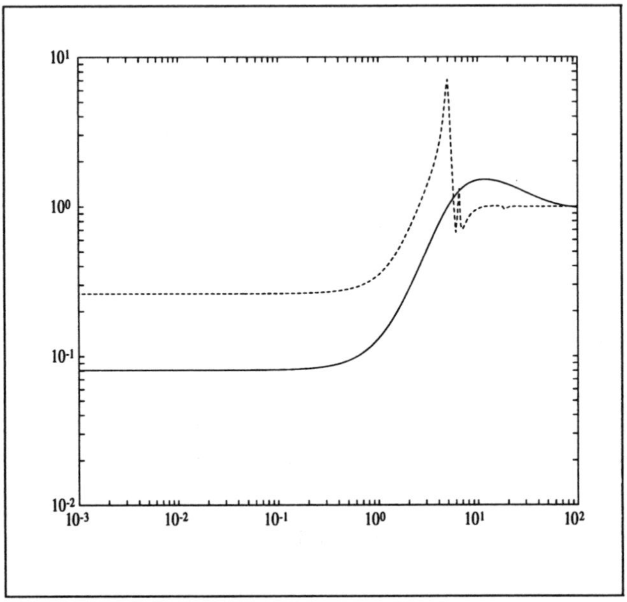

Figure 7.8: Bode magnitude plots: optimal analog controller (solid), optimal discrete-time controller (dash); $h = 0.5$.

large deterioration in performance because h is so large. Figure 7.9 shows the results for $h = 0.01$: The solid line is for the optimal analog controller, the dashed line is for its discretization (as in Figure 3.4), and the dotted line is for the optimal discrete-time controller; all three methods give essentially the same result.

This example is continued in Example 8.4.3.

Exercises

7.1 Compute the $\mathcal{H}_\infty(\mathbb{D})$-norm of

$$\begin{bmatrix} \frac{1}{2+\lambda} & \lambda^2 \\ 1-\lambda & \frac{1}{12-\lambda+\lambda^2} \end{bmatrix}$$

by finding a state model and using Corollary 7.1.1.

7.2 Compute the $\mathcal{H}_\infty(\mathbb{D})$-norm of the preceding matrix by converting to continuous time using the bilinear transformation, then using the MATLAB function *hinfnorm*.

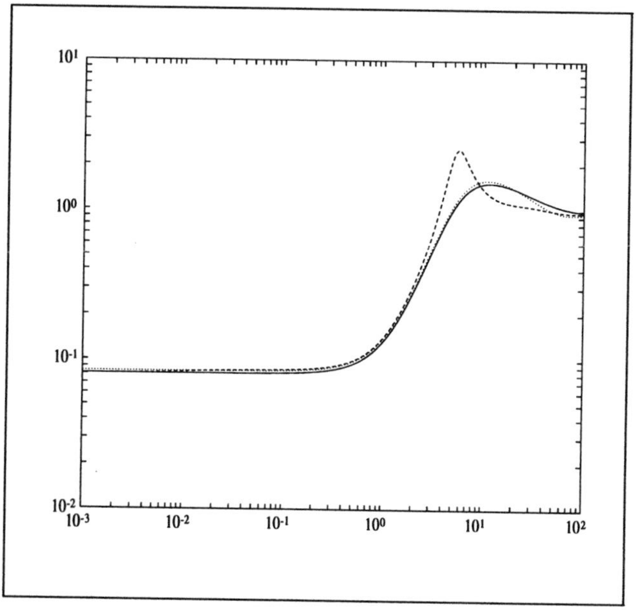

Figure 7.9: Bode magnitude plots: optimal analog controller (solid), discretization of optimal analog controller (dash), optimal discrete-time controller (dot); $h = 0.01$.

Notes and References

The $\mathcal{H}_\infty$-optimal control problem in discrete time is treated in [75], [13], [103], [61], [64], [106], [131], [107], [93], [63], [77], and [156]. Theorem 7.1.1 is adapted from the analogous continuous-time result in [20].

Chapter 8

Fast Discretization of SD Feedback Systems

The methods in Chapters 5, 6, and 7 are to design controllers for the plant discretized at the sampling frequency. This might result in unfortunate intersample ripple (Example 6.6.1). One way to guard against this is to discretize the signals of interest at a rate faster than the sampling frequency. This chapter develops this approach. The same idea also finds application in computer simulation of sampled-data systems.

8.1 Lifting Discrete-Time Signals

The idea of lifting a discrete-time signal has already been introduced in Example 4.4.2. Consider this situation: There is an underlying clock with base period h and there is a discrete-time signal $v(k)$ that is referred to the subperiod h/n of the base period, where n is some positive integer. That is, $v(0)$ occurs at time $t = 0$, $v(1)$ at $t = h/n$, $v(2)$ at $t = 2h/n$, etc. The *lifted signal*, $\underline{v}$, is defined as follows: If

$$v = \{v(0), v(1), v(2), \ldots\},$$

then

$$\underline{v} = \left\{ \begin{bmatrix} v(0) \\ v(1) \\ \vdots \\ v(n-1) \end{bmatrix}, \begin{bmatrix} v(n) \\ v(n+1) \\ \vdots \\ v(2n-1) \end{bmatrix}, \ldots \right\}.$$

Thus the dimension of $\underline{v}(k)$ equals n times that of $v(k)$ (hence the term "lifting") and $\underline{v}$ is regarded as referred to the base period; that is, $\underline{v}(k)$ occurs at time $t = kh$. The *lifting operator* L is defined to be the map $v \mapsto \underline{v}$ and is depicted by the block diagram

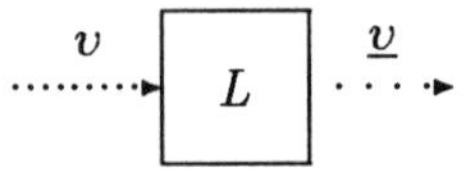

The slow-rate signal is shown with slow-frequency dots; the fast-rate signal with high-frequency dots. The vector representation of the equation $\underline{v} = Lv$ when $n = 2$ is

$$\begin{bmatrix} \underline{v}(0) \\ \underline{v}(1) \\ \underline{v}(2) \\ \vdots \end{bmatrix} = \left[\begin{array}{c|c|c|c|c|c} I & 0 & 0 & 0 & 0 & \cdots \\ 0 & I & 0 & 0 & 0 & \cdots \\ \hline 0 & 0 & I & 0 & 0 & \cdots \\ 0 & 0 & 0 & I & 0 & \cdots \\ \hline 0 & 0 & 0 & 0 & I & \cdots \\ 0 & 0 & 0 & 0 & 0 & \cdots \\ \hline \vdots & \vdots & \vdots & \vdots & \vdots & \end{array}\right] \begin{bmatrix} v(0) \\ v(1) \\ v(2) \\ \vdots \end{bmatrix}.$$

For the partition shown, $[L]$ is neither lower-triangular nor Toeplitz; therefore, as a system L is non-causal and time-varying.

It can be arranged that L is norm-preserving. Let us see this for the case of ℓ_2-norms:

$$\begin{aligned} v &\in \ell_2(\mathbb{Z}_+, \mathbb{R}^m) \\ \underline{v} &\in \ell_2(\mathbb{Z}_+, \mathbb{R}^{nm}) \\ \|v\|_2^2 &= v(0)'v(0) + v(1)'v(1) + \cdots \\ \|\underline{v}\|_2^2 &= \underline{v}(0)'\underline{v}(0) + \underline{v}(1)'\underline{v}(1) + \cdots \\ &= \begin{bmatrix} v(0) \\ \vdots \\ v(n-1) \end{bmatrix}' \begin{bmatrix} v(0) \\ \vdots \\ v(n-1) \end{bmatrix} \\ &\quad + \begin{bmatrix} v(n) \\ \vdots \\ v(2n-1) \end{bmatrix}' \begin{bmatrix} v(n) \\ \vdots \\ v(2n-1) \end{bmatrix} + \cdots \\ &= v(0)'v(0) + v(1)'v(1) + \cdots \\ &= \|v\|_2^2. \end{aligned}$$

Thus $\|Lv\|_2 = \|v\|_2$.

The inverse of lifting, L^{-1}, is defined as follows: If

$$\psi = \left\{ \begin{bmatrix} \psi_1(0) \\ \vdots \\ \psi_n(0) \end{bmatrix}, \begin{bmatrix} \psi_1(1) \\ \vdots \\ \psi_n(1) \end{bmatrix}, \ldots \right\}$$

and

$$v = L^{-1}\psi,$$

then

$$v = \{\psi_1(0), \cdots, \psi_n(0), \psi_1(1), \cdots, \psi_n(1), \cdots\}.$$

The corresponding matrix for $n = 2$ is

$$[L^{-1}] = \left[\begin{array}{cc|cc|cc|c} I & 0 & 0 & 0 & 0 & 0 & \cdots \\ \hline 0 & I & 0 & 0 & 0 & 0 & \cdots \\ \hline 0 & 0 & I & 0 & 0 & 0 & \cdots \\ \hline 0 & 0 & 0 & I & 0 & 0 & \cdots \\ \hline 0 & 0 & 0 & 0 & I & 0 & \cdots \\ \hline \vdots & \vdots & \vdots & \vdots & \vdots & \vdots & \end{array}\right].$$

Clearly, L^{-1} is causal but time-varying.

8.2 Lifting Discrete-Time Systems

For a discrete-time FDLTI system G_d with underlying period h/n, lifting the input and output signals so that the lifted signals correspond to the base period h results in a *lifted system*, $\underline{G}_d := LG_dL^{-1}$, as depicted in Figure 8.1. It is not hard to show that $\underline{G}_d$ is LTI too. Given $\hat{g}_d$ in terms of state-space

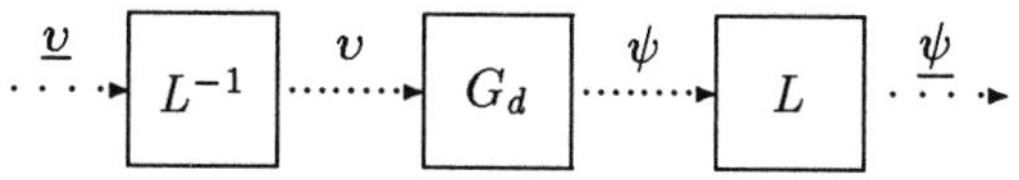

Figure 8.1: The lifted system.

data,

$$\hat{g}_d(\lambda) = \left[\begin{array}{c|c} A & B \\ \hline C & D \end{array}\right],$$

how can one compute the transfer function $\hat{\underline{g}}_d$ for $\underline{G}_d$?

Theorem 8.2.1 *The lifted system $\underline{G}_d$ is FDLTI; in fact,*

$$\hat{\underline{g}}_d(\lambda) = \left[\begin{array}{c|cccc} A^n & A^{n-1}B & A^{n-2}B & \cdots & B \\ \hline C & D & 0 & \cdots & 0 \\ CA & CB & D & \cdots & 0 \\ \vdots & \vdots & \vdots & & \vdots \\ CA^{n-1} & CA^{n-2}B & CA^{n-3}B & \cdots & D \end{array}\right].$$

Proof We shall prove the result when $n = 2$; the general case follows similarly. Looking at the system matrices, we have

$$\begin{aligned}
[\underline{G}_d] &= [L]\,[G_d]\,[L^{-1}] \\
&= [L] \begin{bmatrix} D & 0 & 0 & 0 & \cdots \\ CB & D & 0 & 0 & \cdots \\ CAB & CB & D & 0 & \cdots \\ CA^2B & CAB & CB & D & \cdots \\ \vdots & \vdots & \vdots & \vdots & \end{bmatrix} [L^{-1}] \\
&= \left[\begin{array}{cc|cc|c} D & 0 & 0 & 0 & \cdots \\ CB & D & 0 & 0 & \cdots \\ \hline CAB & CB & D & 0 & \cdots \\ CA^2B & CAB & CB & D & \cdots \\ \hline \vdots & \vdots & \vdots & \vdots & \end{array}\right].
\end{aligned}$$

(Note that $[\underline{G}_d]$ looks the same as $[G_d]$ except for a repartition of the blocks.) The transfer matrix corresponding to the latter matrix is

$$\left[\begin{array}{c|cc} A^2 & AB & B \\ \hline C & D & 0 \\ CA & CB & D \end{array}\right].$$

This is therefore $\hat{\underline{g}}_d$ (when $n = 2$). With such a state-space model, $\underline{G}_d$ is clearly FDLTI and causal. ■

If A is stable, so is A^n. Since the lifting operation preserves norms, it follows that norms of the two transfer functions $\hat{g}_d$ and $\hat{\underline{g}}_d$ satisfy: $\|\hat{g}_d\|_2^2 = \|\hat{\underline{g}}_d\|_2^2/n$ and $\|\hat{g}_d\|_\infty = \|\hat{\underline{g}}_d\|_\infty$. This is a very useful property.

8.3 Fast Discretization of a SD System

We now turn to the standard SD system in Figure 8.2. Choose an integer n, introduce fast sample and hold operators, S_f and H_f, with period h/n, and consider the discretized system in Figure 8.3. This is the *fast discretization* of the system in Figure 8.2. We would expect the latter to emulate the former if n is sufficiently large, and thus fast discretization is useful for both analysis and design. The design techniques in Chapters 5, 6 had $n = 1$.

To make the system in Figure 8.3 amenable to analysis, first move the samplers and holds in Figure 8.3 into G to get Figure 8.4, where

$$P := \begin{bmatrix} S_f & 0 \\ 0 & S \end{bmatrix} G \begin{bmatrix} H_f & 0 \\ 0 & H \end{bmatrix}.$$

This is a discrete-time system, alright, but at two sampling rates. Figure 8.4 therefore represents a time-varying discrete-time system.

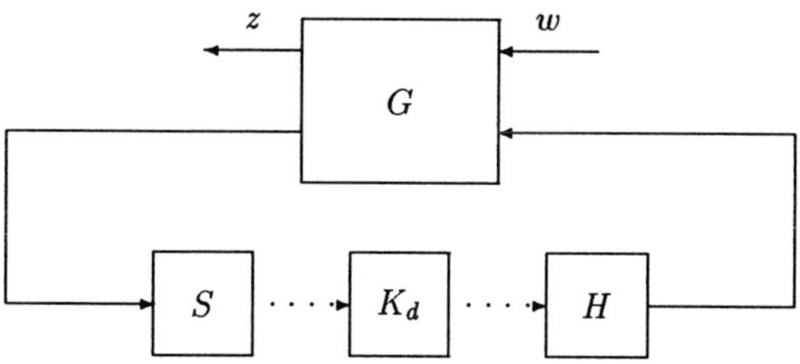

Figure 8.2: The standard SD system.

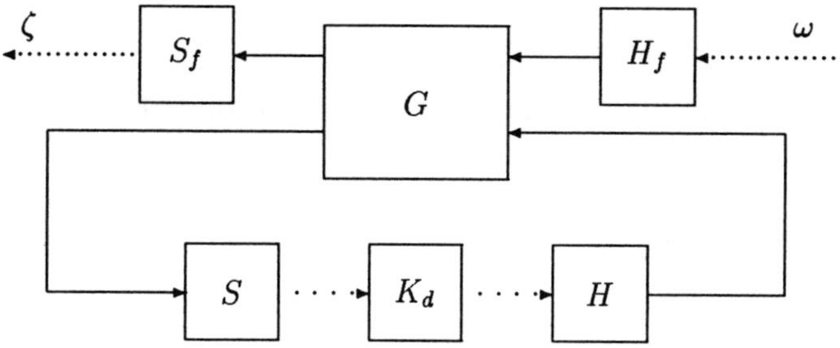

Figure 8.3: Fast discretization.

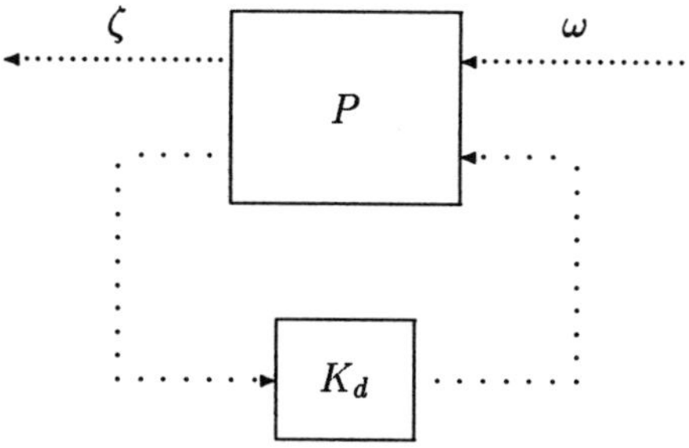

Figure 8.4: Two-rate discrete-time system.

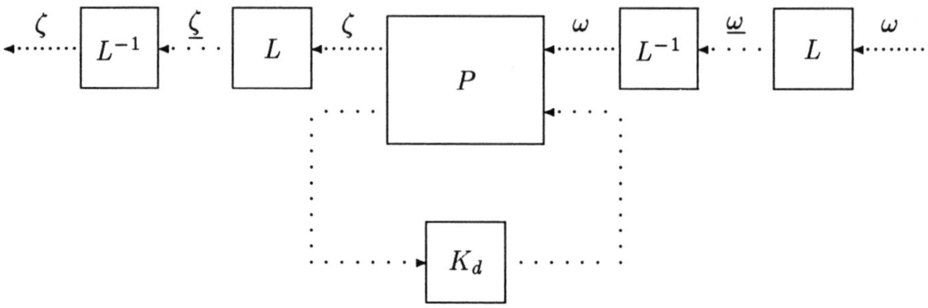

Figure 8.5: Two-rate system with lifting.

Introduce lifting in Figure 8.4 as shown in Figure 8.5. Absorb L and L^{-1} into P to get Figure 8.6, where

$$\underline{P} := \begin{bmatrix} L & 0 \\ 0 & I \end{bmatrix} P \begin{bmatrix} L^{-1} & 0 \\ 0 & I \end{bmatrix}.$$

Finally, if we focus on the lifted signals $\underline{\zeta}$ and $\underline{\omega}$ instead of ζ and ω, we arrive

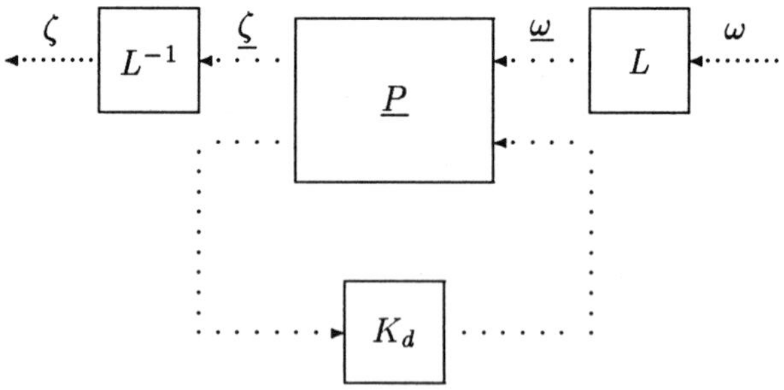

Figure 8.6: Two-rate system with lifted generalized plant.

at Figure 8.7, a single-rate discrete-time system.

The advantage of Figure 8.7 is that $\underline{P}$ is time-invariant, as we shall see. Back in Figure 8.2 partition G as

$$G = \begin{bmatrix} G_{11} & G_{12} \\ G_{21} & G_{22} \end{bmatrix}.$$

Then

$$\underline{P} = \begin{bmatrix} \underline{P}_{11} & \underline{P}_{12} \\ \underline{P}_{21} & \underline{P}_{22} \end{bmatrix} = \begin{bmatrix} LS_f G_{11} H_f L^{-1} & LS_f G_{12} H \\ SG_{21} H_f L^{-1} & SG_{22} H \end{bmatrix}.$$

Take any realization of G:

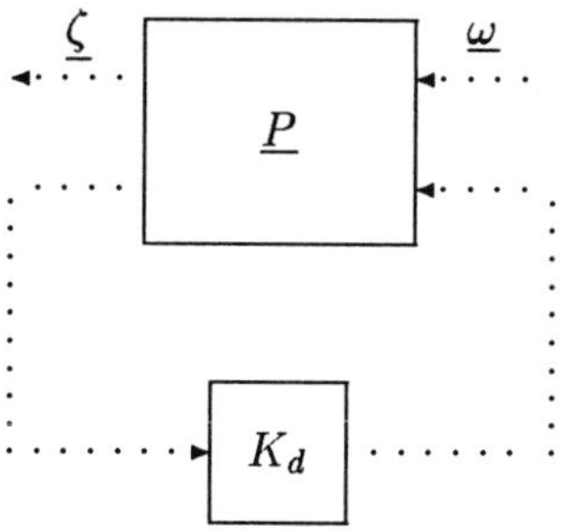

Figure 8.7: Single-rate lifted system.

$$\hat{g}(s) = \left[\begin{array}{c|cc} A & B_1 & B_2 \\ \hline C_1 & D_{11} & D_{12} \\ C_2 & D_{21} & D_{22} \end{array}\right].$$

Discretize at the slow rate

$$[A_d, B_{2d}] := c2d\ (A, B_2, h)$$

and at the fast rate:

$$[A_f, [B_{1f}, B_{2f}]] := c2d\ (A, [B_1, B_2], h/n).$$

We shall derive a state model for $\underline{P}$ based on these matrices; but first let us look at each of the four entries in $\underline{P}$ in turn.

Transfer Function for $\underline{P}_{11}$

Note that $S_f G_{11} H_f$ is the fast-rate discretization of G_{11}; the corresponding transfer function is

$$\left[\begin{array}{c|c} A_f & B_{1f} \\ \hline C_1 & D_{11} \end{array}\right].$$

The transfer function for $\underline{P}_{11}$ follows directly from Theorem 8.2.1: Define

$$\begin{aligned} \underline{B}_1 &= \left[\begin{array}{cccc} A_f^{n-1}B_{1f} & A_f^{n-2}B_{1f} & \cdots & B_{1f} \end{array}\right] \\ \underline{C}_1 &= \left[\begin{array}{c} C_1 \\ C_1 A_f \\ \vdots \\ C_1 A_f^{n-1} \end{array}\right] \\ \underline{D}_{11} &= \left[\begin{array}{cccc} D_{11} & 0 & \cdots & 0 \\ C_1 B_{1f} & D_{11} & \cdots & 0 \\ \vdots & \vdots & & \vdots \\ C_1 A_f^{n-2} B_{1f} & C_1 A_f^{n-3} B_{1f} & \cdots & D_{11} \end{array}\right]. \end{aligned}$$

Then

$$\hat{\underline{p}}_{11}(\lambda) = \left[\begin{array}{c|c} A_d & \underline{B}_1 \\ \hline \underline{C}_1 & \underline{D}_{11} \end{array}\right].$$

(Note that $A_d = A_f^n$.)

Transfer Function for $\underline{P}_{12}$

The definition of $\underline{P}_{12}$ is

$$\underline{P}_{12} = LS_f G_{12} H.$$

It is easy to check that $H_f S_f H = H$ (Exercise 8.5), that is, when acting on the output of H, $H_f S_f$ equals the identity system. Thus

$$\underline{P}_{12} = L(S_f G_{12} H_f) S_f H.$$

Again, $S_f G_{12} H_f$ is the fast-rate discretization of G_{12}, its transfer function being

$$\left[\begin{array}{c|c} A_f & B_{2f} \\ \hline C_1 & D_{12} \end{array}\right].$$

Also, the matrix representation of $S_f H$ is

$$[S_f H] = \begin{bmatrix} \left.\begin{matrix} I & 0 \\ \vdots & \vdots \\ I & 0 \end{matrix}\right\} n & \\ \left.\begin{matrix} 0 & I \\ \vdots & \vdots \\ 0 & I \end{matrix}\right\} n & \\ & \ddots \end{bmatrix}.$$

From this and $[L]$ it can be inferred that

$$LS_f H = \begin{bmatrix} I \\ \vdots \\ I \end{bmatrix}, \quad (n \text{ blocks})$$

that is,

$$S_f H = L^{-1} \begin{bmatrix} I \\ \vdots \\ I \end{bmatrix}.$$

Thus

$$\underline{P}_{12} = L(S_f G_{12} H_f) L^{-1} \begin{bmatrix} I \\ \vdots \\ I \end{bmatrix}.$$

The transfer function for $L(S_f G_{12} H_f)L^{-1}$ can be obtained again by Theorem 8.2.1. In this way, we get the transfer function for $\underline{P}_{12}$:

$$\begin{aligned}
\hat{\underline{p}}_{12}(\lambda) &= \left[\begin{array}{c|cccc} A_d & A_f^{n-1}B_{2f} & A_f^{n-2}B_{2f} & \cdots & B_{2f} \\ \hline C_1 & D_{12} & 0 & \cdots & 0 \\ C_1 A_f & C_1 B_{2f} & D_{12} & \cdots & 0 \\ \vdots & \vdots & \vdots & & \vdots \\ C_1 A_f^{n-1} & C_1 A_f^{n-2} B_{2f} & C_1 A_f^{n-3} B_{2f} & \cdots & D_{12} \end{array}\right] \begin{bmatrix} I \\ I \\ \vdots \\ I \end{bmatrix} \\
&= \left[\begin{array}{c|c} A_d & (A_f^{n-1} + A_f^{n-2} + \cdots + I)B_{2f} \\ \hline C_1 & D_{12} \\ C_1 A_f & C_1 B_{2f} + D_{12} \\ \vdots & \vdots \\ C_1 A_f^{n-1} & C_1 A_f^{n-2} B_{2f} + \cdots + C_1 B_{2f} + D_{12} \end{array}\right].
\end{aligned}$$

Define $\underline{C}_1$ as before and define

$$\underline{D}_{12} = \begin{bmatrix} D_{12} \\ C_1 B_{2f} + D_{12} \\ \vdots \\ C_1 A_f^{n-2} B_{2f} + \cdots + C_1 B_{2f} + D_{12} \end{bmatrix}.$$

It follows from the equation

$$\int_0^h e^{\tau A} d\tau = \int_0^{h/n} e^{\tau A} d\tau + \cdots + \int_{(n-1)h/n}^h e^{\tau A} d\tau$$

that

$$(A_f^{n-1} + \cdots + A_f + I)B_{2f} = B_{2d}.$$

Thus

$$\hat{\underline{p}}_{12}(\lambda) = \left[\begin{array}{c|c} A_d & B_{2d} \\ \hline \underline{C}_1 & \underline{D}_{12} \end{array}\right].$$

Transfer Function for $\underline{P}_{21}$

It is an easy-to-prove fact that $S = SH_f S_f$ (Exercise 8.5). Also, the matrix representation of SH_f is

$$[SH_f] = \left[\begin{array}{cc} \overbrace{I\,0\,\ldots\,0}^{n} & \overbrace{0\,0\,\ldots\,0}^{n} \\ 0\,0\,\ldots\,0 & I\,0\,\ldots\,0 \\ & & \ddots \end{array} \right],$$

from which it follows that

$$SH_f = \left[\begin{array}{cccc} I & 0 & \cdots & 0 \end{array}\right] L.$$

Thus

$$\begin{aligned} \underline{P}_{21} &= SG_{21}H_fL^{-1} \\ &= SH_f(S_fG_{21}H_f)L^{-1} \\ &= SH_fL^{-1}L(S_fG_{21}H_f)L^{-1} \\ &= \left[\begin{array}{cccc} I & 0 & \cdots & 0 \end{array}\right] L(S_fG_{21}H_f)L^{-1}. \end{aligned}$$

Thus a state model can be derived,

$$\hat{\underline{p}}_{21}(\lambda) = \left[\begin{array}{c|c} A_d & \underline{B}_1 \\ \hline C_2 & \underline{D}_{21} \end{array}\right],$$

where $\underline{B}_1$ was defined earlier and

$$\underline{D}_{21} := \left[\begin{array}{cccc} D_{21} & 0 & \cdots & 0 \end{array}\right].$$

Transfer Function for $\underline{P}_{22}$

This is simply the slow-rate discretization of G_{22}:

$$\hat{\underline{p}}_{22}(\lambda) = \left[\begin{array}{c|c} A_d & B_{2d} \\ \hline C_2 & D_{22} \end{array}\right].$$

Transfer Function for $\underline{P}$

The realizations for the four entries in $\underline{P}$ fit nicely together to form a realization for $\underline{P}$:

$$\hat{\underline{p}}(\lambda) = \left[\begin{array}{c|cc} A_d & \underline{B}_1 & B_{2d} \\ \hline \underline{C}_1 & \underline{D}_{11} & \underline{D}_{12} \\ C_2 & \underline{D}_{21} & D_{22} \end{array}\right].$$

In detail, the underlined matrices on the right-hand side is

$$\underline{B}_1 = \left[\begin{array}{cccc} A_f^{n-1}B_{1f} & A_f^{n-2}B_{1f} & \cdots & B_{1f} \end{array}\right]$$

$$\underline{D}_{11} = \left[\begin{array}{cccc} D_{11} & 0 & \cdots & 0 \\ C_1B_{1f} & D_{11} & \cdots & 0 \\ \vdots & \vdots & & \vdots \\ C_1A_f^{n-2}B_{1f} & C_1A_f^{n-3}B_{1f} & \cdots & D_{11} \end{array}\right]$$

$$\underline{D}_{21} = \begin{bmatrix} D_{21} & 0 & \cdots & 0 \end{bmatrix}$$

$$\underline{C}_1 = \begin{bmatrix} C_1 \\ C_1 A_f \\ \vdots \\ C_1 A_f^{n-1} \end{bmatrix}, \quad \underline{D}_{12} = \begin{bmatrix} D_{12} \\ C_1 B_{2f} + D_{12} \\ \vdots \\ \sum_{i=0}^{n-2} C_1 A_f^i B_{2f} + D_{12} \end{bmatrix}.$$

Thus $\underline{P}$ is time-invariant, as mentioned before.

In summary, fast discretization of the standard SD system leads to a two-rate discrete-time system which can be lifted to a single-rate discrete-time system.

8.4 Design Examples

In this section we look at design of three examples via fast discretization.

Example 8.4.1 Bilateral hybrid telerobot (cont'd). Let us use the fast discretization method for the telerobot problem of Example 6.5.1. Recall that the telerobot under sampled-data control can be configured to the standard form shown in Figure 8.8. The input is the 2-vector

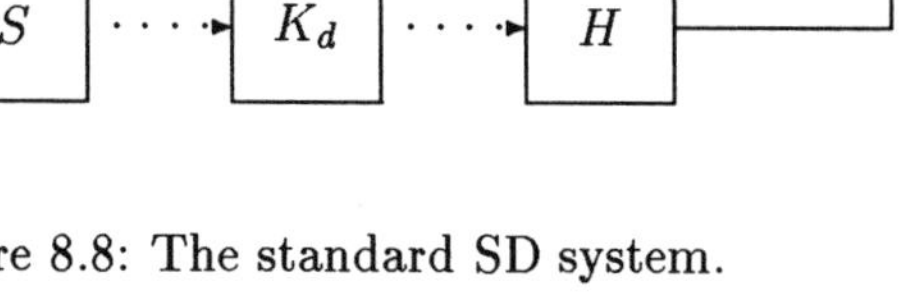

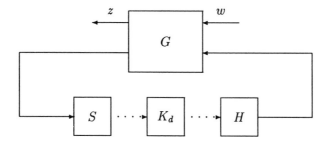

Figure 8.8: The standard SD system.

$$w = \begin{bmatrix} w_h \\ w_e \end{bmatrix}$$

and the output is the 4-vector

$$z = \begin{bmatrix} \alpha_v(v_m - v_s) \\ \alpha_c(f_h - v_m) \\ \alpha_f(f_m - f_e) \\ \alpha_s f_s \end{bmatrix}.$$

The state model for G has the form

$$\left[\begin{array}{c|cc} A & B_1 & B_2 \\ \hline C_1 & 0 & D_{12} \\ C_2 & 0 & 0 \end{array}\right].$$

Let T_{zw} denote the SD system from w to z in Figure 8.8. Let us choose to fast-sample at twice the base rate: Let S_f denote the sampler with period $h/2$. Letting e_1, e_2 denote the standard basis in $\mathbb{R}^2$, we choose to minimize

$$\sum_i \|S_f T_{zw} \delta e_i\|_2^2.$$

(Recall that in Example 6.5.1 we minimized $\sum_i \|ST_{zw}\delta e_i\|_2^2$.) Since lifting is norm-preserving, we can equivalently minimize

$$\sum_i \|LS_f T_{zw} \delta e_i\|_2^2.$$

The relevant block diagram for this optimization problem is Figure 8.9. Bringing L, S_f, S, and H into G gives the system in Figure 8.10 (com-

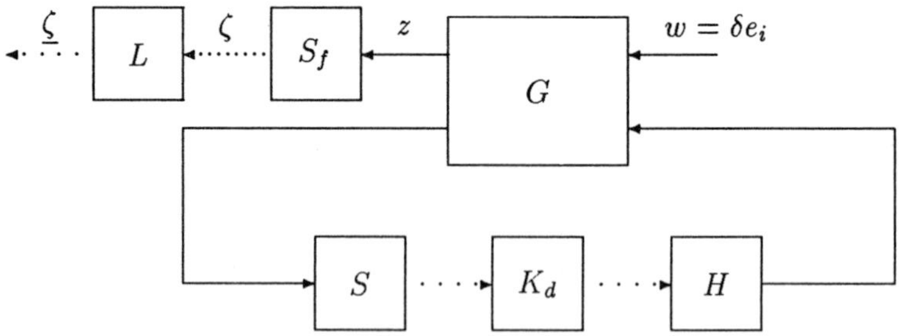

Figure 8.9: Fast sample and lift z.

pare with Figure 6.2). Finally, as in the derivation of Figure 6.3, Figure 8.10 can be converted to the equivalent discrete-time system in Figure 8.11, where

$$\hat{g}_{dis}(\lambda) = \left[\begin{array}{c|cc} A_d & A_d B_1 & B_{2d} \\ \hline C_1 & C_1 B_1 & D_{12} \\ C_1 A_f & C_1 A_f B_1 & C_1 B_{2f} + D_{12} \\ & & \\ C_2 & C_2 B_1 & 0 \end{array}\right]. \tag{8.1}$$

(The derivation is left as Exercise 8.11.)

To summarize, let $T_{\zeta\omega}$ denote the LTI discrete-time system from ω to ζ in Figure 8.11. Then

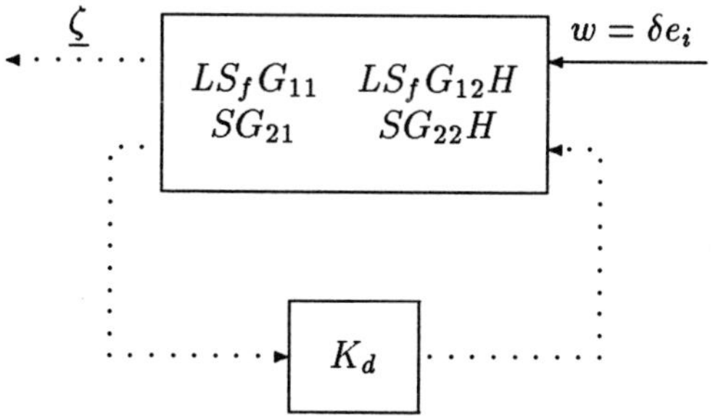

Figure 8.10: Discretized system with continuous-time input w.

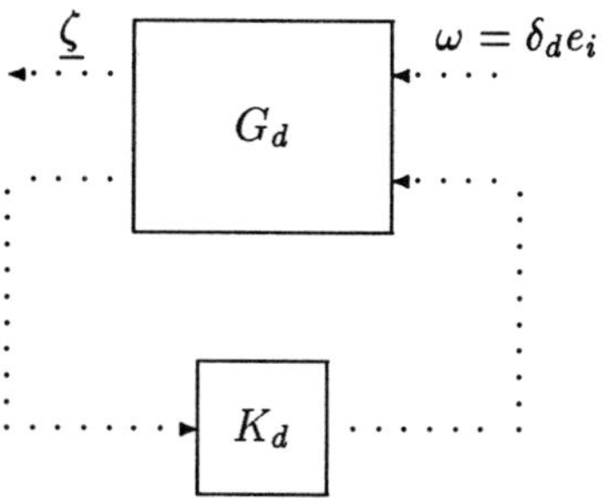

Figure 8.11: Equivalent discrete-time system.

$$\sum_i \|S_f T_{zw} \delta e_i\|_2^2 \text{ in Figure 8.8} = \|\hat{t}_{\zeta w}\|_2^2 \text{ in Figure 8.11.}$$

In this way we arrive at a discrete-time $\mathcal{H}_2$ problem, namely, the minimization of $\|\hat{t}_{\zeta w}\|_2$ in Figure 8.11.

For computations, we take h, α_v, α_c, α_f, and α_s as in Example 6.5.1. The same regularization is performed, namely, 1) the poles of $\hat{g}_m(s)$ and $\hat{g}_s(s)$ at $s = 0$ are perturbed to $s = 10^{-3}$, 2) the matrix in the D_{21}-location of G_{dis} is perturbed to

$$\begin{bmatrix} 0 & 0 & 0 & 20 \\ 0.1 & 0 & 0 & 0 \\ 0 & 0.1 & 0 & 0 \end{bmatrix},$$

and 3) the matrices in the D_{11}- and B_1-locations are padded with two zero columns.

The resulting responses are shown in Figures 8.12 and 8.13. In comparing these plots with Figures 6.4 and 6.5, one sees only a small improvement, namely, the error between f_m and f_e converges to zero a little faster. Presumably, some additional improvement could be obtained by fast sampling faster than twice the base frequency. We shall return to this example in Chapter 12.

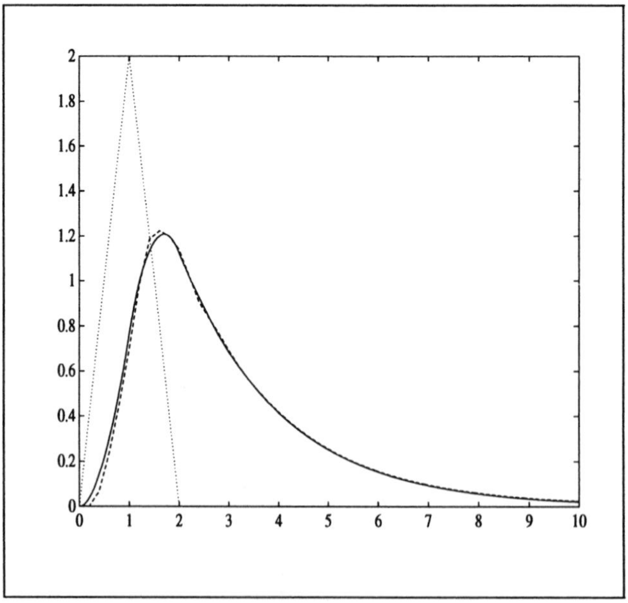

Figure 8.12: Design by fast discretization: v_s (solid), v_m (dash), and f_h (dot).

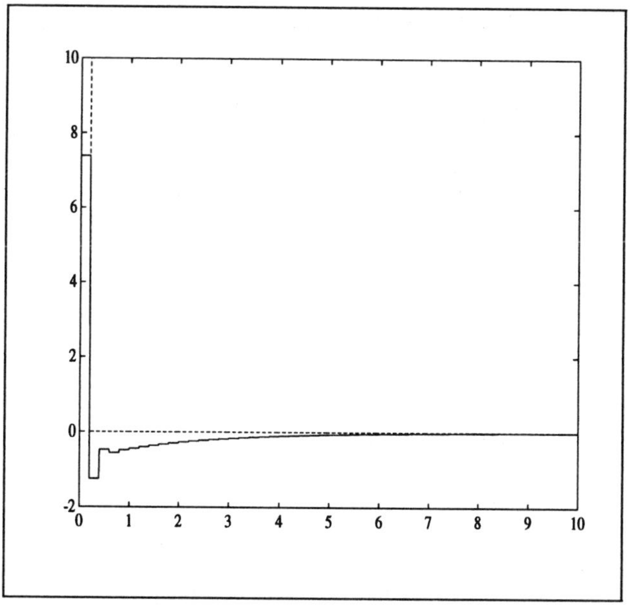

Figure 8.13: Design by fast discretization: f_m (solid) and f_e (dash).

The next example involves step tracking. Let us develop a general procedure for doing this. Start with the system in Figure 8.14. In this figure, the

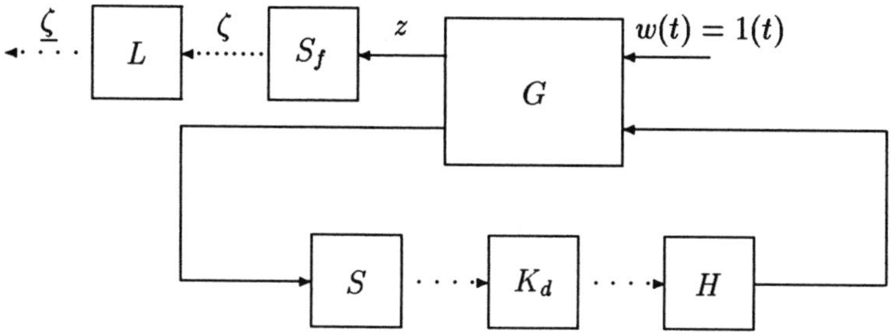

Figure 8.14: SD system with step input; z is fast-sampled and lifted.

input w is the 1-dimensional unit step function $1(t)$ (for a multidimensional step of the form $w(t) = 1(t)\omega_0$, the constant vector ω_0 can be absorbed into G). It is desired to minimize the continuous-time error $\|z\|_2$. This is approximated by $\|S_f z\|_2$, which equals $\|LS_f z\|_2$. In this way, our goal is to design K_d to minimize $\|\underline{\zeta}\|_2$ in Figure 8.14.

A step enjoys the property that it can be the output of a hold operator. In Figure 8.14 write the continuous-time step $1(t)$ as the response of H to the discrete-time step $1_d(k)$; this gives Figure 8.15. Finally, convert Figure 8.15

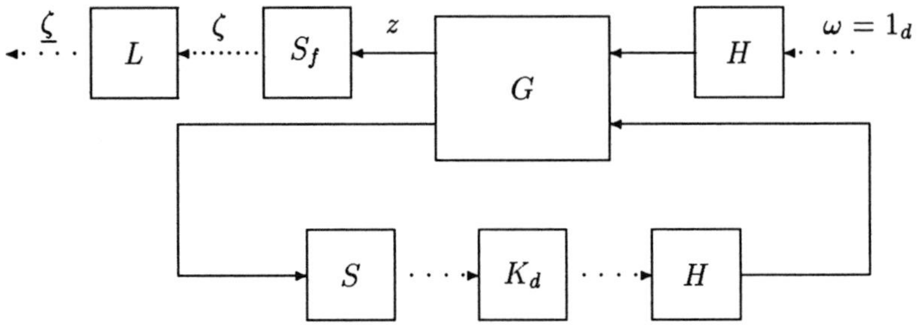

Figure 8.15: Discrete-time step input.

into Figure 8.16 by defining

$$\begin{aligned} G_{dis} &= \begin{bmatrix} LS_f & 0 \\ 0 & S \end{bmatrix} \begin{bmatrix} G_{11} & G_{12} \\ G_{21} & G_{22} \end{bmatrix} \begin{bmatrix} H & 0 \\ 0 & H \end{bmatrix} \\ &= \begin{bmatrix} LS_f G_{11} H & LS_f G_{12} H \\ SG_{21} H & SG_{22} H \end{bmatrix}. \end{aligned}$$

In terms of state models, if

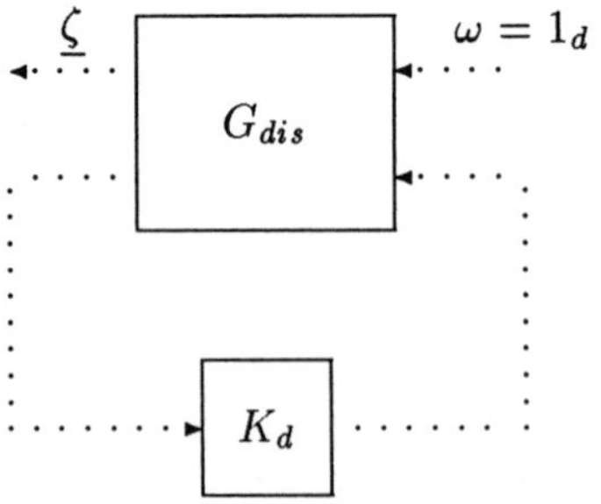

Figure 8.16: Equivalent discrete-time system with step input.

$$\hat{g}(s) = \left[\begin{array}{c|cc} A & B_1 & B_2 \\ \hline C_1 & D_{11} & D_{12} \\ C_2 & D_{21} & 0 \end{array} \right],$$

then

$$\hat{g}_{dis}(\lambda) =$$

$$
\left[\begin{array}{c|cc}
A_d & B_{1d} & B_{2d} \\
\hline
C_1 & D_{11} & D_{12} \\
C_1 A_f & D_{11} + C_1 B_{1f} & D_{12} + C_1 B_{2f} \\
\vdots & \vdots & \vdots \\
C_1 A_f^{n-1} & D_{11} + \cdots + C_1 A_f^{n-2} B_{1f} & D_{12} + \cdots + C_1 A_f^{n-2} B_{2f} \\
 & & \\
C_2 & D_{21} & 0
\end{array}\right].
$$

The problem depicted in Figure 8.16 is a standard $\mathcal{H}_2$-optimal step-tracking problem as studied in Section 6.6.

Example 8.4.2 Let us reconsider Example 6.6.1, where we got severe inter-sample ripple by naively discretizing the plant. The block diagram is shown in Figure 8.17. The plant transfer function is

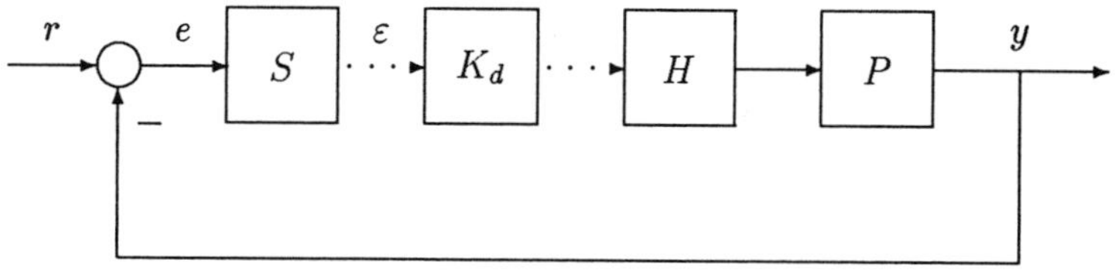

Figure 8.17: A sampled-data tracking system.

$$
\hat{p}(s) = \frac{1}{(10s+1)(25s+1)},
$$

the reference input r is the unit step, and the sampling period is $h = 1$. Instead of minimizing the ℓ_2-norm of ε, as we did in Example 6.6.1, let us minimize the ℓ_2-norm of $\zeta := S_f e$, as shown in Figure 8.18, for $n = 2$.

The computation of the optimal K_d can be performed by transforming Figure 8.18 into Figure 8.14 and applying the general procedure. The solution is

$$
\hat{k}_d(\lambda) = \frac{-488.85(\lambda - 1.1052)(\lambda - 1.0408)}{(\lambda + 1.3955)(\lambda - 1)}.
$$

The sampled-data system of Figure 8.17 with this controller was simulated for a step input. Figure 8.19 shows the plot of $y(t)$ versus t in solid. For comparison, the dashed line is for the controller of Example 6.6.1. The improvement of the new method is obvious.

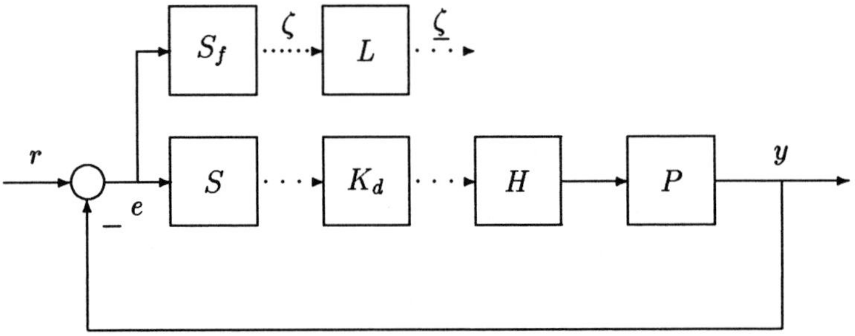

Figure 8.18: A tracking system with fast sampling of the error.

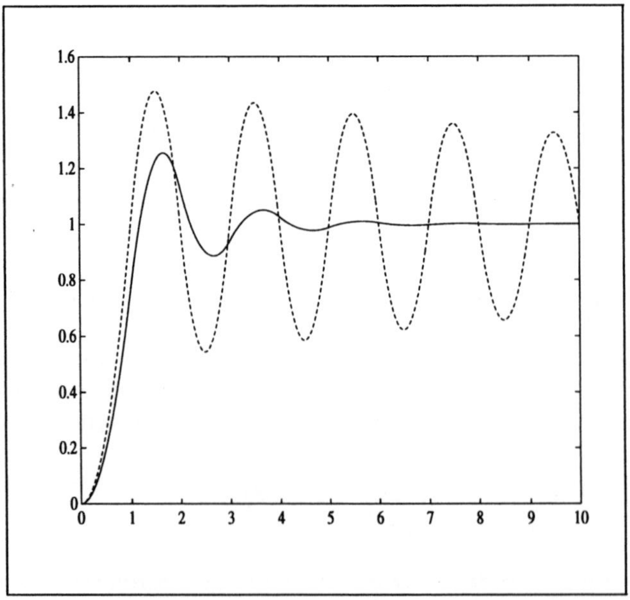

Figure 8.19: Step-response of example: design by fast discretization (solid) and by slow discretization (dash).

Example 8.4.3 Let us reconsider Example 7.2.1. There we designed an optimal $\mathcal{H}_\infty$-controller by discretizing at the sampling frequency of the controller. An alternative procedure is to discretize the problem at a faster rate than the sampling frequency of the controller. That is, instead of Figure 7.7 we could discretize as in Figure 8.20. Lifting ζ and ω as in Section 8.3, we

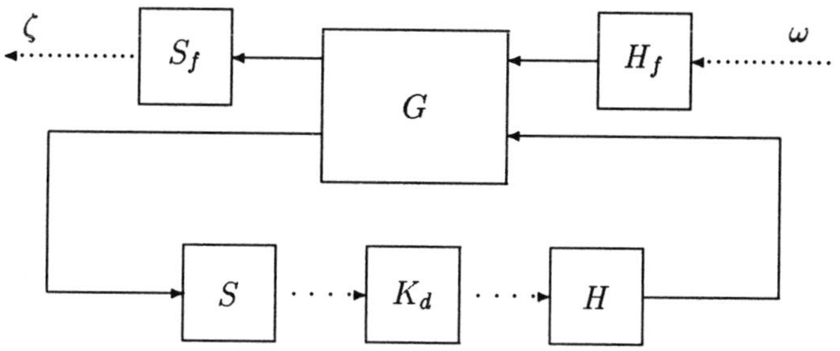

Figure 8.20: Fast discretization.

arrive again at Figure 7.4.

Figure 8.21 shows the results of this fast-discretization method for $h = 0.5$ and the sampling period for S_f and H_f equal to $h/2$. The solid line is the Bode magnitude plot of $1/(1+\hat{p}\hat{k}\hat{f})$, where $\hat{k}$ is the continuous-time controller obtained by $\mathcal{H}_\infty$-optimization in continuous time; the dashed line is for $1/(1+\hat{p}\hat{r}\hat{k}_d\hat{f})$, where $\hat{k}_d$ is the discrete-time controller obtained by $\mathcal{H}_\infty$-optimization in discrete time; and the dotted line is for the discrete-time controller obtained by $\mathcal{H}_\infty$-optimization after fast discretization. The fast-discretization method is a little better than the slow-discretization one in that the peak is smaller. Figure 8.22 shows the results for $h = 0.15$, where the fast-discretization method is quite superior on the operating band even though h is still quite large.

8.5 Simulation of SD Systems

Fast discretization provides an effective way to simulate the standard SD system of Figure 8.2. A simulation procedure takes the data

a model for G

a model for K_d

the sampling period h

the input $w(t)$ over some period $0 \le t \le t_f$

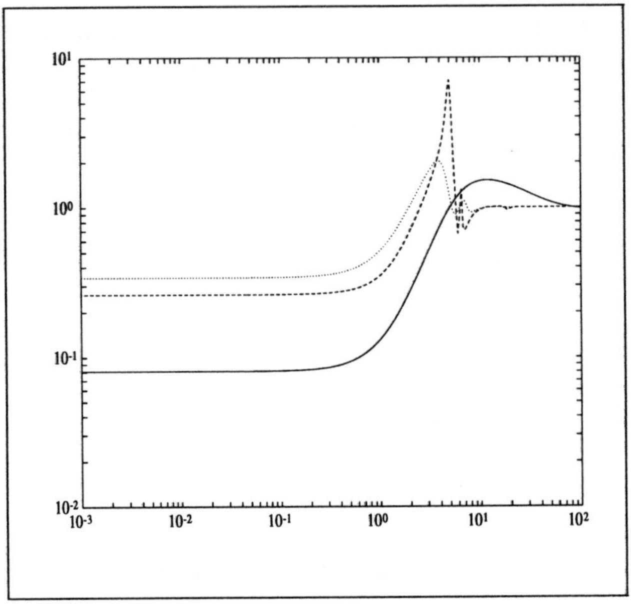

Figure 8.21: Bode magnitude plots: optimal analog controller (solid), optimal discrete-time controller (dash), optimal discrete-time controller via fast discretization (dot); $h = 0.5$.

and computes

the output $z(t)$ over the period $0 \leq t \leq t_f$.

The continuous-time input w can be closely approximated by $H_f S_f w$ provided n is large enough. Define $\omega = S_f w$. Since only a finite amount of data can be computed, it makes sense for the procedure to compute $S_f z$ instead of z. In this way we arrive at Figure 8.3.

For simplicity, suppose the simulation time interval $[0, t_f]$ consists of an integral number, m, of slow sampling intervals, that is, $t_f = mh$. Relative to Figure 8.3, the simulation procedure is to input the data

a model for G

a model for K_d

integers m and n and the sampling period h

the input $\omega(k)$ for $0 \leq k \leq mn$

and compute

the output $\zeta(k)$ for $0 \leq k \leq mn$.

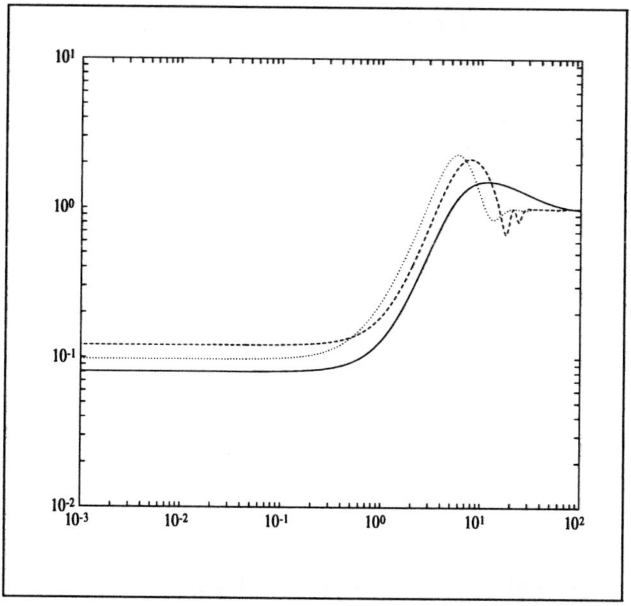

Figure 8.22: Bode magnitude plots: optimal analog controller (solid), optimal discrete-time controller (dash), optimal discrete-time controller via fast discretization (dot); $h = 0.15$.

A simple simulation procedure is therefore as follows:

Step 1 Lift: $\underline{\omega} = L\omega$.

Step 2 Simulate the discrete-time system in Figure 8.7.

Step 3 Inverse lift: $\zeta = L^{-1}\underline{\zeta}$.

Step 4 Compute the times corresponding to the sampled values $\zeta(k)$, namely, $t = kh/n$ $(k = 1, \ldots, mn)$.

This procedure can readily be implemented in MATLAB.

Exercises

8.1 This problem looks at the lifting operation in the frequency domain. Let $v \in \ell_2(\mathbb{Z}_+)$ and $\underline{v} = Lv$, where L is the lifting operator for $n = 2$. Find the relationship between $\hat{v}$ and $\hat{\underline{v}}$. Extend your result for general n.

8.2 Let G be an LTI discrete-time system and let L denote the lifting operator for $n = 2$. Write the transfer matrix $\hat{g}_d(\lambda)$ as follows:

$$\hat{g}(\lambda) \quad = \quad g(0) + \lambda g(1) + \lambda^2 g(2) + \cdots$$

$$
\begin{aligned}
&= \left[g(0) + \lambda^2 g(2) + \cdots\right] + \lambda\left[g(1) + \lambda^2 g(3) + \cdots\right] \\
&=: \hat{g}_0(\lambda^2) + \lambda \hat{g}_1(\lambda^2). \qquad (8.2)
\end{aligned}
$$

Prove that the transfer matrix of LGL^{-1} equals

$$
\begin{bmatrix} \hat{g}_0(\lambda) & \lambda\hat{g}_1(\lambda) \\ \hat{g}_1(\lambda) & \hat{g}_0(\lambda) \end{bmatrix}.
$$

Extend this result for general n.

8.3 Let G be a continuous-time LTI system and consider the discrete-time system $S_f GH$ obtained by fast-sampling the output and slow-holding the input. Is it LTI? Causal? Repeat for SGH_f.

8.4 Let P_d be a linear discrete-time system. Recall from Chapter 4 that P_d is time-invariant if $U^* P_d U = P_d$, where U and U^* are unit delay and unit advance respectively. For a positive integer n, let us define P_d to be *n-periodic* if $(U^*)^n P_d U^n = P_d$. This means shifting the input by n time units corresponds to shifting the output by n units. For the lifting setup in Figure 8.1, show that $\underline{G}_d$ is time-invariant iff G_d is n-periodic. As a special case, if G_d is LTI, so is $\underline{G}_d$.

8.5 Prove that $H_f S_f H = H$ and $S = SH_f S_f$.

8.6 Derive the matrix representations of $S_f H$, SH_f, and $LS_f H$.

8.7 This question concerns causality of G_d and $\underline{G}_d$ in Figure 8.1. Assume G_d is linear and n-periodic. Show that G_d is causal iff $\underline{G}_d$ is causal and $\underline{g}_d(0)$ is (block) lower-triangular, where $\underline{g}_d$ is the impulse response of $\underline{G}_d$.

8.8 Repeat the design in Example 8.4.2 for $n = 4$ and compare the step response with that for $n = 2$.

8.9 Write a MATLAB function *sd_sim.m* for SD simulation. Its arguments should be the state matrices for G (assuming $D_{22} = 0$),

$$A, B_1, B_2, C_1, C_2, D_{11}, D_{12}, D_{21},$$

the state matrices for K_d,

$$A_K, B_K, C_K, D_K,$$

the numbers h, m, n, and the input vector ω. It should compute the output vector ζ, together with the time vector, the elements of which are the simulation instants.

As a simple test of your program, simulate the following system:

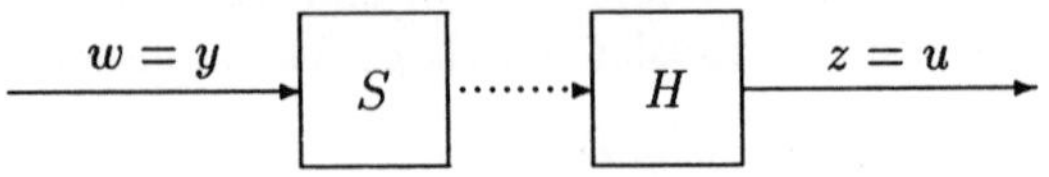

(Choose your own values for h, m, n.)

8.10 Consider the analog control system

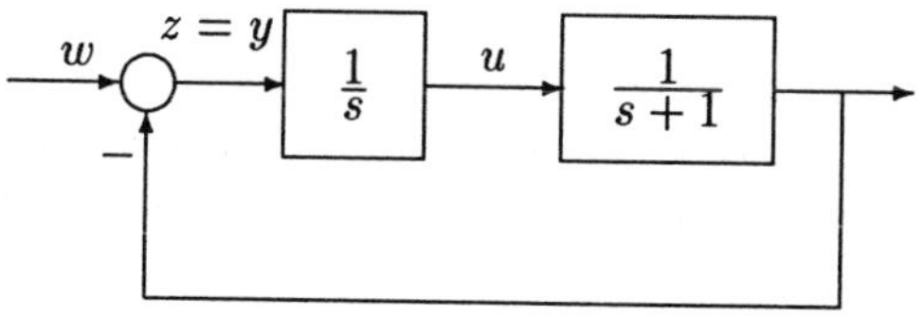

and a sampled-data implementation of the controller, as shown here:

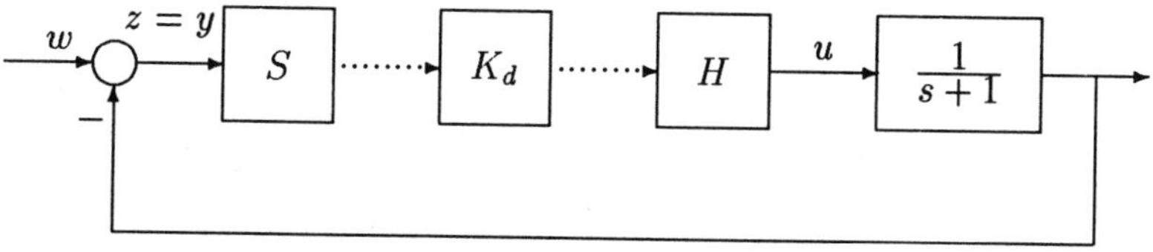

Using your program, simulate the step response of the latter system, taking $h = 0.1, m = 50, n = 10$. Plot the output z versus time and overlay the step response of the analog system for comparison. Repeat with $h = 1, m = 50, n = 10$. Observe the effect of the sampling period on the quality of the digital implementation (remember that the objective is to recover the analog design).

8.11 Derive equation (8.1).

8.12 Let G be a continuous-time system with transfer function

$$\hat{g}(s) = \left[\begin{array}{c|c} A & B \\ \hline C & D \end{array}\right].$$

Let H denote the hold with period h and S_f the fast sampler with period $h/3$. Give a procedure to compute the induced norm

$$\sup_{\|v\|_2 \leq 1} \|S_f G H v\|_2.$$

Notes and References

The technique of lifting is basic to the early work of Kranc on multirate sampled-data systems [94]. He used a method called *switch decomposition* to convert a multirate system into a single-rate one. The idea is as follows, described for the case $n = 2$ for simplicity. Consider the switch system

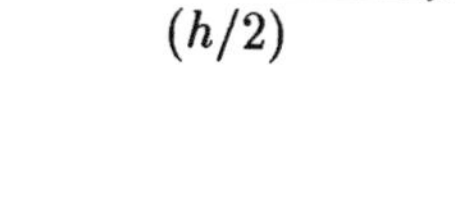

The switch closes periodically with period $h/2$, so that

$$y(t) = \begin{cases} u(t), & t = kh/2, \quad k \in \mathbb{Z} \\ 0, & \text{else.} \end{cases}$$

Thus the switch is like the fast sampler S_f except its output is continuous-time. This switch system is input-output equivalent to the following system, where the fast switch has been decomposed into two parallel slow ones:

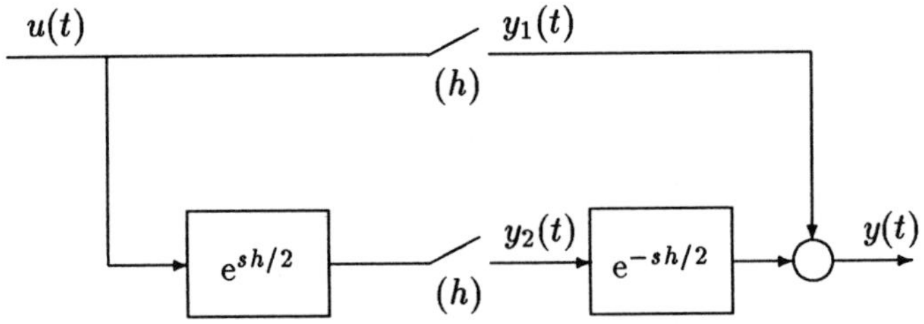

In a certain sense $\begin{bmatrix} y_1 \\ y_2 \end{bmatrix}$ is y lifted. More precisely, if we let $\psi(k) = y(kh/2)$ and $\psi_1(k) = y_1(kh)$, $\psi_2(k) = y_2(kh)$, then $\begin{bmatrix} \psi_1 \\ \psi_2 \end{bmatrix}$ is precisely ψ lifted. So Kranc's switch decomposition is essentially lifting.

Friedland [56] developed lifting explicitly to convert a periodic discrete-time system into a time-invariant one. Later, Davis [35] did the same for stability analysis of the feedback connection of an LTI plant and a periodic (memoryless) controller. The idea of lifting is similar to the digital signal processing idea of *blocking*—taking a discrete-time signal, regarding it as a vector, and subdividing the vector into blocks. This is used, for example, in block convolution [116] [112]. The decomposition in (8.2) is called a *polyphase decomposition*. Introduced by Davis [35], it is a very important tool in multirate signal processing [144].

The lifting framework of Section 8.1 was developed by Khargonekar, Poolla, and Tannenbaum [90]. They showed how closed-loop zeros could be assigned by periodic controllers, concluding, for example, that the system's gain margin could be increased without bound (even if there is a bound for time-invariant controllers). Discrete lifting was also used by Araki and Yamamoto [7] in their study of multirate SD systems.

The fast sampling idea for approximating continuous-time performance is due to Keller and Anderson [87].

The simulation procedure in Section 8.5 is new. Several software packages can simulate nonlinear sampled-data systems.

Part II

Direct SD Design

Chapter 9

Properties of S and H

In this chapter we view S and H as linear transformations and derive some of their properties. We also study performance recovery—how analog performance specifications can be recovered when an analog controller is implemented digitally and the sampling period tends to zero.

9.1 Review of Input-Output Stability of LTI Systems

Sampled-data systems are time-varying. Before we look at stability of such systems, it is instructive to review the case of LTI systems.

Definitions

1. For $1 \le p \le \infty$, $\mathcal{L}_p(\mathbf{R}_+, \mathbf{R}^n)$ denotes the space of piecewise-continuous functions from $\mathbf{R}_+$ to $\mathbf{R}^n$ such that the following norm is finite:

$$\|u\|_p = \begin{cases} \left[\int_0^\infty \|u(t)\|^p dt\right]^{1/p}, & 1 \le p < \infty \\ \sup_t \|u(t)\|, & p = \infty. \end{cases}$$

   Here the norm on $\mathbf{R}^n$ is the Euclidean norm. If n is irrelevant, we write just $\mathcal{L}_p(\mathbf{R}_+)$. We are primarily interested in the cases $p = 2, \infty$, for then the p-norm has a physical significance—$\|u\|_2^2$ is energy and $\|u\|_\infty$ is maximum value.

2. $\mathcal{C}(\mathbf{R}_+, \mathbf{R}^n)$, or just $\mathcal{C}(\mathbf{R}_+)$, denotes the space of continuous functions from $\mathbf{R}_+$ to $\mathbf{R}^n$.

Now we summarize some stability theory for LTI systems. Let G be an LTI, causal, continuous-time system, that is, it has a representation of the form

$$y(t) = g_0 u(t) + \int_0^t g_1(t-\tau)u(\tau)d\tau,$$

where g_0 is a constant matrix and $g_1(t) = 0$ for $t < 0$. The impulse response function is therefore $\delta(t)g_0 + g_1(t)$.

Theorem 9.1.1 *The following are equivalent:*

1. $G : \mathcal{L}_1(\mathbb{R}_+) \longrightarrow \mathcal{L}_1(\mathbb{R}_+)$ *is bounded.*
2. $G : \mathcal{L}_\infty(\mathbb{R}_+) \longrightarrow \mathcal{L}_\infty(\mathbb{R}_+)$ *is bounded.*
3. $G : \mathcal{L}_p(\mathbb{R}_+) \longrightarrow \mathcal{L}_p(\mathbb{R}_+)$ *is bounded for every* p.
4. *Each element of* g_1 *is in* $\mathcal{L}_1(\mathbb{R}_+)$.
5. *Assume* G *is finite-dimensional and let* $\left[\begin{array}{c|c} A & B \\ \hline C & D \end{array}\right]$ *be a minimal realization of the transfer matrix. Then* A *is stable (all eigs in the open left half-plane).*

So any one of these five conditions could qualify as a definition of stability of G.

Continuing with such G, define the induced norm

$$M_p := \sup_{\|u\|_p \le 1} \|Gu\|_p.$$

This is the gain of the system from $\mathcal{L}_p(\mathbb{R}_+)$ to $\mathcal{L}_p(\mathbb{R}_+)$. To get an upper bound on this gain, introduce a matrix N as follows: Take the ij^{th} element of the impulse response matrix and write it as $g_0\delta(t) + g_1(t)$; then the ij^{th} element of N is defined as $|g_0| + \|g_1\|_1$.

Theorem 9.1.2 *Assume* G *is stable. Then* $M_p \le \sigma_{\max}(N)$. *Also,* $M_2 = \|\hat{g}\|_\infty$.

Now assume G is LTI, strictly causal ($g_0 = 0$), and stable. If we want to sample its output, we are interested in knowing that its output is continuous. It is not hard to prove that $G\mathcal{L}_p(\mathbb{R}_+) \subset \mathcal{C}(\mathbb{R}_+)$.

9.2 M. Riesz Convexity Theorem

We shall want to be able to infer $\mathcal{L}_2(\mathbb{R}_+)$-stability from $\mathcal{L}_1(\mathbb{R}_+)$- and $\mathcal{L}_\infty(\mathbb{R}_+)$-stability. Theorem 9.1.1 provides this for LTI systems. Since sampled-data systems are not time-invariant, we need a more powerful result, the M. Riesz convexity theorem.

Let $\mathcal{L}_{1e}(\mathbb{R}_+)$ denote the extended $\mathcal{L}_1(\mathbb{R}_+)$ space, the space of all piecewise-continuous functions from $\mathbb{R}_+$ to $\mathbb{R}^n$ such that $\|u(t)\|$ is integrable on every finite time interval.

Theorem 9.2.1 *Suppose G is a linear system with the property that $Gu \in \mathcal{L}_{1e}(\mathbb{R}_+)$ for every input u in $\mathcal{L}_{1e}(\mathbb{R}_+)$. If*

$$G : \mathcal{L}_1(\mathbb{R}_+) \longrightarrow \mathcal{L}_1(\mathbb{R}_+) \text{ is bounded, with induced norm } M_1, \text{ and}$$

$$G : \mathcal{L}_\infty(\mathbb{R}_+) \longrightarrow \mathcal{L}_\infty(\mathbb{R}_+) \text{ is bounded, with induced norm } M_\infty,$$

then for every p

$$G : \mathcal{L}_p(\mathbb{R}_+) \longrightarrow \mathcal{L}_p(\mathbb{R}_+) \text{ is bounded, with induced norm } M_p \leq M_1^{\frac{1}{p}} M_\infty^{1-\frac{1}{p}}.$$

Why is this called a convexity theorem? Note that $M_p \leq M_1^{\frac{1}{p}} M_\infty^{1-\frac{1}{p}}$ iff

$$\ln M_p \leq \frac{1}{p} \ln M_1 + \left(1 - \frac{1}{p}\right) \ln M_\infty,$$

or, with $x := 1/p$, $f(x) := \ln M_p$,

$$f(x) \leq x f(1) + (1-x) f(0).$$

This says that f is a convex function on the interval $[0,1]$.

9.3 Boundedness of S and H

Now we have the mathematical machinery to begin our study of S and H. The (perhaps unexpected) fact is that sample-and-hold is not a bounded operator on, for example, $\mathcal{L}_2(\mathbb{R}_+)$.

Theorem 9.3.1

1. *For any $\psi \in \ell_p(\mathbb{Z}_+)$, $\|H\psi\|_p = h^{1/p}\|\psi\|_p$. Thus $H : \ell_p(\mathbb{Z}_+) \to \mathcal{L}_p(\mathbb{R}_+)$ is bounded and of norm $h^{1/p}$.*

2. *$S : \mathcal{C}(\mathbb{R}_+) \cap \mathcal{L}_\infty(\mathbb{R}_+) \to \ell_\infty(\mathbb{Z}_+)$ is bounded and of norm 1.*

3. *$S : \mathcal{C}(\mathbb{R}_+) \cap \mathcal{L}_p(\mathbb{R}_+) \to \ell_p(\mathbb{Z}_+)$ is not bounded for any $1 \leq p < \infty$.*

Proof The first and second statements are immediate. Now we show the third.

Define u to be a series of ever-narrowing triangular pulses as follows:

$$u(t) = \sum_{k=1}^{\infty} v_k(t),$$

$$v_k(t) = \begin{cases} 1 - \dfrac{2k^2}{h}|t - kh|, & \text{if } |t - kh| < \dfrac{h}{2k^2} \\ 0, & \text{otherwise.} \end{cases}$$

Thus v_k is a triangular pulse centered at $t = kh$ with height 1 and base width h/k^2. Then $(Su)(k) = 1$ for all $k \geq 1$, so $Su \notin \ell_p(\mathbb{Z}_+)$. Yet

$$\|v_k\|_p^p = 2\int_{kh}^{kh+\frac{h}{2k^2}} \left[1 - \frac{2k^2}{h}(t-kh)\right]^p dt = \frac{h}{(p+1)k^2},$$

so

$$\|u\|_p^p = \sum_{k=1}^{\infty} \|v_k\|_p^p = \frac{h}{p+1}\sum_{k=1}^{\infty}\frac{1}{k^2} < \infty.$$

Therefore u belongs to $\mathcal{C}(\mathbb{R}_+) \cap \mathcal{L}_p(\mathbb{R}_+)$. ∎

The hold operator has the nice property that by proper scaling it is norm-preserving from $\ell_p(\mathbb{Z}_+)$ to $\mathcal{L}_p(\mathbb{R}_+)$. But the sampling operator is unbounded from $\mathcal{C}(\mathbb{R}_+) \cap \mathcal{L}_p(\mathbb{R}_+)$ to $\ell_p(\mathbb{Z}_+)$ for all $1 \leq p < \infty$, which unfortunately includes the case $p = 2$ that we shall focus on. The counterexample in the preceding proof shows that the problem is caused by allowing into S a signal of ever-increasing derivative. If the derivative of the input is limited, one might expect S to be bounded. This is indeed the case. It will be proven next that S preceded by a low-pass filter is a bounded operator on $\mathcal{C}(\mathbb{R}_+) \cap \mathcal{L}_p(\mathbb{R}_+)$.

The setup is shown in Figure 9.1—the sampling operator with a prefilter F. Assume F is causal and LTI with impulse response matrix f in $\mathcal{L}_1(\mathbb{R}_+)$, i.e.,

$$\|f\|_1 := \int_0^{\infty} \sigma_{max}[f(t)]dt < \infty,$$

where the norm on $f(t)$ is the maximum singular value of the matrix $f(t)$. Thus F is a bounded operator on $\mathcal{L}_p(\mathbb{R}_+)$ for every p. Define the sequence ϕ

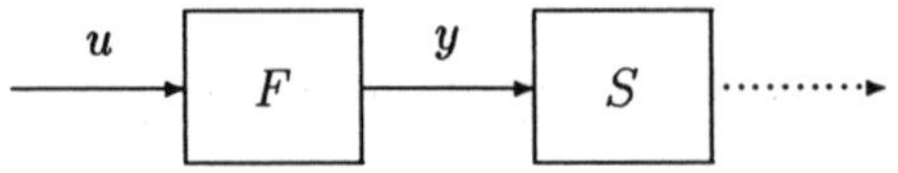

Figure 9.1: S with a prefilter.

by

$$\phi(k) := \sup_{t \in [kh,\,(k+1)h)} \|f(t)\|. \tag{9.1}$$

Theorem 9.3.2 *If $\phi \in \ell_1(\mathbb{Z}_+)$, then $SF : \mathcal{L}_p(\mathbb{R}_+) \to \ell_p(\mathbb{Z}_+)$ is bounded for every $1 \leq p \leq \infty$.*

The proof uses the following basic fact:

Lemma 9.3.1 *If g and u are in $\mathcal{L}_1(\mathbb{R}_+)$, so is their convolution $g * u$; moreover*

$$\|g * u\|_1 \le \|g\|_1 \cdot \|u\|_1.$$

The same result holds for two sequences in $\ell_1(\mathbb{Z}_+)$.

Proof of Theorem 9.3.2 First note from Theorem 9.3.1 that for any p, the operator $SF : \mathcal{L}_p(\mathbb{R}_+) \to \ell_p(\mathbb{Z}_+)$ is bounded iff $HSF : \mathcal{L}_p(\mathbb{R}_+) \to \mathcal{L}_p(\mathbb{R}_+)$ is. Apply Theorem 9.2.1 to the operator HSF: It suffices to show that SF is bounded $\mathcal{L}_\infty(\mathbb{R}_+) \to \ell_\infty(\mathbb{Z}_+)$ and $\mathcal{L}_1(\mathbb{R}_+) \to \ell_1(\mathbb{Z}_+)$. It is easy to see that SF is bounded $\mathcal{L}_\infty(\mathbb{R}_+) \to \ell_\infty(\mathbb{Z}_+)$, with $\|SF\| \le \|f\|_1$.

Now suppose $u \in \mathcal{L}_1(\mathbb{R}_+)$. For $k \ge 0$

$$y(kh) = \int_0^{kh} f(kh-\tau)u(\tau)\,d\tau = \sum_{i=1}^{k} \int_{(i-1)h}^{ih} f(kh-\tau)u(\tau)\,d\tau$$

So

$$\begin{aligned}
\|y(kh)\| &\le \sum_{i=1}^{k} \int_{(i-1)h}^{ih} \|f(kh-\tau)\| \cdot \|u(\tau)\|\,d\tau \\
&\le \sum_{i=1}^{k} \phi(k-i) \int_{(i-1)h}^{ih} \|u(\tau)\|\,d\tau
\end{aligned}$$

Define the sequence v by $v(i) := \int_{(i-1)h}^{ih} \|u(\tau)\|\,d\tau$. Then the right-hand side of the above is the convolution of ϕ and v. Note that $\phi \in \ell_1(\mathbb{Z}_+)$ by the hypothesis and $v \in \ell_1(\mathbb{Z}_+)$ since $\|v\|_1 = \|u\|_1 < \infty$. Invoke Lemma 9.3.1 to get

$$\sum_k \|y(kh)\| \le \|\phi\|_1 \cdot \|v\|_1 = \|\phi\|_1 \cdot \|u\|_1$$

Hence the operator SF is bounded $\mathcal{L}_1(\mathbb{R}_+) \to \ell_1(\mathbb{Z}_+)$ with $\|SF\| \le \|\phi\|_1$. ∎

It follows from the proof that upper bounds for the $\mathcal{L}_1(\mathbb{R}_+)$- and $\mathcal{L}_\infty(\mathbb{R}_+)$-induced norms of HSF are $h\|\phi\|_1$ and $\|f\|_1$, respectively. Thus by Theorem 9.2.1 an upper bound for the $\mathcal{L}_p(\mathbb{R}_+)$-induced norm is $(h\|\phi\|_1)^{1/p}(\|f\|_1)^{1/q}$ where $\frac{1}{p} + \frac{1}{q} = 1$. Since $\|f\|_1 \le h\|\phi\|_1$, we obtain that $h\|\phi\|_1$ is an upper bound for $\|HSF\|$ on every $\mathcal{L}_p(\mathbb{R}_+)$ space.

Corollary 9.3.1 *If F is FDLTI, stable, and strictly causal, then*

$$SF : \mathcal{L}_p(\mathbb{R}_+) \to \ell_p(\mathbb{Z}_+)$$

is bounded for every $1 \le p \le \infty$ and for every sampling period $h > 0$.

Proof Such a filter F admits a state-space representation

$$\hat{F}(s) = \left[\begin{array}{c|c} A & B \\ \hline C & 0 \end{array}\right]$$

with A stable. Then $f(t) = Ce^{tA}B1(t)$. It can be verified that the corresponding ϕ belongs to $\ell_1(\mathbb{Z}_+)$ for any $h > 0$. ■

So far, F has been a filter with impulse-response function in $\mathcal{L}_1(\mathbb{R})$, such as a stable, strictly causal, FDLTI filter. What about an ideal low-pass filter: $\hat{f}(j\omega) = 1$ up to some cutoff frequency, $\hat{f}(j\omega) = 0$ at higher frequency? This is not causal and its impulse-response function is the well-known sinc function, $a\sin(bt)/t$. This function is *not* absolutely integrable, that is, not in $\mathcal{L}_1(\mathbb{R})$. The next result handles such a filter.

Lemma 9.3.2 *Let F_{id} be an ideal bandpass filter with passband $\Omega := [\omega_1, \omega_2]$ and F an LTI filter such that $\hat{f}(j\omega)$ is invertible for every $\omega \in \Omega$. If T is a linear map such that TF is bounded on $\mathcal{L}_2(\mathbb{R})$, then TF_{id} is bounded on $\mathcal{L}_2(\mathbb{R})$ and*

$$||TF_{id}|| \leq \max_{\omega\in\Omega} ||\hat{f}(j\omega)^{-1}|| \cdot ||TF||.$$

Proof It is a fact that F_{id} is bounded on $\mathcal{L}_2(\mathbb{R})$ with norm 1. Thus by the hypothesis, the map TFF_{id} is bounded on $\mathcal{L}_2(\mathbb{R})$ and hence on the subspace $\mathcal{B} := F_{id}\mathcal{L}_2(\mathbb{R})$, namely, the space of $\mathcal{L}_2(\mathbb{R})$ functions that are bandlimited to Ω. The operator FF_{id} maps $\mathcal{B}$ to $\mathcal{B}$ and the restriction $FF_{id}|\mathcal{B}$ has a bounded inverse since $\hat{f}(j\omega)$ is invertible over the frequency band Ω. Let $T|_{\mathcal{B}}$ denote the map T when the domain space is restricted to $\mathcal{B}$. Then

$$T|_{\mathcal{B}} = (TFF_{id})(FF_{id}|_{\mathcal{B}})^{-1}$$

So $TF_{id} = T|_{\mathcal{B}}F_{id}$ is bounded on $\mathcal{L}_2(\mathbb{R})$ and

$$||TF_{id}|| \leq ||T|_{\mathcal{B}}|| \leq ||(FF_{id}|_{\mathcal{B}})^{-1}|| \cdot ||TFF_{id}|| \leq \max_{\omega\in\Omega} ||\hat{f}(j\omega)^{-1}|| \cdot ||TF||.$$

■

The preceding result is applied as follows: Given any ideal bandpass filter F_{id}, invoke Corollary 9.3.1 [with, say, $\hat{f}(s) = 1/(s+1)$] and Lemma 9.3.2 to conclude that HSF_{id} is bounded on $\mathcal{L}_2(\mathbb{R})$ for every sampling frequency (not just those greater than the Nyquist frequency). However, since the impulse-response function of F_{id} does not belong to $\mathcal{L}_1(\mathbb{R})$, the map F_{id} is bounded on neither $\mathcal{L}_\infty(\mathbb{R})$ nor $\mathcal{L}_1(\mathbb{R})$. Hence HSF_{id} is not bounded on $\mathcal{L}_\infty(\mathbb{R})$ or $\mathcal{L}_1(\mathbb{R})$.

Now we turn to the question of whether or not $HS \approx I$ as the sampling period tends to zero; more precisely, does $\lim_{h\to 0} ||I - HS|| = 0$? To see that the answer is no, check that the norm of $I - HS$ on $\mathcal{L}_\infty(\mathbb{R}_+)$ equals 2 no matter how small h is! Again, low-pass prefiltering rectifies the situation.

Let F be causal and LTI with impulse-response function f in $\mathcal{L}_1(\mathbb{R}_+)$. Define

$$f_h(t) := \sup_{a\in(0,h)} \|f(t) - f(t-a)\|. \tag{9.2}$$

From (9.1)

$$\phi(0) = \sup_{t\in[0,h)} \|f(t)\|.$$

Theorem 9.3.3 *If $\phi(0)$ is finite for some $h > 0$ and $\lim_{h\to 0} \|f_h\|_1 = 0$, then $(I - HS)F$ converges to zero as h tends to zero in the $\mathcal{L}_p(\mathbb{R}_+)$-induced norm for every $1 \le p \le \infty$.*

Proof Again by Theorem 9.2.1 it suffices to prove that $\lim_{h\to 0} \|(I-HS)F\| = 0$ for the $\mathcal{L}_1(\mathbb{R}_+)$- and $\mathcal{L}_\infty(\mathbb{R}_+)$-induced norms.

First let $u \in \mathcal{L}_\infty(\mathbb{R}_+)$. For any $t > 0$, choose k such that $kh < t \le (k+1)h$. Then since $y = Fu$,

$$\begin{aligned}[(I - HS)Fu](t) &= y(t) - y(kh)\\ &= \int_0^{kh} [f(t-\tau) - f(kh-\tau)]u(\tau)d\tau + \int_{kh}^t f(t-\tau)u(\tau)d\tau.\end{aligned}$$

Hence

$$\begin{aligned}&\|[(I - HS)Fu](t)\|\\ &\le \int_0^{kh} f_h(t-\tau)\|u(\tau)\|d\tau + \int_{kh}^t \|f(t-\tau)\| \cdot \|u(\tau)\|d\tau \qquad (9.3)\\ &\le \{\|f_h\|_1 + h\phi(0)\}\|u\|_\infty.\end{aligned}$$

The quantity in parentheses is independent of t, so it is an upper bound for the $\mathcal{L}_\infty(\mathbb{R}_+)$-induced norm of the operator $(I - HS)F$. By our hypothesis this upper bound tends to zero as h tends to zero.

Next suppose $u \in \mathcal{L}_1(\mathbb{R}_+)$. Again, for any $t > 0$, choose k such that $kh < t \le (k+1)h$. From inequality (9.3)

$$\begin{aligned}\|[(I - HS)Fu](t)\| &\le \int_0^t f_h(t-\tau)\|u(\tau)\|\, d\tau +\\ &\qquad \int_{kh}^t \|f(t-\tau)\| \cdot \|u(\tau)\|\, d\tau\\ &\le (f_h * \|u\|)(t) + \phi(0) \int_{kh}^{(k+1)h} \|u(\tau)\|\, d\tau\end{aligned}$$

Apply Lemma 9.3.1 to get

$$\|(I - HS)Fu\|_1 \le \|f_h\|_1 \cdot \|u\|_1 + h\phi(0)\|u\|_1$$

Thus an upper bound for the $\mathcal{L}_1(\mathbb{R}_+)$-induced norm of the operator $(I - HS)F$ is again $\|f_h\|_1 + h\phi(0)$, which tends to zero as h tends to zero by our hypothesis. ■

In Theorem 9.3.3, there is no assumption that ϕ belongs to $\ell_1(\mathbb{Z}_+)$; hence, there is no implication that HSF is bounded [only $(I - HS)F$ is bounded for small enough h].

Corollary 9.3.2 *If F is FDLTI, stable, and strictly causal, then HSF converges to F as h tends to zero in the $\mathcal{L}_p(\mathbb{R}_+)$-induced norm for every $1 \le p \le \infty$.*

Proof Bring in a realization for F as in the proof of Corollary 9.3.1 to get $f(t) = Ce^{tA}B1(t)$, where A is stable. It is clear that $\phi(0)$ is bounded for any $h > 0$. From (9.2)

$$\begin{aligned} f_h(t) &\le \|B\| \cdot \|C\| \cdot \|e^{tA}\| \sup_{a \in (0,h)} \|I - e^{-aA}\| \\ &\le \|B\| \cdot \|C\| \cdot \|e^{tA}\|(e^{h\|A\|} - 1) \end{aligned}$$

Since A is stable, $\int_0^\infty \|e^{tA}\| \, dt$ is finite. It follows that $\|f_h\|_1 \to 0$ at least as fast as $h \to 0$. ■

Corollary 9.3.2 and Lemma 9.3.2 allow us to deduce that for any ideal bandpass filter F_{id}, HSF_{id} converges to F_{id} as h tends to zero in the $\mathcal{L}_2(\mathbb{R}_+)$-induced norm.

9.4 Performance Recovery

This section looks at an application of the preceding properties: We will see that internal stability of the digital implementation of an analog system is recovered as the sampling period tends to zero. The same can be proved for other types of performance specifications. This is comforting to know, but it isn't always useful because frequently the sampling period is not designable.

Start with the setup in Figure 9.2. Assume P and K are FDLTI and strictly causal. Moreover, assume the system is *internally stable* in the sense that the mapping

$$\begin{bmatrix} I & P \\ -K & I \end{bmatrix}^{-1} : \begin{bmatrix} r \\ d \end{bmatrix} \mapsto \begin{bmatrix} z \\ u \end{bmatrix} : \mathcal{L}_2(\mathbb{R}_+) \longrightarrow \mathcal{L}_2(\mathbb{R}_+)$$

is bounded.

Now do a digital implementation as in Figure 9.3. This isn't set up quite right for $\mathcal{L}_2(\mathbb{R}_+)$ stability because the input to the left-hand S isn't filtered.

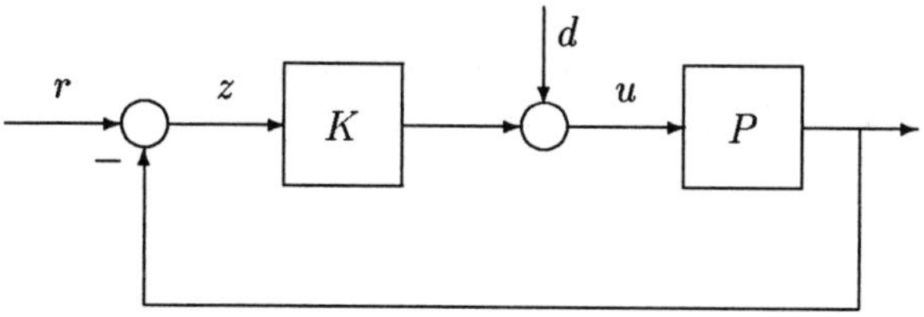

Figure 9.2: Multivariable analog feedback system.

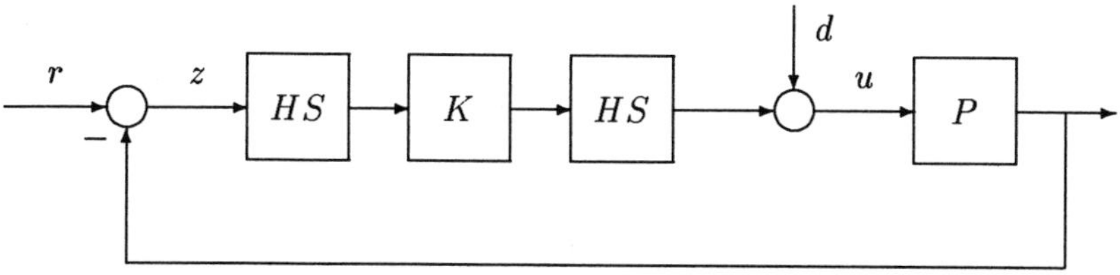

Figure 9.3: Implementation of controller via *c2d*.

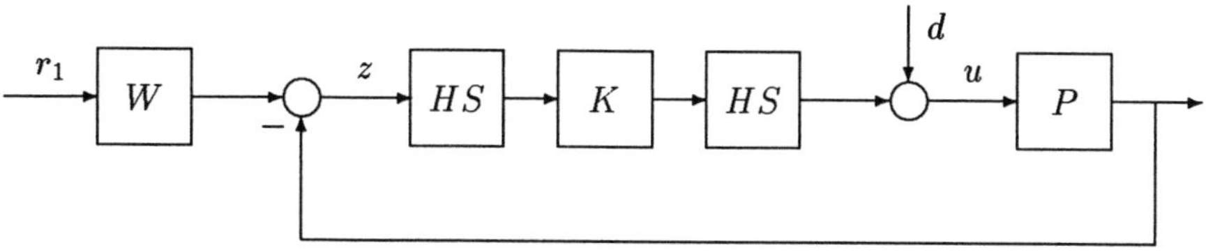

Figure 9.4: Inclusion of filter on reference input.

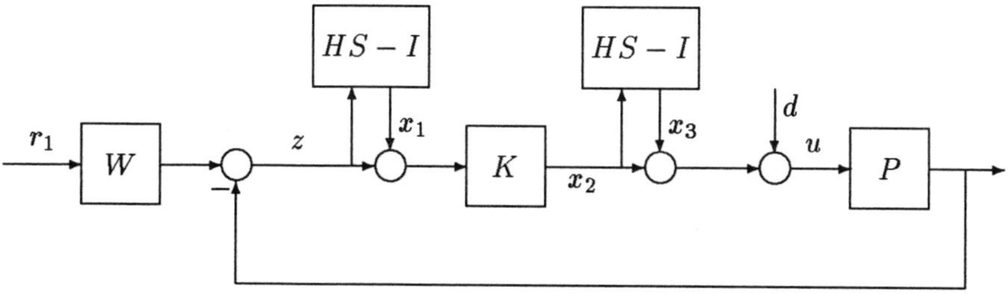

Figure 9.5: Drawing the SD system as a perturbation of the analog system.

Instead, we'll study stability of the system in Figure 9.4, where W is FDLTI, strictly causal, and stable.

Reconfigure the preceding figure to look like Figure 9.5. This in turn can be viewed as in Figure 9.6, where

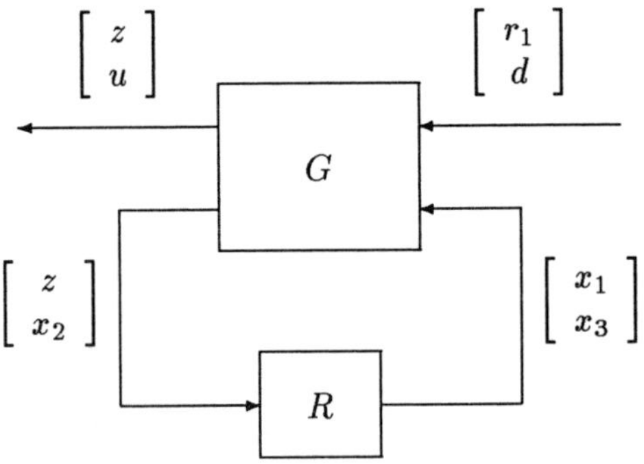

Figure 9.6: Linear-fractional representation.

$$
\begin{aligned}
G &= \begin{bmatrix} G_{11} & G_{12} \\ G_{21} & G_{22} \end{bmatrix} \\
G_{11} &= \begin{bmatrix} (I+PK)^{-1}W & -P(I+KP)^{-1} \\ K(I+PK)^{-1}W & (I+KP)^{-1} \end{bmatrix} \\
G_{12} &= \begin{bmatrix} -P(I+KP)^{-1}K & -P(I+KP)^{-1} \\ (I+KP)^{-1}K & (I+KP)^{-1} \end{bmatrix} \\
G_{21} &= \begin{bmatrix} (I+PK)^{-1}W & -P(I+KP)^{-1} \\ K(I+PK)^{-1}W & -KP(I+KP)^{-1} \end{bmatrix} \\
G_{22} &= \begin{bmatrix} -P(I+KP)^{-1}K & -P(I+KP)^{-1} \\ (I+KP)^{-1}K & -KP(I+KP)^{-1} \end{bmatrix}
\end{aligned}
$$

and

$$R = \begin{bmatrix} HS-I & 0 \\ 0 & HS-I \end{bmatrix}.$$

Thus G is FDLTI and stable and R is time-varying. Notice that R is a perturbation of the original analog system.

The mapping from $\begin{bmatrix} r_1 \\ d \end{bmatrix}$ to $\begin{bmatrix} z \\ u \end{bmatrix}$ is

$$G_{11} + G_{12}(I - RG_{22})^{-1}RG_{21}.$$

Since G_{21} and G_{22} are strictly causal, we have from Theorem 9.3.2 that RG_{21} and RG_{22} are bounded on $\mathcal{L}_2(\mathbb{R}_+)$; also, from Theorem 9.3.3 $\|RG_{22}\|$ tends to zero as $h \longrightarrow 0$. Thus by the small-gain theorem $(I - RG_{22})^{-1}$ is bounded for small enough h.

For reference, one version of the *small-gain theorem* is as follows: Let $\mathcal{X}$ be a Banach space and let T be a bounded linear operator on $\mathcal{X}$. If the induced norm of T is less than 1, then $(I-T)^{-1}$ is a bounded linear operator on $\mathcal{X}$. In summary:

Theorem 9.4.1 *If the analog system Figure 9.2 is internally stable and if h is small enough—namely, if $\|RG_{22}\| < 1$—then the sampled-data system Figure 9.4 is stable [bounded on $\mathcal{L}_2(\mathbb{R}_+)$].*

Exercises

9.1 Consider a SISO, LTI system, $y = g * u$, with $g \in \mathcal{L}_1(\mathbb{R}_+)$ and satisfying $g(t) \geq 0$ for all t. Prove that

$$\sup_{\|u\|_2 \leq 1} \|y\|_2 = \sup_{\|u\|_\infty \leq 1} \|y\|_\infty .$$

9.2 Consider a continuous-time system with transfer matrix

$$\begin{bmatrix} \frac{1}{s+1} & \frac{s}{s+2} \\ \frac{1}{10s+1} & 4 \end{bmatrix} .$$

Compute an upper bound for M_3, the induced norm on $\mathcal{L}_3(\mathbb{R}_+)$.

9.3 Let $\hat{f}(s) = 1/(s+1)$. Corollary 9.3.1 concludes that $SF : \mathcal{L}_2(\mathbb{R}_+) \rightarrow \ell_2(\mathbb{Z}_+)$ is bounded. Show that $S : F\mathcal{L}_2(\mathbb{R}_+) \rightarrow \ell_2(\mathbb{Z}_+)$ is not bounded. Hint: S is not bounded iff

$$(\forall M)(\exists y)\ y \in F\mathcal{L}_2(\mathbb{R}_+), \|y\|_2 \leq 1, \|Sy\|_2 > M.$$

Note also that $y \in F\mathcal{L}_2(\mathbb{R}_+)$ iff $y, \dot{y} \in \mathcal{L}_2(\mathbb{R}_+)$.

9.4 Consider a SISO, FDLTI, stable, strictly causal, continuous-time system G. Is $SG : \mathcal{L}_2(\mathbb{R}_+) \rightarrow \ell_\infty(\mathbb{Z}_+)$ bounded? If so, how can its norm be computed?

9.5 Repeat Exercise 4 for the operator $GH : \ell_2(\mathbb{Z}_+) \rightarrow \mathcal{L}_\infty(\mathbb{R}_+)$.

9.6 Let $G : \mathcal{L}_\infty(\mathbb{R}_+) \rightarrow \mathcal{L}_\infty(\mathbb{R}_+)$ be a SISO system with state model

$$\hat{g}(s) = \left[\begin{array}{c|c} A & B \\ \hline C & 0 \end{array}\right],$$

where A is stable.

1. Prove that $\|G\| = \|g\|_1$.

2. State the discrete-time counterpart.

3. Let $G_d = SGH : \ell_\infty(\mathbb{Z}_+) \to \ell_\infty(\mathbb{Z}_+)$. Show that if $g(t)$ does not change sign for $t \geq 0$, then $\|G\| = \|G_d\|$. (All first-order systems have this property.)

9.7 Consider the analog control system in Figure 9.7. Assume W, P, and

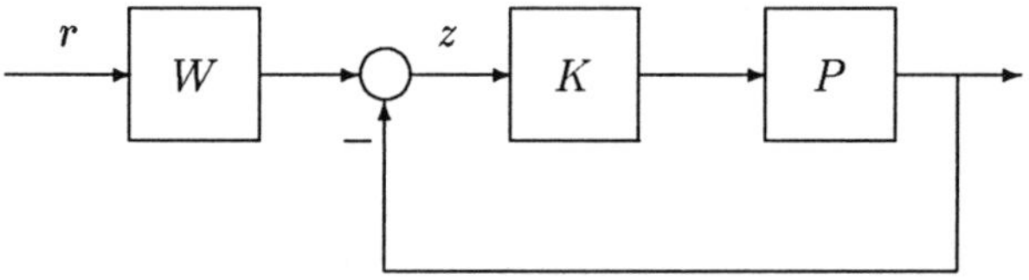

Figure 9.7: Analog control system.

K are FDLTI, strictly causal, with W stable. Assume this feedback system is internally stable and that the following performance specification is satisfied:

$$\sup_{\|r\|_\infty \leq 1} \|z\|_\infty < \epsilon. \tag{9.4}$$

Suppose now that the controller K is replaced by the sampled-data controller HK_dS, where $K_d = SKH$. Prove that if h is sufficiently small, then the performance specification (9.4) is recovered. What if K_d is obtained from K by bilinear transformation?

9.8 Consider the SD system K_dSP, where

$$\hat{p}(s) = \frac{1}{s}, \quad \hat{k}_d(\lambda) = 1 - \lambda.$$

Find a continuous-time LTI system G such that $SG = K_dSP$. Conclude that $K_dSP : \mathcal{L}_2(\mathbb{R}_+) \to \ell_2(\mathbb{Z}_+)$ is bounded, even though P is not bounded on $\mathcal{L}_2(\mathbb{R}_+)$.

Notes and References

Theorems 9.1.1 and 9.1.2 are from [145] (Section 6.4.1, Theorem 45 and Lemma 46). For the M. Riesz convexity theorem, see [129]. Sections 9.3 and 9.4 are based on [28]. It was observed at least as far back as 1970 [127] that S is unbounded on $\mathcal{L}_p$ ($1 \leq p < \infty$) and that lowpass filtering the input rectifies this situation. Recent results on the boundedness of S are contained in [86]. For extensions of the results of this chapter to time-varying systems, see [76].

Chapter 10

Continuous Lifting

In Chapter 8 we saw how to lift certain time-varying discrete-time systems into time-invariant systems. The basic idea was to extend the input and output spaces properly. The advantage of doing so is obvious: After lifting, we get LTI systems.

In this chapter we shall introduce a construction to "lift" a continuous-time signal into a discrete-time one. This construction will also be used to associate a time-invariant discrete-time system to a continuous-time SD one. The lifting technique will be the main tool for the development of this chapter and several chapters to follow.

As an application of the continuous lifting technique, in this chapter we shall also study induced norms of several special SD systems. Induced norms measure the input-output gain of the system, say from disturbance input to plant output; the concept of induced norms for systems is central in the philosophy of modern analog control design methods, such as $\mathcal{H}_\infty$ and $\mathcal{L}_1$ optimization.

10.1 Lifting Continuous-Time Signals

In this section we shall develop a construction to lift a continuous-time signal into a discrete-time one. The fact that SD systems are periodic warrants that this construction can be used to associate a time-invariant discrete-time system to a continuous-time SD one.

Lifting can be done in a fairly general framework, but for the first pass let us do the concrete case of $\mathcal{L}_2$. Start with a signal u in the extended space $\mathcal{L}_{2e}(\mathbb{R})$ [i.e., $\mathcal{L}_{2e}(\mathbb{R}, \mathbb{R}^n)$], that is,

$$\int_0^T u(t)'u(t)dt < \infty \quad \forall T > 0.$$

Now think of the pieces of u on the sampling intervals

$$\ldots, [-h, 0), [0, h), [h, 2h), \ldots.$$

Denote by u_k the piece in the kth sampling interval $[kh, (k+1)h)$ translated to the interval $[0, h)$; that is,

$$u_k(t) := u(kh + t), \quad 0 \le t < h.$$

Each u_k therefore lives in the space $\mathcal{K} := \mathcal{L}_2[0, h)$ [i.e., $\mathcal{L}_2([0, h), \mathbb{R}^n)$], which is an infinite-dimensional Hilbert space with inner product

$$\langle v, w \rangle = \int_0^h v(t)' w(t) dt$$

and norm

$$\|v\| = \left[\int_0^h v(t)' v(t) dt \right]^{1/2}.$$

We think of u_k as the kth component of u. It is then natural to introduce the discrete-time signal $\underline{u}$:

$$\underline{u} = \left[\begin{array}{c} \vdots \\ u_{-2} \\ u_{-1} \\ \hline u_0 \\ u_1 \\ \vdots \end{array} \right].$$

The horizontal lines in the latter array separate the time intervals $\{k < 0\}$ and $\{k \ge 0\}$. This discrete-time signal, $\underline{u}$, lives in the space $\ell(\mathbb{Z}, \mathcal{K})$, defined as all sequences of the form

$$v = \left[\begin{array}{c} \vdots \\ v(-2) \\ v(-1) \\ \hline v(0) \\ v(1) \\ \vdots \end{array} \right],$$

where each $v(k)$ lives in $\mathcal{K}$.

Let L denote the *lifting operator* mapping u in $\mathcal{L}_{2e}(\mathbb{R})$ to $\underline{u}$ in $\ell(\mathbb{Z}, \mathcal{K})$. It can be depicted by the block diagram

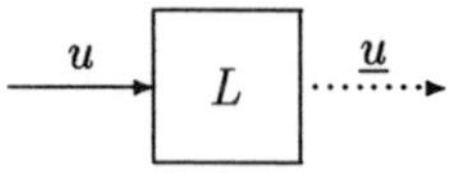

If u has finite 2-norm, that is, if u lives in $\mathcal{L}_2(\mathbb{R})$, then $\underline{u}$ lives in $\ell_2(\mathbb{Z}, \mathcal{K})$, the subspace of $\ell(\mathbb{Z}, \mathcal{K})$ of all square-summable sequences with values in $\mathcal{K}$. This is a Hilbert space with inner product

$$\langle v, \psi \rangle = \sum_{-\infty}^{\infty} \langle v(k), \psi(k) \rangle,$$

the right-hand inner product being the one on $\mathcal{K}$.

The following computation shows that L as a mapping from $\mathcal{L}_2(\mathbb{R})$ to $\ell_2(\mathbb{Z}, \mathcal{K})$ preserves inner products:

$$u, v \in \mathcal{L}_2(\mathbf{R}) \Rightarrow$$

$$\begin{aligned}
\langle u, v \rangle &= \int_{-\infty}^{\infty} u(t)'v(t)dt \\
&= \sum_{k=-\infty}^{\infty} \int_{kh}^{(k+1)h} u(t)'v(t)dt \\
&= \sum_{k=-\infty}^{\infty} \int_{0}^{h} u_k(t)'v_k(t)dt \\
&= \sum_{k=-\infty}^{\infty} \langle u_k, v_k \rangle \\
&= \langle \underline{u}, \underline{v} \rangle \\
&= \langle Lu, Lv \rangle.
\end{aligned}$$

So L is norm-preserving as well. Since L is surjective, it is an isomorphism of Hilbert spaces.

10.2 Lifting Open-Loop Systems

Now we look at what lifting means for certain systems. In what follows, G is a FDLTI system. Its input, state, and output evolve in finite-dimensional Euclidean spaces. Because the dimensions of these spaces will be irrelevant, they will all be denoted by $\mathcal{E}$. We will not initially assume that G is stable. Thus G is regarded as a linear operator on the extended space $\mathcal{L}_{2e}(\mathbb{R}, \mathcal{E})$. Let A, B, C, D denote parameters of some state realization and $x(t)$ the corresponding state vector.

Lifting G

We begin by lifting G itself. If the input-output equation is $y = Gu$, then the relation between the lifted input, $\underline{u} = Lu$, and the lifted output, $\underline{y} = Ly$, is $\underline{y} = \underline{G}\,\underline{u}$, where $\underline{G} := LGL^{-1}$ is the *lifted system*. The block diagram is

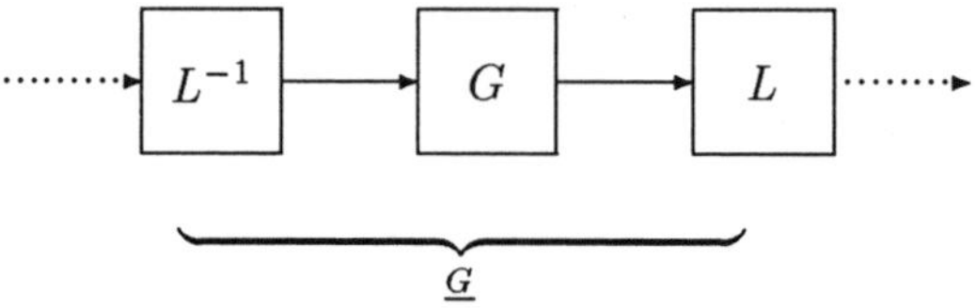

The lifted system, $\underline{G}$, acts on the discrete-time space $\ell_{2e}(\mathbb{Z}, \mathcal{K})$ and consequently it has a matrix representation of the form

$$\left[\begin{array}{cc|cccc} & \vdots & \vdots & \vdots & \vdots & \\ \cdots & \underline{g}(-1,-1) & 0 & 0 & 0 & \cdots \\ \hline \cdots & \underline{g}(0,-1) & \underline{g}(0,0) & 0 & 0 & \cdots \\ \cdots & \underline{g}(1,-1) & \underline{g}(1,0) & \underline{g}(1,1) & 0 & \cdots \\ & \vdots & \vdots & \vdots & \vdots & \end{array}\right].$$

The horizontal and vertical lines again separate the time intervals $\{k < 0\}$ and $\{k \geq 0\}$. Each block, $\underline{g}(i, j)$, is a linear transformation from $\mathcal{K}$ to $\mathcal{K}$. Because G is time-invariant, so is $\underline{G}$, so this matrix is Toeplitz:

$$\left[\begin{array}{cc|cccc} & \vdots & \vdots & \vdots & \vdots & \\ \cdots & \underline{g}(0) & 0 & 0 & 0 & \cdots \\ \hline \cdots & \underline{g}(1) & \underline{g}(0) & 0 & 0 & \cdots \\ \cdots & \underline{g}(2) & \underline{g}(1) & \underline{g}(0) & 0 & \cdots \\ & \vdots & \vdots & \vdots & \vdots & \end{array}\right].$$

The computation below will show that this matrix in fact has the form

$$\left[\begin{array}{cc|cccc} & \vdots & \vdots & \vdots & \vdots & \\ \cdots & \underline{D} & 0 & 0 & 0 & \cdots \\ \hline \cdots & \underline{C}\underline{B} & \underline{D} & 0 & 0 & \cdots \\ \cdots & \underline{C}\underline{A}\underline{B} & \underline{C}\underline{B} & \underline{D} & 0 & \cdots \\ & \vdots & \vdots & \vdots & \vdots & \end{array}\right], \tag{10.1}$$

so $\underline{G}$ could be modeled by the discrete-time equations

$$\begin{aligned} \xi(k+1) &= \underline{A}\xi(k) + \underline{B}u_k, \\ y_k &= \underline{C}\xi(k) + \underline{D}u_k. \end{aligned}$$

Here u_k and y_k are the kth components of the lifted input and output of G.

To determine these four new operators, $\underline{A}, \ldots, \underline{D}$, apply an input to G having support in $[0, h)$:

$$u(t) = \begin{cases} 0, & t < 0 \\ u_0(t), & 0 \leq t < h \\ 0, & t \geq h. \end{cases}$$

The output is then

$$y(t) = \begin{cases} 0, & t < 0 \\ Du_0(t) + \int_0^t Ce^{(t-\tau)A}Bu_0(\tau)d\tau, & 0 \le t < h \\ \int_0^h Ce^{(t-\tau)A}Bu_0(\tau)d\tau, & t \ge h. \end{cases}$$

The corresponding input and output of $\underline{G}$ are

$$\underline{u} = Lu = \begin{bmatrix} \vdots \\ 0 \\ \hline u_0 \\ 0 \\ \vdots \end{bmatrix}, \quad \underline{y} = Ly = \begin{bmatrix} \vdots \\ 0 \\ \hline y_0 \\ y_1 \\ \vdots \end{bmatrix}$$

where

$$\begin{aligned} y_0(t) &:= Du_0(t) + \int_0^t Ce^{(t-\tau)A}Bu_0(\tau)d\tau \\ y_1(t) &:= y(t+h) \\ &= Ce^{tA}\int_0^h e^{(h-\tau)A}Bu_0(\tau)d\tau \\ y_2(t) &:= y(t+2h) \\ &= Ce^{tA}e^{hA}\int_0^h e^{(h-\tau)A}Bu_0(\tau)d\tau \end{aligned}$$

etc.

Defining linear transformations

$$\begin{aligned} &\underline{A}: \ \mathcal{E} \to \mathcal{E}, \quad \underline{A}x = e^{hA}x \\ &\underline{B}: \ \mathcal{K} \to \mathcal{E}, \quad \underline{B}u = \int_0^h e^{(h-\tau)A}Bu(\tau)d\tau \\ &\underline{C}: \ \mathcal{E} \to \mathcal{K}, \quad (\underline{C}x)(t) = Ce^{tA}x \\ &\underline{D}: \ \mathcal{K} \to \mathcal{K}, \quad (\underline{D}u)(t) = Du(t) + \int_0^t Ce^{(t-\tau)A}Bu(\tau)d\tau, \end{aligned}$$

we have

$$\begin{aligned} y_0 &= \underline{D}u_0 \\ y_1 &= \underline{C}\,\underline{B}u_0 \\ y_2 &= \underline{C}\,\underline{A}\,\underline{B}u_0 \end{aligned}$$

etc.

as required for matrix (10.1).

The important point to observe is how finite-dimensionality of G is manifest in $\underline{G}$, namely, $\underline{A}$ acts on the same state-space as does the original A: Its matrix is $A_d := e^{hA}$, which would appear in a discretization of G using sample and hold. It is customary to identify the linear transformation $\underline{A}$ with its matrix representation A_d. Then the matrix representation of $\underline{G}$ is given by

$$\left[\begin{array}{cc|cccc} & \vdots & \vdots & \vdots & \vdots & \\ \cdots & \underline{D} & 0 & 0 & 0 & \cdots \\ \hline \cdots & \underline{C}\,\underline{B} & \underline{D} & 0 & 0 & \cdots \\ \cdots & \underline{C}A_d\underline{B} & \underline{C}\,\underline{B} & \underline{D} & 0 & \cdots \\ & \vdots & \vdots & \vdots & \vdots & \end{array}\right]$$

and the state model by

$$\left[\begin{array}{c|c} A_d & \underline{B} \\ \hline \underline{C} & \underline{D} \end{array}\right]. \tag{10.2}$$

Note that operators $\underline{B}$ and $\underline{C}$ have finite rank—the co-domain of $\underline{B}$ and the domain of $\underline{C}$ are $\mathcal{E}$. Operator $\underline{D}$ is the compression of G to $\mathcal{K}$.

It is not hard to see that the corresponding state vector for $\underline{G}$ in (10.2) is ξ defined via $\xi(k) = x(kh)$.

Lifting SG

A more interesting system to lift is SG, which maps continuous time to discrete time. The resulting system maps discrete time to discrete time, and consequently is a simpler object of study.

We shall lift SG, where G is as before except with $D = 0$. Then SG maps $\mathcal{L}_{2e}(\mathbb{R}, \mathcal{E})$ to $\ell_{2e}(\mathbb{Z}, \mathcal{E})$. The output from SG is already discrete-time, so we need lift only the input. The *lifted system*, $\underline{SG} := SGL^{-1}$, acts from $\ell_{2e}(\mathbb{Z}, \mathcal{K})$ to $\ell_{2e}(\mathbb{Z}, \mathcal{E})$. The block diagram is

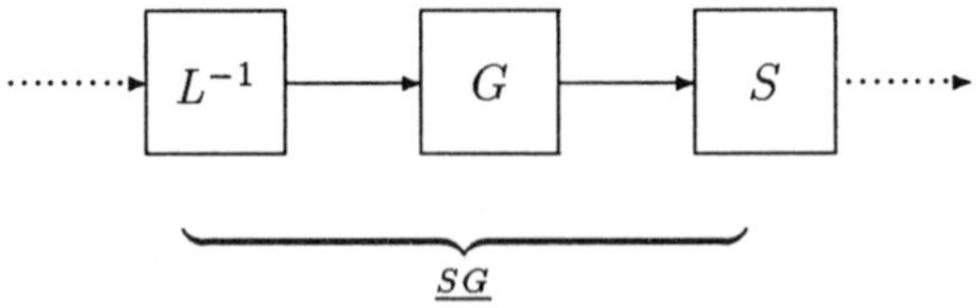

and the matrix is easily derived to be

$$\left[\begin{array}{cc|cccc} & \vdots & \vdots & \vdots & \vdots & \\ \cdots & 0 & 0 & 0 & 0 & \cdots \\ \hline \cdots & C\underline{B} & 0 & 0 & 0 & \cdots \\ \cdots & CA_d\underline{B} & C\underline{B} & 0 & 0 & \cdots \\ & \vdots & \vdots & \vdots & \vdots & \end{array}\right], \tag{10.3}$$

with $\underline{B}$ as above. Note that the other linear transformation is C (i.e., $x \mapsto Cx$), not $\underline{C}$.

Again, the state model for $\underline{SG}$ is

$$\left[\begin{array}{c|c} A_d & \underline{B} \\ \hline C & 0 \end{array}\right]$$

with the same state vector ξ as above.

Lifting GH

Finally, we shall lift GH. This is an operator from $\ell_{2e}(\mathbb{Z}, \mathcal{E})$ to $\mathcal{L}_{2e}(\mathbb{R}, \mathcal{E})$. The input to GH is already discrete-time, so we need lift only the output. The *lifted system*, $\underline{GH} := LGH$, acts from $\ell_{2e}(\mathbb{Z}, \mathcal{E})$ to $\ell_{2e}(\mathbb{Z}, \mathcal{K})$. The block diagram is

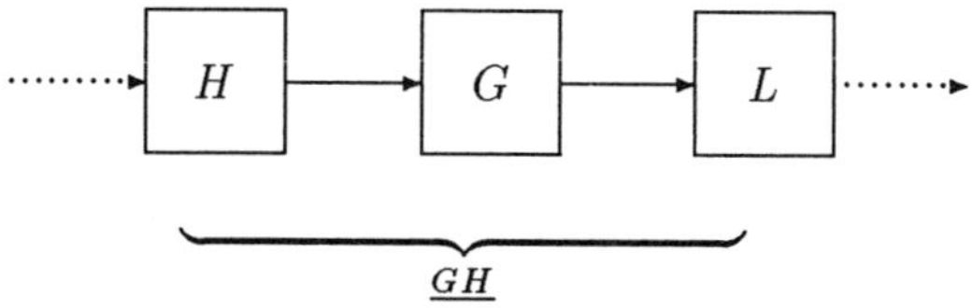

and the matrix is

$$\left[\begin{array}{cc|cccc} & \vdots & \vdots & \vdots & \vdots & \\ \cdots & \underline{D}_{res} & 0 & 0 & 0 & \cdots \\ \hline \cdots & \underline{C}B_d & \underline{D}_{res} & 0 & 0 & \cdots \\ \cdots & \underline{C}A_d B_d & \underline{C}B_d & \underline{D}_{res} & 0 & \cdots \\ & \vdots & \vdots & \vdots & \vdots & \end{array}\right],$$

where $B_d := \int_0^h e^{\tau A} d\tau B$ and $\underline{D}_{res}$ denotes the restriction of $\underline{D}$ to $\mathcal{E}$, that is,

$$\underline{D}_{res} : \mathcal{E} \to \mathcal{K}, \quad (\underline{D}_{res} v)(t) = \left[D + \int_0^t C e^{\tau A} d\tau B\right] v.$$

A state model for $\underline{GH}$, with the same state vector ξ, is

$$\left[\begin{array}{c|c} A_d & B_d \\ \hline \underline{C} & \underline{D}_{res} \end{array}\right].$$

10.3 Lifting SD Feedback Systems

In this section we shall put the formulas in the preceding section together to get a lifted model for the standard SD system shown in Figure 10.1 with

$$\hat{g}(s) = \left[\begin{array}{c|cc} A & B_1 & B_2 \\ \hline C_1 & D_{11} & D_{12} \\ C_2 & 0 & 0 \end{array}\right].$$

We have taken D_{21} to be zero so that the signal is lowpass filtered prior to the sampler. The state vector for this model is $x_G(t)$.

In the preceding section we lifted a continuous-time LTI system G into a discrete-time LTI system $\underline{G} := LGL^{-1}$. For $\underline{G}$ to be time-invariant, it is necessary and sufficient that G be h-periodic in continuous time. To see this, suppose that G is h-periodic, i.e.,

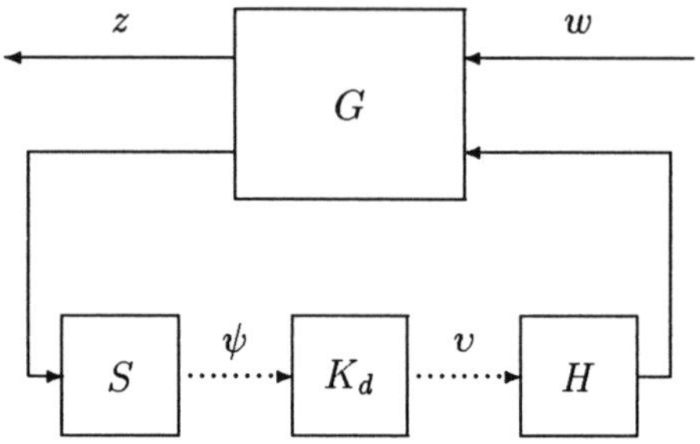

Figure 10.1: The standard SD system.

$$D_h^* G D_h = G,$$

where D_h and D_h^* are time delay and time advance by h in continuous time respectively. Let U and U^* be unit time delay and time advance operators on $\ell_{2e}(\mathbb{Z}, \mathcal{K})$, defined in the obvious way. It is readily verified that

$$U^* L = L D_h^*, \quad L^{-1} U = D_h L^{-1}.$$

Thus

$$\begin{aligned} U^* \underline{G} U &= U^* L G L^{-1} U \\ &= L D_h^* G D_h L^{-1} \\ &= L G L^{-1} \\ &= \underline{G} \end{aligned}$$

This means that $\underline{G}$ is time-invariant. The converse is true by reversing the argument.

Now T_{zw}, being the map $w \mapsto z$ in Figure 10.1, is h-periodic. So we lift this operator to get $L T_{zw} L^{-1}$. This corresponds to do the following in Figure 10.1: Move S and H into the generalized plant and introduce the lifting operators L and L^{-1} appropriately to get Figure 10.2, the lifted SD system, where $\underline{G}$, the lifted generalized plant, is given by

$$\begin{aligned} \underline{G} &= \begin{bmatrix} L & 0 \\ 0 & S \end{bmatrix} G \begin{bmatrix} L^{-1} & 0 \\ 0 & H \end{bmatrix} \\ &= \begin{bmatrix} L G_{11} L^{-1} & L G_{12} H \\ S G_{21} L^{-1} & G_{22d} \end{bmatrix}. \end{aligned}$$

Thus the lifted operator $L T_{zw} L^{-1}$ is exactly the map $\underline{w} \mapsto \underline{z}$ in Figure 10.2.

We stress that Figure 10.2 is a discrete-time LTI system with $\underline{w} = Lw$ and $\underline{z} = Lz$ both living in $\ell_{2e}(\mathbb{Z}, \mathcal{K})$. The lifted plant $\underline{G}$ maps $\ell_{2e}(\mathbb{Z}, \mathcal{K}) \oplus$

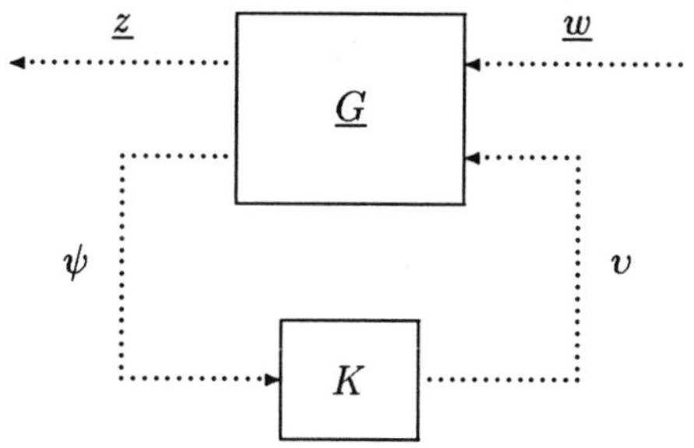

Figure 10.2: The lifted SD system.

$\ell_{2e}(\mathbb{Z},\mathcal{E})$ to $\ell_{2e}(\mathbb{Z},\mathcal{K}) \oplus \ell_{2e}(\mathbb{Z},\mathcal{E})$ and has four blocks: The (2, 2)-block is the discretization of G_{22} and the other blocks were studied in Section 10.2. Given the state model of G, we can obtain state models for all the four blocks and then put them together to obtain a state model for $\underline{G}$,

$$\left[\begin{array}{c|cc} A_d & \underline{B}_1 & B_{2d} \\ \hline \underline{C}_1 & \underline{D}_{11} & \underline{D}_{12} \\ C_2 & 0 & 0 \end{array}\right],$$

where $(A_d,\ B_{2d})$ is obtained via *c2d* of $(A,\ B_2)$ and the operator-valued entries are

$$\begin{aligned} \underline{B}_1 &: \quad \mathcal{K} \to \mathcal{E}, \quad \underline{B}_1 w = \int_0^h e^{(h-\tau)A} B_1 w(\tau)\, d\tau \\ \underline{C}_1 &: \quad \mathcal{E} \to \mathcal{K}, \quad (\underline{C}_1 x)(t) = C_1 e^{tA} x \\ \underline{D}_{11} &: \quad \mathcal{K} \to \mathcal{K}, \quad (\underline{D}_{11} w)(t) = D_{11} w(t) + C_1 \int_0^t e^{(t-\tau)A} B_1 w(\tau)\, d\tau \\ \underline{D}_{12} &: \quad \mathcal{E} \to \mathcal{K}, \quad (\underline{D}_{12} v)(t) = D_{12} v + C_1 \int_0^t e^{\tau A}\, d\tau\, B_2 v. \end{aligned}$$

The corresponding state vector for this state model of $\underline{G}$ is ξ_G defined via $\xi_G(k) = x_G(kh)$.

In summary, SD feedback systems can be lifted into time-invariant discrete-time systems with infinite-dimensional input and output spaces. This procedure will be the main tool for analysis and synthesis of SD systems in later chapters.

10.4 Adjoint Operators

Adjoints are fundamental to the study of operators on Hilbert space. Let us begin with the familiar case.

Example 10.4.1 Consider a matrix A in $\mathbb{R}^{n\times m}$. The corresponding linear transformation, denoted by, say, T, is multiplication by A:

$$T : \mathbb{R}^m \to \mathbb{R}^n, \quad Tx = Ax.$$

(Normally the same symbol is used for both A and T, but for clarity we use two different symbols here.) The transpose matrix, A', satisfies the equation

$$(Ax)'y = x'(A'y), \quad x \in \mathbb{R}^m, \ y \in \mathbb{R}^n,$$

or in inner-product notation

$$\langle Ax, y\rangle = \langle x, A'y\rangle.$$

The linear transformation

$$T^* : \mathbb{R}^n \to \mathbb{R}^m, \quad T^*y = A'y$$

is called the *adjoint* of T; it is the unique linear transformation satisfying the equation

$$\langle Tx, y\rangle = \langle x, T^*y\rangle.$$

The next example, of a discrete-time system, is similar.

Example 10.4.2 Consider a causal, LTI discrete-time system G that is bounded as an operator from $\ell_2(\mathbb{Z})$ to $\ell_2(\mathbb{Z})$. Its matrix representation then has the form

$$\left[\begin{array}{cc|cccc} & \vdots & \vdots & \vdots & \vdots & \\ \cdots & G_0 & 0 & 0 & 0 & \cdots \\ \hline \cdots & G_1 & G_0 & 0 & 0 & \cdots \\ \cdots & G_2 & G_1 & G_0 & 0 & \cdots \\ & \vdots & \vdots & \vdots & \vdots & \end{array}\right].$$

The adjoint G^* of G turns out to be the operator from $\ell_2(\mathbb{Z})$ to $\ell_2(\mathbb{Z})$ whose matrix representation is the transpose of the preceding one:

$$\left[\begin{array}{cc|ccc} & \vdots & \vdots & \vdots & \\ \cdots & G_0' & G_1' & G_2' & \cdots \\ \hline \cdots & 0 & G_0' & G_1' & \cdots \\ \cdots & 0 & 0 & G_0' & \cdots \\ & \vdots & \vdots & \vdots & \end{array}\right].$$

Let us consider a more interesting example.

Example 10.4.3 Consider the hold operator $H : \ell_2(\mathbb{Z}) \to \mathcal{L}_2(\mathbb{R})$:

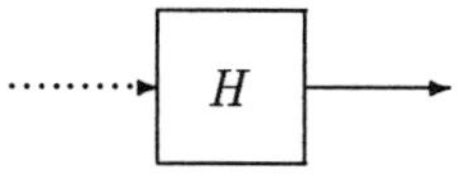

Its adjoint, H^*, maps in the reverse direction, $H^* : \mathcal{L}_2(\mathbb{R}) \to \ell_2(\mathbb{Z})$:

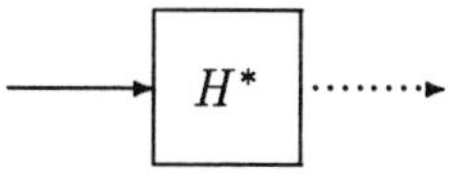

It is uniquely determined by the equation

$$\langle Hv, y\rangle = \langle v, H^* y\rangle .$$

We can find the action of H^* by evaluating both sides of this equation. First, define $x = Hv$. Then the left-hand side equals

$$\begin{aligned} \int x(t)'y(t)dt &= \sum_k \int_{kh}^{(k+1)h} x(t)'y(t)dt \\ &= \sum_k v(k)' \int_{kh}^{(k+1)h} y(t)dt. \end{aligned}$$

Similarly, if $\psi := H^* y$, the right-hand side equals

$$\sum v(k)'\psi(k).$$

It follows that

$$\psi(k) = \int_{kh}^{(k+1)h} y(t)dt.$$

To recap, H^* is defined as follows:

$$H^* : \mathcal{L}_2(\mathbb{R}) \to \ell_2(\mathbb{Z}), \quad (H^* y)(k) = \int_{kh}^{(k+1)h} y(t)dt.$$

Thus the value of $H^* y$ at time k equals the integral of y over the kth sampling interval.

Let us continue and find the operator $H^*H : \ell_2(\mathbb{Z}) \to \ell_2(\mathbb{Z})$. Note that this operates solely in the discrete-time domain:

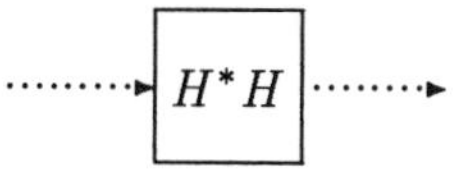

Apply an arbitrary input v to H^*H and let ψ denote the output. From the two equations

$$\begin{aligned} y &= Hv \\ \psi &= H^*y, \end{aligned}$$

it follows that $\psi = hv$. Therefore, $H^*H = hI$, where I denotes the identity operator.

The concept of an adjoint is general: For every bounded operator T from a Hilbert space $\mathcal{X}$ to another $\mathcal{Y}$, there is a unique bounded operator T^* from $\mathcal{Y}$ to $\mathcal{X}$ satisfying the equation

$$\langle Tx, y \rangle = \langle x, T^*y \rangle.$$

For two Hilbert space operators $T : \mathcal{X} \to \mathcal{Y}$ and $G : \mathcal{Y} \to \mathcal{Z}$, it is easily verified that $(GT)^* = T^*G^*$.

We conclude this section with a final example.

Example 10.4.4 In the preceding section, we introduced several new operators associated with a state model (A, B, C, D) for a FDLTI system G. In particular, we had

$$\begin{aligned} &\underline{B}: \ \mathcal{K} \to \mathcal{E}, \quad \underline{B}u = \int_0^h e^{(h-\tau)A} Bu(\tau)d\tau \\ &\underline{C}: \ \mathcal{E} \to \mathcal{K}, \quad (\underline{C}x)(t) = Ce^{tA}x. \end{aligned}$$

The adjoints of these operators can be calculated to be

$$\begin{aligned} &\underline{B}^*: \ \mathcal{E} \to \mathcal{K}, \quad (\underline{B}^*x)(t) = B'e^{(h-t)A'}x \\ &\underline{C}^*: \ \mathcal{K} \to \mathcal{E}, \quad \underline{C}^*v = \int_0^h e^{tA'}C'v(t)dt. \end{aligned}$$

From these formulas we then get that $\underline{B}\,\underline{B}^*$ and $\underline{C}^*\underline{C}$ both act on $\mathcal{E}$, that is, they are (equivalent to) matrices. Specifically,

$$\begin{aligned} \underline{B}\,\underline{B}^* &= \int_0^h e^{tA}BB'e^{tA'}dt \\ \underline{C}^*\underline{C} &= \int_0^h e^{tA'}C'Ce^{tA}dt. \end{aligned}$$

10.5 The Norm of SG

We return now to the study of SG and the computation of its induced norm. So that SG is indeed a bounded operator from $\mathcal{L}_2(\mathbb{R}, \mathcal{E})$ to $\ell_2(\mathbb{Z}, \mathcal{E})$, we shall assume that G has a strictly causal state model $(A, B, C, 0)$ with A stable. The induced norm of SG is defined to be

$$\|SG\| = \sup_{\|u\|_2 \le 1} \|SGu\|_2.$$

The lifted system $\underline{SG} = SGL^{-1}$ is then bounded from $\ell_2(\mathbb{Z}, \mathcal{K})$ to $\ell_2(\mathbb{Z}, \mathcal{E})$. Since L is norm-preserving, we have that $\|SG\| = \|\underline{SG}\|$. Now it is a general fact that for any bounded Hilbert space operator T,

$$\|T\| = \|T^*\| = \|T^*T\|^{1/2} = \|TT^*\|^{1/2}.$$

Therefore

$$\|SG\| = \|\underline{SG}\,(\underline{SG})^*\|^{1/2}.$$

This equation motivates us to determine the operator $\underline{SG}\,(\underline{SG})^* : \ell_2(\mathbb{Z}, \mathcal{E}) \to \ell_2(\mathbb{Z}, \mathcal{E})$ explicitly.

The matrix representation of $\underline{SG}$ in (10.3) can be written as the product

$$\left[\begin{array}{cc|ccc} & \vdots & \vdots & \vdots & \\ \cdots & 0 & 0 & 0 & \cdots \\ \hline \cdots & C & 0 & 0 & \cdots \\ \cdots & CA_d & C & 0 & \cdots \\ & \vdots & \vdots & \vdots & \end{array}\right] \left[\begin{array}{cc|ccc} & \vdots & \vdots & \vdots & \\ \cdots & \underline{B} & 0 & 0 & \cdots \\ \hline \cdots & 0 & \underline{B} & 0 & \cdots \\ \cdots & 0 & 0 & \underline{B} & \cdots \\ & \vdots & \vdots & \vdots & \end{array}\right].$$

Note that the left-hand matrix, denoted $[M_1]$, represents a discrete system $M_1 : \ell_2(\mathbb{Z}, \mathcal{E}) \to \ell_2(\mathbb{Z}, \mathcal{E})$, its transfer matrix being

$$\hat{m}_1(\lambda) = \left[\begin{array}{c|c} A_d & I \\ \hline C & 0 \end{array}\right].$$

Therefore, the matrix representation of $(\underline{SG})^*$ is

$$\left[\begin{array}{cc|ccc} & \vdots & \vdots & \vdots & \\ \cdots & \underline{B}^* & 0 & 0 & \cdots \\ \hline \cdots & 0 & \underline{B}^* & 0 & \cdots \\ \cdots & 0 & 0 & \underline{B}^* & \cdots \\ & \vdots & \vdots & \vdots & \end{array}\right] [M_1]'.$$

Multiplying these two matrices, we get that the matrix representation of $\underline{SG}\,(\underline{SG})^*$ is

$$[M_1] \left[\begin{array}{cc|ccc} & \vdots & \vdots & \vdots & \\ \cdots & \underline{B}\,\underline{B}^* & 0 & 0 & \cdots \\ \hline \cdots & 0 & \underline{B}\,\underline{B}^* & 0 & \cdots \\ \cdots & 0 & 0 & \underline{B}\,\underline{B}^* & \cdots \\ & \vdots & \vdots & \vdots & \end{array}\right] [M_1]'. \qquad (10.4)$$

The center matrix represents a pure gain from $\ell_2(\mathbb{Z},\mathcal{E})$ to $\ell_2(\mathbb{Z},\mathcal{E})$—we calculated the matrix $\underline{B}\,\underline{B}^*$ in the preceding section.

For convenience, define

$$J := \underline{B}\,\underline{B}^* = \int_0^h e^{tA}BB'e^{tA'}dt$$

and bring in a matrix B_J satisfying $B_J B_J' = J$ (for example, $B_J = J^{1/2}$). Also, introduce the discrete-time system $M : \ell_2(\mathbb{Z},\mathcal{E}) \to \ell_2(\mathbb{Z},\mathcal{E})$ with transfer matrix

$$\begin{aligned}\hat{m}(\lambda) &= \hat{m}_1(\lambda)B_J \\ &= \left[\begin{array}{c|c} A_d & B_J \\ \hline C & 0 \end{array}\right].\end{aligned}$$

It follows then from (10.4) that

$$\underline{SG}\,(\underline{SG})^* = MM^*.$$

We conclude that $\|SG\| = \|M\|$. The advantage of this is that M is a FDLTI system, so $\|M\|$ equals the norm in $\mathcal{H}_\infty(\mathbb{D})$ of $\hat{m}$.

Let us summarize with the following procedure to compute $\|SG\|$:

Step 1 Start with state parameters $(A, B, C, 0)$ of G, A stable.

Step 2 Compute

$$J = \int_0^h e^{tA}BB'e^{tA'}dt.$$

Step 3 Compute B_J satisfying $B_J B_J' = J$ (Cholesky factorization).

Step 4 Define

$$\hat{m}(\lambda) = \left[\begin{array}{c|c} A_d & B_J \\ \hline C & 0 \end{array}\right] \quad (A_d = e^{hA}).$$

Step 5 Then the induced norm of $SG : \mathcal{L}_2(\mathbb{R},\mathcal{E}) \to \ell_2(\mathbb{Z},\mathcal{E})$ equals $\|\hat{m}\|_\infty$.

Example 10.5.1 For the simplest case, take

$$\hat{g}(s) = \frac{1}{s+1} = \left[\begin{array}{c|c} -1 & 1 \\ \hline 1 & 0 \end{array}\right].$$

Then we calculate that

$$
\begin{aligned}
J &= \int_0^h e^{-2t} dt \\
&= \frac{1-e^{-2h}}{2} \\
\hat{m}(\lambda) &= \left[\begin{array}{c|c} e^{-h} & \sqrt{J} \\ \hline 1 & 0 \end{array}\right] \\
&= \frac{\lambda\sqrt{J}}{1-e^{-h}\lambda} \\
\|SG\| &= \|\hat{m}\|_\infty \\
&= \frac{\sqrt{J}}{1-e^{-h}} \\
&= \left[\frac{1+e^{-h}}{2(1-e^{-h})}\right]^{1/2}.
\end{aligned}
$$

Continuing, recall that the hold operator has the property that $\|Hv\|_2 = \sqrt{h}\|v\|_2$ for every v in $\ell_2(\mathbb{Z}, \mathcal{E})$. Thus, for this example

$$
\|HSG\| = \sqrt{h}\|SG\| = \left[\frac{h(1+e^{-h})}{2(1-e^{-h})}\right]^{1/2}.
$$

In particular,

$$
\lim_{h\to 0} \|HSG\| = 1.
$$

Also,

$$
\|G\| = \|\hat{g}\|_\infty = 1.
$$

This verifies for this example the general fact that

$$
\lim_{h\to 0} \|HSG\| = \|G\|.
$$

In the example, computing J was trivial (A was a scalar); but it requires some work in the matrix case. Now we look at how to do Step 2 in general using matrix exponentials. The key lemma serving this purpose is as follows.

Lemma 10.5.1 *Let A_{11} and A_{22} both be square and define*

$$
\begin{bmatrix} F_{11}(t) & F_{12}(t) \\ 0 & F_{22}(t) \end{bmatrix} := \exp\left\{t\begin{bmatrix} A_{11} & A_{12} \\ 0 & A_{22} \end{bmatrix}\right\}, \quad t \geq 0. \tag{10.5}
$$

Then $F_{11}(t) = e^{tA_{11}}$, $F_{22}(t) = e^{tA_{22}}$, and

$$
F_{12}(t) = \int_0^t e^{(t-\tau)A_{11}} A_{12} e^{\tau A_{22}}\, d\tau.
$$

Proof Since the matrices are block upper-triangular, we easily get

$$F_{11}(t) = e^{tA_{11}}, \qquad F_{22}(t) = e^{tA_{22}}.$$

Differentiate (10.5) to get

$$\frac{d}{dt}\begin{bmatrix} F_{11}(t) & F_{12}(t) \\ 0 & F_{22}(t) \end{bmatrix} = \begin{bmatrix} A_{11} & A_{12} \\ 0 & A_{22} \end{bmatrix}\begin{bmatrix} F_{11}(t) & F_{12}(t) \\ 0 & F_{22}(t) \end{bmatrix}.$$

Thus

$$\frac{d}{dt}F_{12}(t) = A_{11}F_{12}(t) + A_{12}F_{22}(t).$$

Solve this differential equation, noting $F_{22}(t) = e^{tA_{22}}$ and $F_{12}(0) = 0$:

$$F_{12}(t) = \int_0^t e^{(t-\tau)A_{11}} A_{12} e^{\tau A_{22}}\, d\tau.$$

■

To compute

$$J = \int_0^h e^{tA} BB' e^{tA'}\, dt$$

in Step 2, define

$$\begin{bmatrix} P_{11} & P_{12} \\ 0 & P_{22} \end{bmatrix} = \exp\left\{ h \begin{bmatrix} -A & BB' \\ 0 & A' \end{bmatrix} \right\}.$$

Then by Lemma 10.5.1

$$\begin{aligned} P_{22} &= e^{hA'} \\ P_{12} &= \int_0^h e^{(\tau-h)A} BB' e^{\tau A'}\, d\tau \\ &= e^{-hA}\int_0^h e^{\tau A} BB' e^{\tau A'}\, d\tau. \end{aligned}$$

So

$$J = P_{22}' P_{12}.$$

This gives a way to evaluate J via computing a matrix exponential, which is easy in MATLAB.

In conclusion, in this section we saw that SG, a hybrid operator mapping continuous to discrete time, has the same norm as a certain purely discrete-time system M. The induced norm of M is readily calculated as the $\mathcal{H}_\infty(\mathbb{D})$-norm of its transfer matrix.

10.6 The Norm of GH

A companion to SG is the operator GH, to which we now turn. So that GH is a bounded operator from $\ell_2(\mathbb{Z},\mathcal{E})$ to $\mathcal{L}_2(\mathbb{R},\mathcal{E})$, we shall assume that G has a state model (A,B,C,D) with A stable (strict properness is not necessary).

Again, the main ingredient in the computation is a lift: We have

$$\|GH\| = \|\underline{GH}\| = \|(\underline{GH})^*\underline{GH}\|^{1/2}.$$

The operator $(\underline{GH})^*\underline{GH} : \ell_2(\mathbb{Z},\mathcal{E}) \to \ell_2(\mathbb{Z},\mathcal{E})$ is again FDLTI, but noncausal. The next step is to introduce a discrete-time system $N : \ell_2(\mathbb{Z},\mathcal{E}) \to \ell_2(\mathbb{Z},\mathcal{E})$ such that

$$(\underline{GH})^*\underline{GH} = N^*N.$$

To compute N, we observe that the matrix representation of $\underline{GH}$ has the following factorization

$$\left[\begin{array}{cc|ccc}
 & \vdots & \vdots & \vdots & \\
\cdots & [\,\underline{C}\;\; \underline{D}_{res}\,] & 0 & 0 & \cdots \\
\hline
\cdots & 0 & [\,\underline{C}\;\; \underline{D}_{res}\,] & 0 & \cdots \\
\cdots & 0 & 0 & [\,\underline{C}\;\; \underline{D}_{res}\,] & \cdots \\
 & \vdots & \vdots & \vdots &
\end{array}\right][N_1],$$

where $[N_1]$ is the matrix representation of the discrete-time system N_1 with transfer matrix

$$\hat{n}_1(\lambda) = \left[\begin{array}{c|c} A_d & B_d \\ \hline I & 0 \\ 0 & I \end{array}\right].$$

Thus similar analysis leads first to computing matrices C_d, D_d satisfying the equation

$$\begin{bmatrix} C_d' \\ D_d' \end{bmatrix}\begin{bmatrix} C_d & D_d \end{bmatrix} = \begin{bmatrix} \underline{C}^* \\ \underline{D}_{res}^* \end{bmatrix}\begin{bmatrix} \underline{C} & \underline{D}_{res} \end{bmatrix} \tag{10.6}$$

and then taking N to have transfer matrix

$$\begin{aligned}\hat{n}(\lambda) &= \begin{bmatrix} C_d & D_d \end{bmatrix}\hat{n}_1(\lambda)\\ &= \left[\begin{array}{c|c} A_d & B_d \\ \hline C_d & D_d \end{array}\right].\end{aligned}$$

It remains to compute the right-hand side of (10.6), denoted J:

$$\begin{aligned} J &= \begin{bmatrix} \underline{C}^*\underline{C} & \underline{C}^*\underline{D}_{res} \\ \underline{D}_{res}^*\underline{C} & \underline{D}_{res}^*\underline{D}_{res} \end{bmatrix}, \\ \underline{C}^*\underline{C} &= \int_0^h e^{tA'}C'Ce^{tA}\,dt, \end{aligned}$$

$$\begin{aligned}
\underline{C}^* \underline{D}_{res} &= \int_0^h e^{tA'} C' \left[D + C \int_0^t e^{\tau A} d\tau B \right] dt, \\
\underline{D}_{res}^* \underline{D}_{res} &= \int_0^h \left[D + C \int_0^t e^{\tau A} d\tau B \right]' \left[D + C \int_0^t e^{\tau A} d\tau B \right] dt.
\end{aligned}$$

The matrix J can be computed again via matrix exponentials. Define the square matrix

$$\underline{A} := \begin{bmatrix} A & B \\ 0 & 0 \end{bmatrix}.$$

Then by Lemma 10.5.1

$$\begin{aligned}
e^{t\underline{A}} &= \begin{bmatrix} e^{tA} & \int_0^t e^{(t-\tau)A} B \, d\tau \\ 0 & I \end{bmatrix} \\
&= \begin{bmatrix} e^{tA} & \int_0^t e^{\tau A} \, d\tau B \\ 0 & I \end{bmatrix}.
\end{aligned}$$

It is straightforward to check that

$$J = \int_0^h e^{t\underline{A}'} \begin{bmatrix} C & D \end{bmatrix}' \begin{bmatrix} C & D \end{bmatrix} e^{t\underline{A}} \, dt.$$

This type of integral was studied in the preceding section and can be computed using Lemma 10.5.1.

Let us summarize the procedure as follows:

Step 1 Start with state parameters (A, B, C, D) of G, A stable.

Step 2 Compute the square matrices

$$\underline{A} = \begin{bmatrix} A & B \\ 0 & 0 \end{bmatrix}, \quad Q = \begin{bmatrix} C & D \end{bmatrix}' \begin{bmatrix} C & D \end{bmatrix}$$

and

$$\begin{bmatrix} P_{11} & P_{12} \\ 0 & P_{22} \end{bmatrix} = \exp\left\{ h \begin{bmatrix} -\underline{A}' & Q \\ 0 & \underline{A} \end{bmatrix} \right\}.$$

Then $J = P_{22}' P_{12}$.

Step 3 Compute C_d and D_d satisfying

$$\begin{bmatrix} C_d' \\ D_d' \end{bmatrix} \begin{bmatrix} C_d & D_d \end{bmatrix} = J$$

(Cholesky factorization).

Step 4 Define

$$\hat{n}(\lambda) = \left[\begin{array}{c|c} A_d & B_d \\ \hline C_d & D_d \end{array}\right] \quad \left(A_d = e^{hA},\ B_d = \int_0^h e^{\tau A} d\tau B\right).$$

Here A_d and B_d could be read from P_{22}:

$$P_{22} = \left[\begin{array}{cc} A_d & B_d \\ 0 & I \end{array}\right].$$

Step 5 Then the induced norm of $GH : \ell_2(\mathbb{Z}, \mathcal{E}) \to \mathcal{L}_2(\mathbb{R}, \mathcal{E})$ equals $\|\hat{n}\|_\infty$.

Notice that

$$\begin{aligned} \|GH\| &\leq \|G\|\|H\| \\ &= \|\hat{g}\|_\infty \sqrt{h}. \end{aligned}$$

Example 10.6.1 Figure 10.3 shows the magnitude Bode plot of $h^{-1/2}\hat{n}$ for

$$\hat{g}(s) = \frac{1}{s+1},$$

for $h = 0.1$ and $h = 1$. The peak magnitude equals 1 in each case, so $\|GH\| = \sqrt{h}$.

Figure 10.4 is for

$$\hat{g}(s) = \frac{1}{s^2+s+1}.$$

The peak magnitudes give

$$\frac{1}{\sqrt{h}}\|GH\| = \begin{cases} 1.1544, & h = 0.1 \\ 1.1315, & h = 1. \end{cases}$$

10.7 Analysis of Sensor Noise Effect

This section applies the formulas in the preceding section to the analysis of the effect of sensor noise on the controlled output in a SD system. The system to be studied is shown in Figure 10.5. The sensor noise, or a combination of sensor noise and quantization error, is additive at the output of the sampler and is modelled as $W_d\omega$, where W_d is a fixed discrete-time weighting system and is assumed to be FDLTI and stable and ω has bounded $\ell_2(\mathbb{Z})$-norm but is otherwise unknown. Typically, *a priori* frequency-domain information of the noise, e.g., bandwith, is incorporated into the model W_d. Assuming the command input r is zero, we would like to analyze the worst-case effect of ω on the plant output y via the induced norm:

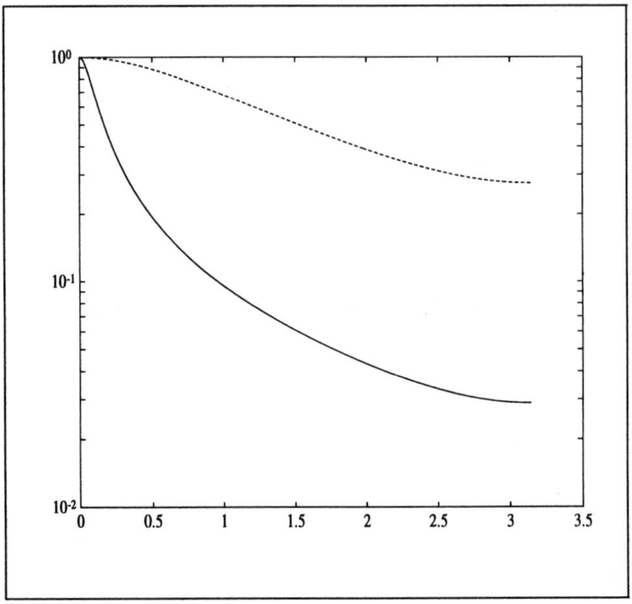

Figure 10.3: Magnitude Bode Plot of $h^{-1/2}\hat{n}$ for $\hat{g}(s) = \frac{1}{s+1}$: $h = 0.1$ solid line, $h = 1$ dash line.

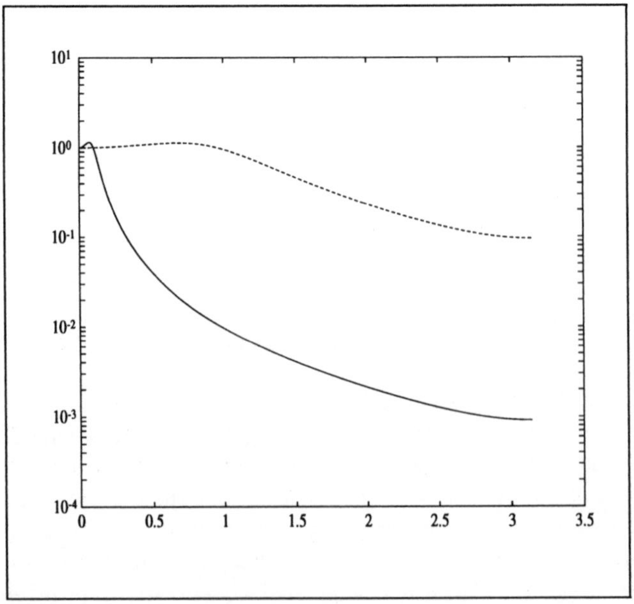

Figure 10.4: Magnitude Bode Plot of $h^{-1/2}\hat{n}$ for $\hat{g}(s) = \frac{1}{s^2+s+1}$: $h = 0.1$ solid line, $h = 1$ dash line.

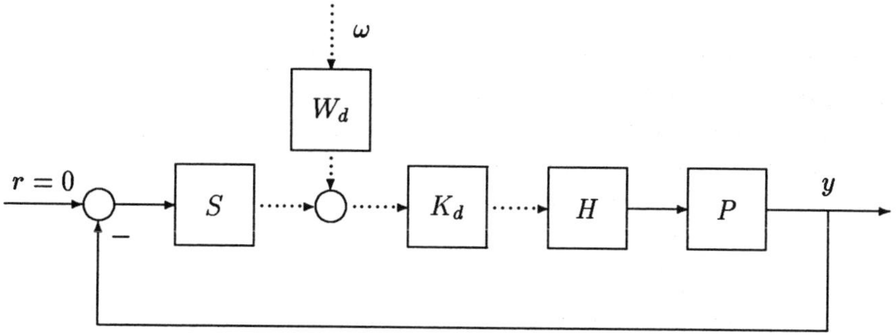

Figure 10.5: A SD system with sensor noise.

> Given P, K_d, W_d, and h, compute the $\ell_2(\mathbb{Z})$ to $\mathcal{L}_2(\mathbb{R})$ norm of the map $\omega \mapsto y$.

Let $T_{y\omega}$ be the map $\omega \mapsto y$ in Figure 10.5. It can be derived that

$$T_{y\omega} = PH(I + K_d P_d)^{-1} K_d W_d,$$

where $P_d := SPH$. As usual, we assume that the continuous P and discrete K_d are both FDLTI and causal and K_d internally stabilizes P_d. Then the discrete-time system

$$T_d := (I + K_d P_d)^{-1} K_d W_d$$

is stable and $T_{y\omega} = PHT_d$. Assume, for simplicity, that P is stable and bring in a FDLTI discrete-time system N_d such that

$$(\underline{PH})^* \underline{PH} = N_d^* N_d.$$

Explicit formulas for N_d in terms of P were derived in the preceding section. For any $\omega \in \ell_2(\mathbb{Z})$, we compute the $\mathcal{L}_2(\mathbb{R})$-norm of $T_{y\omega}\omega$ as follows:

$$\begin{aligned} \|T_{y\omega}\omega\|_2^2 &= \langle \underline{PH}T_d\omega, \underline{PH}T_d\omega \rangle \\ &= \langle T_d^* (\underline{PH})^* \underline{PH} T_d\omega, \omega \rangle \\ &= \langle T_d^* N_d^* N_d T_d \omega, \omega \rangle \\ &= \langle N_d T_d \omega, N_d T_d \omega \rangle \\ &= \|N_d T_d \omega\|_2^2 \end{aligned}$$

This means that for the same input, the SD system $T_{y\omega}$ and the discrete-time system $N_d T_d$ have equal output norm. Thus $\|T_{y\omega}\| = \|N_d T_d\|$, the latter being the $\ell_2(\mathbb{Z})$-induced norm.

We conclude that $\|T_{y\omega}\|$ can be computed via evaluating the $\mathcal{H}_\infty(\mathbb{D})$-norm of a known transfer function:

$$\|T_{yw}\| = \|\hat{n}_d \hat{t}_d\|_\infty.$$

Example 10.7.1 In this example, we look at the noise attenuation properties of the familiar SD tracking system studied in Examples 6.6.1 and 8.4.2. Focusing on the sensor noise effect, we arrive at the block diagram in Figure 10.5. The plant transfer function is

$$\hat{p}(s) = \frac{1}{(10s+1)(25s+1)}$$

and the sampling period is $h = 1$. The sensor noise effect in the SD system will be analyzed for two different controllers: the controller designed in Example 6.6.1 via naive discretization,

$$\hat{k}_d(\lambda) = \frac{-477.1019(\lambda - 1.1052)(\lambda - 1.0408)}{(\lambda + 1.0478)(\lambda - 1)},$$

and the controller designed in Example 8.4.2 via fast discretization,

$$\hat{k}_d(\lambda) = \frac{-488.85(\lambda - 1.1052)(\lambda - 1.0408)}{(\lambda + 1.3955)(\lambda - 1)}.$$

We saw before that the first controller gives severe intersample ripple in the step response but the second does not. What about their sensor noise attenuation properties?

To get noise attenuation properties for the whole frequency range, we take the noise weighting function $\hat{w}_d(\lambda) = 1$. For the two controlled SD systems, we compute the two associated discrete-time systems $\hat{n}_d\hat{t}_d$ and plot their magnitudes versus frequency in Figure 10.6. From the maximum values on these plots, we get that $\|T_{yw}\| = 15.6311$ for the controller designed via naive discretization and $\|T_{yw}\| = 1.0329$ for the controller designed via fast discretization. This shows that the design via fast discretization is superior not only for step response but also for sensor noise attenuation. Additional information can also be obtained from these plots: The two controlled systems have roughly the same capability for rejecting noise at low frequencies, but the second outperforms the first for noise at high frequencies; in fact, the magnitude plot of $\hat{n}_d\hat{t}_d$ for the second controller is almost flat, indicating the noise attenuation capability is quite uniform across the whole frequency range.

Exercises

10.1 Let $m\mathbb{Z}$ denote the set of all integer multiples of m, that is, $\{mk; k \in \mathbb{Z}\}$; let $\ell(m\mathbb{Z})$ denote the set of all sequences $m\mathbb{Z} \longrightarrow \mathbb{R}$; and let $\ell_2(m\mathbb{Z})$ denote the Hilbert space of all square-summable such sequences. The *decimator* is the linear operator

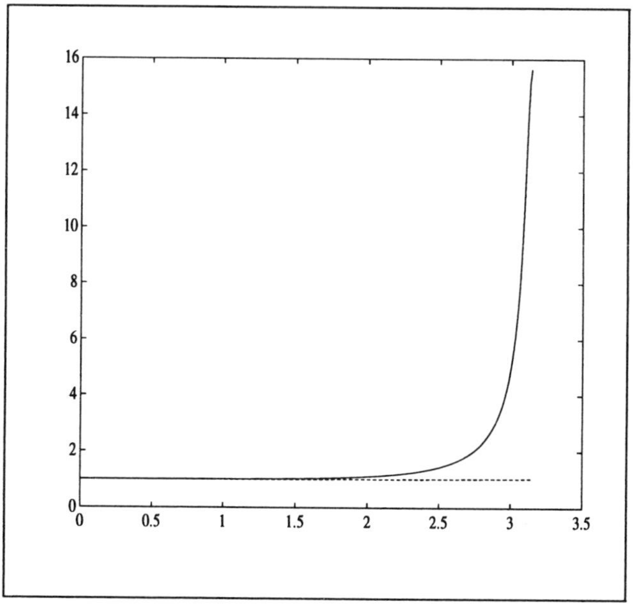

Figure 10.6: Magnitude plots versus frequency of $\hat{n}_d \hat{t}_d$ for the controller designed by naive discretization (solid) and by fast discretization (dash).

$$D : \ell_2(\mathbb{Z}) \longrightarrow \ell_2(m\mathbb{Z})$$

that maps v to ψ, where $\psi(mk) = v(mk)$. Find the adjoint operator D^*.

10.2 Let G be the continuous-time system with transfer function $\hat{g}(s) = 1/(s+5)$. Consider G as an operator $\mathcal{L}_2(\mathbb{R}) \to \mathcal{L}_2(\mathbb{R})$.

1. Find the adjoint operator G^*, that is, the relationship between u and y when $u = G^*y$.

2. Let H be the usual hold, considered as an operator $\ell_2(\mathbb{Z}) \to \mathcal{L}_2(\mathbb{R})$, and let

$$y(t) = \begin{cases} 0, & t < 0 \\ e^{-t}, & t \geq 0. \end{cases}$$

Find $(GH)^*y$ for $h = 1$.

10.3 As in Section 10.2, define the operators

$$\begin{aligned} \underline{B}: \ & \mathcal{K} \to \mathcal{E}, \quad \underline{B}u = \int_0^h e^{(h-\tau)A} Bu(\tau) d\tau \\ \underline{C}: \ & \mathcal{E} \to \mathcal{K}, \quad (\underline{C}x)(t) = Ce^{tA}x. \end{aligned}$$

Compute the following formulas:

$$\underline{B}^* : \ \mathcal{E} \rightarrow \mathcal{K}, \quad (\underline{B}^* x)(t) = B' \mathrm{e}^{(h-t)A'} x$$
$$\underline{C}^* : \ \mathcal{E} \rightarrow \mathcal{K}, \quad \underline{C}^* v = \int_0^h \mathrm{e}^{tA'} C' v(t) dt$$

$$\underline{B}\,\underline{B}^* = \int_0^h \mathrm{e}^{tA} B B' \mathrm{e}^{tA'} dt$$
$$\underline{C}^* \underline{C} = \int_0^h \mathrm{e}^{tA'} C' C \mathrm{e}^{tA} dt.$$

10.4 Let

$$\hat{g}(s) = \begin{bmatrix} \dfrac{1}{s+2} & \dfrac{s-1}{(2s+1)(s+1)} \\ \dfrac{s+1}{3s^2+2s+1} & \dfrac{1}{s^2+2s+3} \end{bmatrix}.$$

Compute $\|SG\|$ and $\|GH\|$ for a range of h.

10.5 Let G be a continuous-time, FDLTI, SISO, stable system with strictly proper transfer function $\hat{g}(s)$. Also, let $\hat{r}(s)$ denote the function

$$\hat{r}(s) = \frac{1 - \mathrm{e}^{-sh}}{sh}.$$

1. Using the results in Section 3.3 and the Cauchy-Schwarz inequality, derive the following bound on the $\mathcal{L}_2(\mathbb{R})$-induced norm of HSG:

$$\|HSG\| \leq$$
$$\sup_\omega \left[\left(\sum_k |\hat{r}(j\omega + jk\omega_s)|^2 \right)^{1/2} \left(\sum_k |\hat{g}(j\omega + jk\omega_s)|^2 \right)^{1/2} \right].$$

2. Prove that

$$\sum_k |\hat{r}(j\omega + jk\omega_s)|^2 = 1 \quad \forall \omega.$$

Thus

$$\|HSG\| \leq \sup_\omega \left[\left(\sum_k |\hat{g}(j\omega + jk\omega_s)|^2 \right)^{1/2} \right].$$

(Actually, equality holds.)

10.6 With reference to Figure 10.5, study the noise attenuation properties of the SD system with

$$\hat{p}(s) = \frac{10}{s(s+1)}, \quad h = 1,$$

and

$$\hat{k}_d(\lambda) = \frac{0.1392\lambda(\lambda - 2.6567)(\lambda - 1.5511)}{(\lambda - 1)(\lambda + 0.6327)(\lambda + 1.2699)}, \quad \hat{w}_d(\lambda) = 1.$$

Note that the plant P here is unstable.

Notes and References

Pioneering work on the $\mathcal{L}_2$-induced norm of a sampled-data system was done by Thompson, Stein, and Athans [138] and by Thompson, Dailey, and Doyle [137]. Perry [119] and Leung, Perry, and Francis [101] treated the same problem, but for bandlimited inputs.

The continuous lifting idea in Section 10.1, which follows naturally from the discrete-time one, was used in the control context independently by Bamieh and Pearson [16], Bamieh et al. [18], Tadmor [136], Toivonen [140], and Yamamoto [151], [154]. Formulas for the norms of SG and GH were derived by Chen and Francis in [26], though not by lifting as here. The first general solution of the problem of computing and optimizing the $\mathcal{L}_2$ induced norm of a SD system was due to Hara and Kabamba [82], [83] and Toivonen [140]. A related reference is [132], which does not assume *ab initio* that the A/D device is a zero-order hold. Induced norms can also be addressed via differential game theory [14]. A more general approach to the $\mathcal{L}_2$-induced norm of a hybrid system is in [126].

The idea of using matrix exponentials to compute integrals involving matrix exponentials, e.g., Lemma 10.5.1, was due to Van Loan [108].

For an extension of the $\mathcal{L}_1$ analog design approach to the SD framework, the $\mathcal{L}_\infty$-induced norm is important. This is treated in [42], [124], [105], and [15].

Chapter 11

Stability and Tracking in SD Systems

In Chapters 9 and 10 we studied several open-loop SD systems considering intersample behaviour; a useful lifting technique was developed in Chapter 10. This chapter continues our study of SD systems, focusing on closed-loop stability and the ability to track command inputs.

11.1 Internal Stability

Section 9.4 showed that input-output stability of a digital implementation is recovered as the sampling period h tends to zero. But what if h is not infinitesimally small? How can we test stability for an arbitrary sampling period?

The system to be studied is shown in Figure 11.1. State models for G and K are denoted as follows:

$$\hat{g}(s) = \left[\begin{array}{c|cc} A & B_1 & B_2 \\ \hline C_1 & D_{11} & D_{12} \\ C_2 & D_{21} & 0 \end{array}\right], \quad \hat{k}(\lambda) = \left[\begin{array}{c|c} A_K & B_K \\ \hline C_K & D_K \end{array}\right].$$

It is *assumed* that (A, B_2) is stabilizable and (C_2, A) is detectable; otherwise, G cannot be stabilized by *any* controller. Let $x_G(t)$ denote the state of G and $\xi_K(k)$ that of K. What should be an appropriate definition of internal stability for the SD feedback system?

Note that the SD system in Figure 11.1 is h-periodic in continuous time. Define the continuous-time vector

$$x_{sd}(t) := \begin{bmatrix} x_G(t) \\ \xi_K(k) \end{bmatrix}, \quad kh \le t < (k+1)h.$$

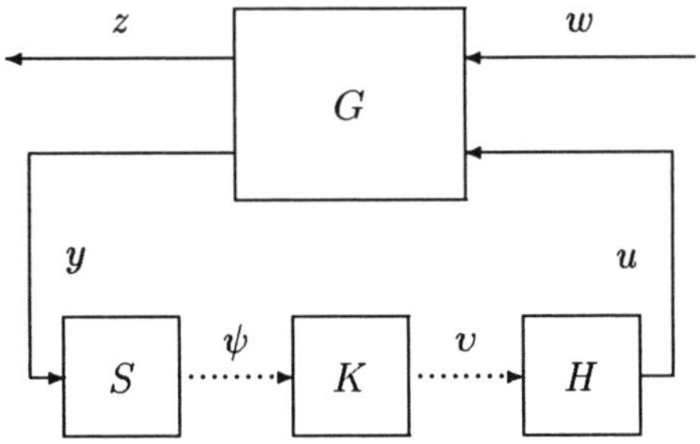

Figure 11.1: Standard SD system.

The autonomous system in Figure 11.1 is *internally stable* if for every initial time t_0, $0 \leq t_0 < h$, and initial state $x_{sd}(t_0)$ we have $x_{sd}(t) \to 0$ as $t \to \infty$. This definition is natural; it means that driven by initial conditions only, the states of the continuous G and discrete K both approach zero asymptotically.

Now we shall relate this notion of stability to the familiar one in discrete time. Look at Figure 11.1 with $w = 0$, that is, Figure 11.2. The realization

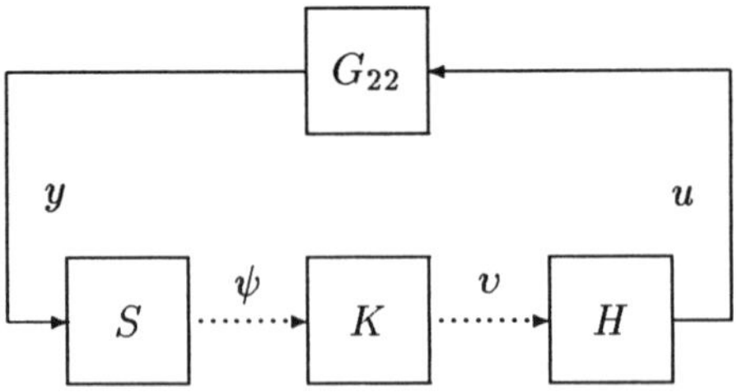

Figure 11.2: SD controller with G_{22}.

of G_{22} is

$$\hat{g}_{22}(s) = \left[\begin{array}{c|c} A & B_2 \\ \hline C_2 & 0 \end{array}\right].$$

Note that its state is still $x_G(t)$.

Now bring S and H around the loop to get Figure 11.3. Here $G_{22d} := SG_{22}H$, having the realization

$$\hat{g}_{22d}(\lambda) = \left[\begin{array}{c|c} A_d & B_{2d} \\ \hline C_2 & 0 \end{array}\right],$$

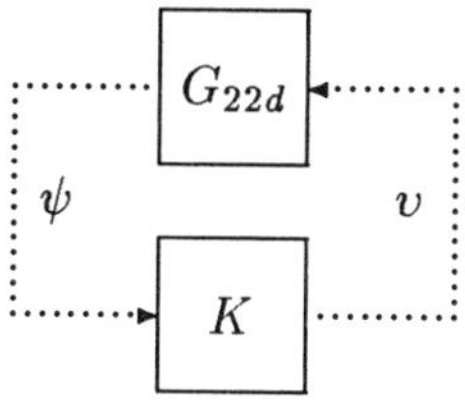

Figure 11.3: Discretization of preceding system.

where (A_d, B_{2d}) is obtained from (A, B_2) via *c2d*. The state of G_{22d} is $\xi_G(k) := x_G(kh)$. The state in Figure 11.3, namely,

$$\xi(k) := \begin{bmatrix} \xi_G(k) \\ \xi_K(k) \end{bmatrix},$$

evolves according to the equation $\dot{\xi} = \underline{A}\xi$, where

$$\underline{A} := \begin{bmatrix} A_d + B_{2d}D_K C_2 & B_{2d}C_K \\ B_K C_2 & A_K \end{bmatrix}.$$

So stability in Figure 11.3 is exactly equivalent to the condition that $\underline{A}$ is stable, that is, $\rho(\underline{A}) < 1$.

Let us observe that for $\underline{A}$ to be stable, it must be true that

$$(A_d, B_{2d}) \text{ is stabilizable and } (C_2, A_d) \text{ is detectable.} \tag{11.1}$$

Proof If $\underline{A}$ is stable, then $\underline{A} - \lambda$ is invertible for every $\lambda \notin \mathbb{D}$. Then we get in turn

$$\begin{bmatrix} A_d + B_{2d}D_K C_2 - \lambda & B_{2d}C_K \\ B_K C_2 & A_K - \lambda \end{bmatrix} \text{ is invertible for all } \lambda \notin \mathbb{D}$$

$$\Longrightarrow \begin{bmatrix} A_d + B_{2d}D_K C_2 - \lambda & B_{2d}C_K \end{bmatrix} \text{ has full row-rank for all } \lambda \notin \mathbb{D}$$

$$\Longrightarrow \begin{bmatrix} A_d + B_{2d}D_K C_2 - \lambda & B_{2d} \end{bmatrix} \text{ has full row-rank for all } \lambda \notin \mathbb{D}$$

$$\Longrightarrow \begin{bmatrix} A_d - \lambda & B_{2d} \end{bmatrix} \text{ has full row-rank for all } \lambda \notin \mathbb{D}.$$

This gives stabilizability of (A_d, B_{2d}). Similarly for detectability. ■

We saw in Chapter 3 that non-pathological sampling is a sufficient condition for (11.1).

It is readily seen from the definition that for the SD system in Figure 11.1 to be internally stable, it is necessary that the matrix $\underline{A}$ be stable. Perhaps it is not surprising that this condition is also sufficient.

Theorem 11.1.1 *The SD system in Figure 11.1 is internally stable iff the matrix $\underline{A}$ is stable.*

Proof Suppose that $\underline{A}$ is stable. For any initial value $x_{sd}(t_0)$, it follows that in Figure 11.3 $\xi_G(k), \xi_K(k) \to 0$ as $k \to \infty$. Thus $v(k) \to 0$ as $k \to \infty$ in Figure 11.3 and hence $u(t) \to 0$ as $t \to \infty$ in Figure 11.2. Now since for $kh \leq t < (k+1)h$,

$$x_G(t) = \mathrm{e}^{(t-kh)A}\xi_G(k) + \int_{kh}^{t} \mathrm{e}^{(t-\tau)A} B_2 u(\tau)\, d\tau,$$

it follows that $x_G(t) \to 0$ as $t \to \infty$; hence internal stability. ■

In summary, internal stability of the SD system in Figure 11.1 is equivalent to stability of the A-matrix of the discretized system. This provides a convenient way of determining the range of sampling periods for which stability is maintained: Plot $\rho(\underline{A})$ versus h and see where $\rho(\underline{A}) < 1$. For numerical reasons, it is actually better to plot

$$d(\underline{A}) := \frac{1}{h}[\rho(\underline{A}) - 1]$$

versus h and see where $d(\underline{A}) < 0$. The reason can be explained as follows.

Let A be a square, real matrix, representing the dynamics of a continuous-time system, and suppose all the eigenvalues of A are in the open left half-plane. The corresponding matrix obtained via *c2d* is $A_d = \mathrm{e}^{hA}$. The eigenvalues of A_d are inside the open unit disk $\mathbb{D}$ for every $h > 0$, but as $h \to 0$, $A_d \to I$, so the eigenvalues of A_d all converge to the point 1, which lies on the boundary of $\mathbb{D}$. One might infer from this that the discretized plant is converging towards an unstable one, which of course is not true. On the other hand, as $h \to 0$, $h^{-1}(A_d - I) \to A$ and $d(A_d)$ converges to the real part of the rightmost eigenvalue of A, a number that is strictly negative. The following example illustrates this point.

Example 11.1.1 Consider the simple SISO feedback system in Figure 11.4, where

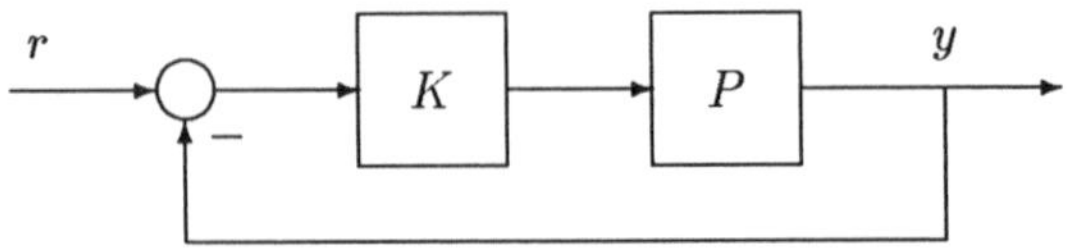

Figure 11.4: Unity feedback loop.

$$\hat{p}(s) = \frac{1}{s^2 - s + 1}.$$

The controller

$$\hat{k}(s) = \frac{10.8215s - 7.3699}{s^2 + 4.5430s + 9.8194}$$

stabilizes the feedback system, putting the closed loop poles at $-0.8409 \pm 0.8409j$, $-0.9306 \pm 0.9306j$. Using MATLAB, we get state models for P and K:

$$\left[\begin{array}{c|c} A_P & B_P \\ \hline C_P & 0 \end{array}\right], \quad \left[\begin{array}{c|c} A_K & B_K \\ \hline C_K & 0 \end{array}\right].$$

Then the closed-loop A-matrix of the continuous-time system is

$$\underline{A}_c := \begin{bmatrix} A_P & B_P C_K \\ -B_K C_P & A_K \end{bmatrix}.$$

Suppose this analog controller is implemented digitally as in Figure 11.5, where K_d is obtained from K via *c2d*. Thus

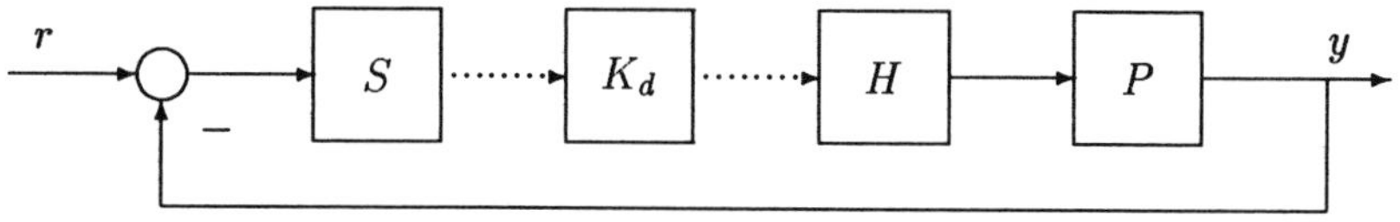

Figure 11.5: Digital implementation of the controller.

$$\hat{k}_d(\lambda) = \left[\begin{array}{c|c} A_{K_d} & B_{K_d} \\ \hline C_K & 0 \end{array}\right].$$

The discretized plant is then

$$\hat{p}_d(\lambda) = \left[\begin{array}{c|c} A_{P_d} & B_{P_d} \\ \hline C_P & 0 \end{array}\right],$$

and the closed-loop A-matrix of the discretized system is

$$\underline{A}_d := \begin{bmatrix} A_{P_d} & B_{P_d} C_{K_d} \\ -B_{K_d} C_{P_d} & A_{K_d} \end{bmatrix}.$$

It can be checked that

$$\underline{A}_d \to I \text{ as } h \to 0,$$

whereas

$$h^{-1}(\underline{A}_d - I) \to \underline{A}_c \text{ as } h \to 0,$$

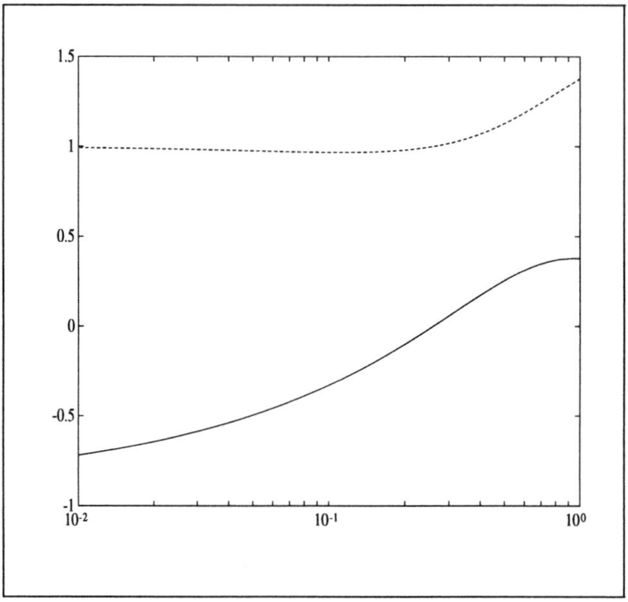

Figure 11.6: $d(\underline{A}_d)$ (solid) and $\rho(\underline{A}_d)$ (dash) versus h for the controller discretized via *c2d*.

Figure 11.6 shows plots of $d(\underline{A}_d)$ and $\rho(\underline{A}_d)$ for h from 0.01 to 1. It can be seen that $d(\underline{A}_d) < 0$ for $h \in [0, 0.26]$. So the SD system in Figure 11.5 is stable for h up to 0.26. Notice in Figure 11.6 that it is much easier to determine where $d(\underline{A}_d) < 0$ than where $\rho(\underline{A}_d) < 1$.

For interest, let us also discretize the analog controller via bilinear transformation; the formulas for this are given in Section 3.4 in terms of state-space data. Figure 11.7 shows plots of $d(\underline{A}_d)$ versus h when the controller is discretized via both *c2d* and bilinear transformation. For bilinear transformation the SD system in Figure 11.5 is stable for h up to 0.47. Thus we may conclude in this example that bilinear transformation gives a larger range of h for stability than does *c2d*.

11.2 Input-Output Stability

Continuing with our discussion and notation in the preceding section, now we want to see what sort of input-output stability in Figure 11.1 follows from its internal stability. Nothing very surprising happens. The first result concerns bounded-input, bounded-output stability in terms of the $\mathcal{L}_\infty$-norm.

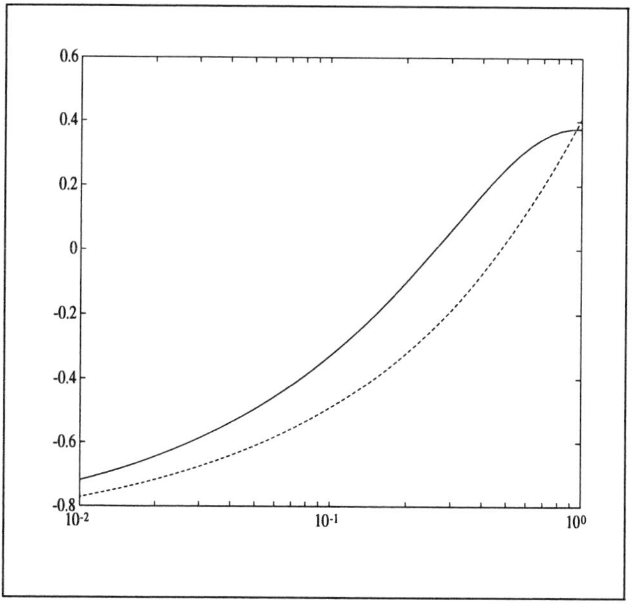

Figure 11.7: $d(\underline{A}_d)$ versus h for discretization via *c2d* (solid) and bilinear transformation (dash).

Theorem 11.2.1 *If the SD system in Figure 11.1 is internally stable, then the map* $w \mapsto z$ *is bounded* $\mathcal{L}_\infty(\mathbb{R}_+) \to \mathcal{L}_\infty(\mathbb{R}_+)$.

The second result is in terms of the $\mathcal{L}_2$-norm. The only modification required is the assumption that $D_{21} = 0$ so that w is lowpass filtered before it enters the sampler.

Theorem 11.2.2 *Assume* $D_{21} = 0$. *If the SD system in Figure 11.1 is internally stable, then the map* $w \mapsto z$ *is bounded* $\mathcal{L}_2(\mathbb{R}_+) \to \mathcal{L}_2(\mathbb{R}_+)$.

The proofs of these theorems are similar; we shall prove the second using continuous lifting.

Let T_{zw} denote the map $w \mapsto z$ in Figure 11.1. As in Section 10.3, we lift this continuous-time, h-periodic operator to get $LT_{zw}L^{-1}$. It follows that T_{zw} is bounded $\mathcal{L}_2(\mathbb{R}_+) \to \mathcal{L}_2(\mathbb{R}_+)$ iff the lifted operator $LT_{zw}L^{-1}$ is bounded $\ell_2(\mathbb{Z}_+, \mathcal{K}) \to \ell_2(\mathbb{Z}_+, \mathcal{K})$. From Section 10.3, $LT_{zw}L^{-1}$ is the map $\underline{w} \mapsto \underline{z}$ in the lifted system of Figure 11.8, where the lifted plant $\underline{G}$ is given by

$$\begin{aligned} \underline{G} &= \begin{bmatrix} L & 0 \\ 0 & S \end{bmatrix} G \begin{bmatrix} L^{-1} & 0 \\ 0 & H \end{bmatrix} \\ &: \quad \ell_{2e}(\mathbb{Z}, \mathcal{K}) \oplus \ell_{2e}(\mathbb{Z}, \mathcal{E}) \to \ell_{2e}(\mathbb{Z}, \mathcal{K}) \oplus \ell_{2e}(\mathbb{Z}, \mathcal{E}) \end{aligned}$$

and has the state model

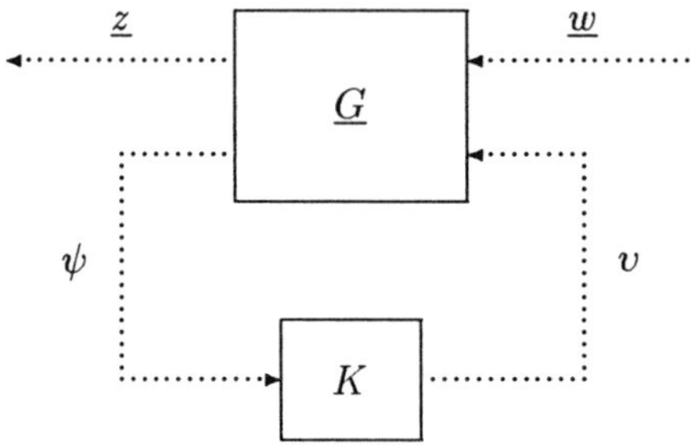

Figure 11.8: The lifted system.

$$\left[\begin{array}{c|cc} A_d & \underline{B}_1 & B_{2d} \\ \hline \underline{C}_1 & \underline{D}_{11} & \underline{D}_{12} \\ C_2 & 0 & 0 \end{array}\right],$$

where the operator-valued entries are

$$\begin{aligned} \underline{B}_1 : &\quad \mathcal{K} \to \mathcal{E}, \quad \underline{B}_1 w = \int_0^h e^{(h-\tau)A} B_1 w(\tau)\, d\tau \\ \underline{C}_1 : &\quad \mathcal{E} \to \mathcal{K}, \quad (\underline{C}_1 x)(t) = C_1 e^{tA} x \\ \underline{D}_{11} : &\quad \mathcal{K} \to \mathcal{K}, \quad (\underline{D}_{11} w)(t) = C_1 \int_0^t e^{(t-\tau)A} B_1 w(\tau)\, d\tau \\ \underline{D}_{12} : &\quad \mathcal{E} \to \mathcal{K}, \quad (\underline{D}_{12} v)(t) = D_{12} v + C_1 \int_0^t e^{(t-\tau)A}\, d\tau\, B_2 v. \end{aligned}$$

These four linear transformations are bounded, because $\mathcal{K} = \mathcal{L}_2[0,h)$ and $h < \infty$. A state model for the closed-loop map $\underline{w} \mapsto \underline{z}$ is therefore

$$\left[\begin{array}{c|c} \underline{A} & \underline{B}_{cl} \\ \hline \underline{C}_{cl} & \underline{D}_{11} \end{array}\right] = \left[\begin{array}{cc|c} A_d + B_{2d} D_K C_2 & B_{2d} C_K & \underline{B}_1 \\ B_K C_2 & A_K & 0 \\ \hline \underline{C}_1 + \underline{D}_{12} D_K C_2 & \underline{D}_{12} C_K & \underline{D}_{11} \end{array}\right],$$

the corresponding state vector being

$$\xi(k) = \begin{bmatrix} \xi_G(k) \\ \xi_K(k) \end{bmatrix}.$$

Note that the A-matrix of this state model is again $\underline{A}$, the A-matrix of the discretized closed-loop system; this important observation is used below to infer input-output stability.

Proof of Theorem 11.2.2 To show that the map $\underline{w} \mapsto \underline{z}$ in Figure 11.8 is bounded $\ell_2(\mathbb{Z}_+, \mathcal{K}) \to \ell_2(\mathbb{Z}_+, \mathcal{K})$, we first look at the lifted state model:

$$\xi(k+1) = \underline{A}\xi(k) + \underline{B}_{cl} \underline{w}_k \tag{11.2}$$

$$\underline{z}_k = \underline{C}_{cl} \xi(k) + \underline{D}_{11} \underline{w}_k. \tag{11.3}$$

Note that $\underline{B}_{cl}$ is a bounded operator mapping $\mathcal{K}$ to $\mathcal{E}$ and so if we define $\rho(k) = \underline{B}_{cl}\underline{w}_k$, we get that the map $\underline{w} \mapsto \rho$ is bounded $\ell_2(\mathbb{Z}_+, \mathcal{K}) \to \ell_2(\mathbb{Z}_+)$ and then (11.2) becomes an equation involving finite-dimensional vectors:

$$\dot{\xi} = \underline{A}\xi + \rho.$$

Since internal stability implies that $\underline{A}$ is stable, we see from this discrete system that the map $\rho \mapsto \xi$ is bounded on $\ell_2(\mathbb{Z}_+)$. Hence the map $\underline{w} \mapsto \xi$ (from input to state) is bounded $\ell_2(\mathbb{Z}_+, \mathcal{K}) \to \ell_2(\mathbb{Z}_+)$. This and boundedness of $\underline{C}_{cl}$ and $\underline{D}_{11}$ in (11.3) imply immediately that $\underline{w} \mapsto \underline{z}$ is bounded. ∎

In summary, to obtain BIBO stability in the sampled-data setup of Figure 11.1 it suffices that the A-matrix of the discretized system be stable. Thus BIBO stability can also be checked easily.

Example 11.2.1 Consider the analog control system in Figure 11.9, where

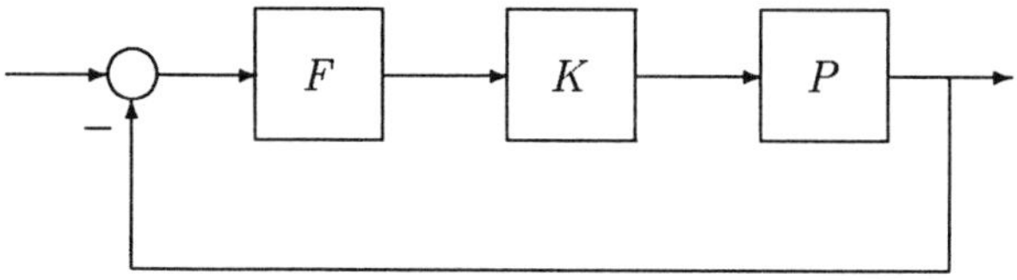

Figure 11.9: Analog control system.

$$\hat{f}(s) = \frac{1}{(0.5/\pi)s + 1}, \quad \hat{k}(s) = 1, \quad \hat{p}(s) = \frac{1}{s}e^{-\tau s}.$$

The filter F represents an antialiasing filter for later digital implementation of the controller K at the sampling period $h = 0.5$ ($\pi/0.5$ is then the Nyquist frequency). The plant, P, has a time delay of τ s. The feedback system is internally stable for $\tau = 0$.

Let us compute the *time-delay margin*, $\tau_{\max}$, the minimum value of τ for which the feedback system is unstable. The feedback system becomes unstable when the Nyquist plot of $\hat{p}\hat{k}\hat{f}$ passes through the critical point, -1, that is, when

$$(\exists\omega > 0)\ (\hat{p}\hat{k}\hat{f})(j\omega) = -1.$$

This latter equation can be written as

$$e^{-\tau j\omega} = 0.1592\omega^2 - j\omega.$$

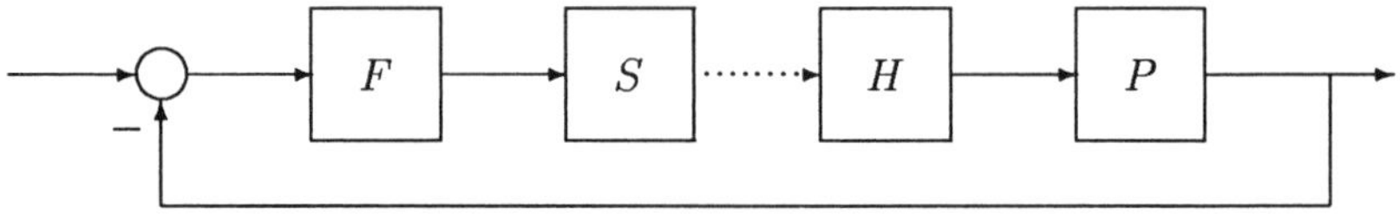

Figure 11.10: Digital implementation of $K = I$.

The unique values of $\omega > 0$ and $\tau > 0$ satisfying this equation are $\omega = 0.9879$ and $\tau = 1.4322$. Thus $\tau_{\max} = 1.4322$.

Consider now a digital implementation of $K = I$ via *c2d*, as shown in Figure 11.10. Let us find $\tau_{\max}$ for this system. The preceding two theorems were developed for a finite-dimensional generalized plant, whereas P in this example has a time delay and is therefore infinite-dimensional. Nevertheless, the theorems can be generalized for such plants. Time-delay systems have the interesting and useful property that their discretizations are finite-dimensional.

Define $G = FP$, that is,

$$\hat{g}(s) = \frac{1}{s(0.1592s+1)} e^{-\tau s}.$$

The system in Figure 11.10 is modelled at the sampling instants by Figure 11.11, where $G_d := SGH$. The discretization of G depends on which

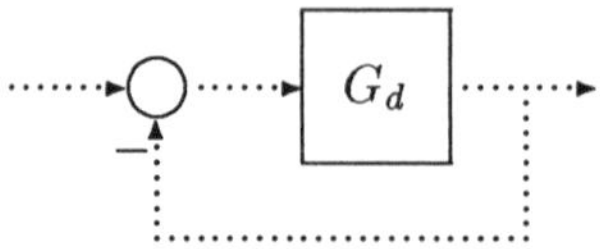

Figure 11.11: Discretized system.

sampling period τ lies in. Suppose $(l-1)h < \tau \leq lh$. Since $\hat{g}(s)$ has the form

$$\hat{g}(s) = \left[\begin{array}{c|c} A & B \\ \hline C & 0 \end{array}\right] e^{-\tau s},$$

it is not hard to derive (Exercise 3.2) that

$$\hat{g}_d(\lambda) = \lambda^l \left[\begin{array}{c|c} A_d & B_d \\ \hline C_d & D_d \end{array}\right],$$

where A_d, B_d are as usual and

$$C_d = Ce^{(lh-\tau)A}, \quad D_d = C\int_0^{lh-\tau} e^{tA}dtB.$$

For $l = 1$ (i.e., $0 < \tau \le h$) the closed-loop A-matrix is

$$\underline{A} = \begin{bmatrix} A_d & B_d \\ -C_d & -D_d \end{bmatrix};$$

for $l = 2$ (i.e., $h < \tau \le 2h$)

$$\underline{A} = \begin{bmatrix} A_d & B_d & 0 \\ 0 & 0 & 1 \\ -C_d & -D_d & 0 \end{bmatrix};$$

and so on.

To be specific, let us take $h = 0.5$. Figure 11.12 shows a plot of $\rho(\underline{A})$ versus τ over the range $0 \le \tau < 1.5$. The spectral radius of $\underline{A}$ reaches 1

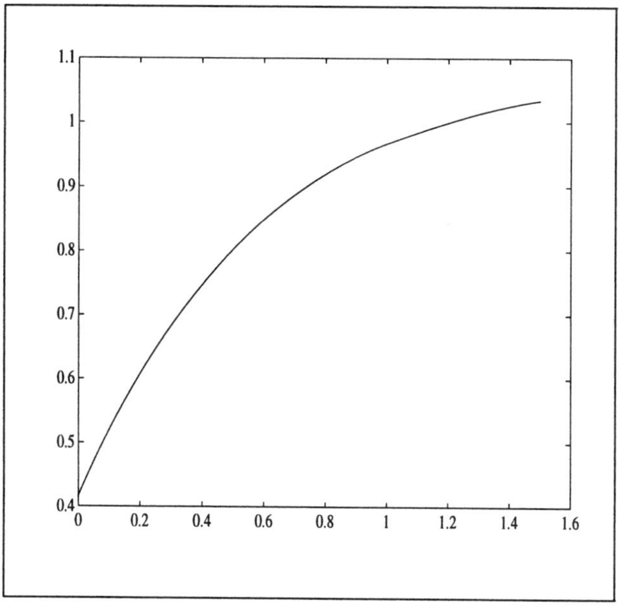

Figure 11.12: $\rho(\underline{A})$ versus τ over the range $0 \le \tau < 1.5$.

when $\tau = 1.2029$. This therefore equals $\tau_{\max}$.

In conclusion, with a sampling period of 0.5 the time-delay margin drops from 1.4322 to 1.2029.

11.3 Robust Stability

In this section we shall see how to compute a stability margin for the representative SD system in Figure 11.13 using the induced norms concept from Chapter 10. The plant P and the filter F are both assumed to be FDLTI,

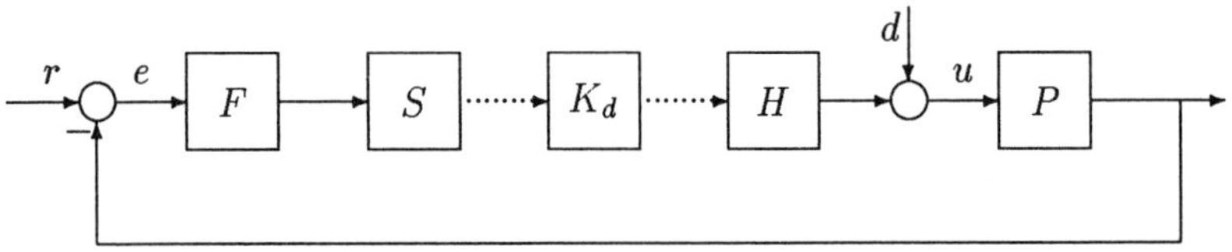

Figure 11.13: Unity-feedback digital control loop.

with P causal and F strictly causal—P is the nominal plant model. Assume that the A-matrix of the discretized system is stable. It follows that the mapping

$$\begin{bmatrix} r \\ d \end{bmatrix} \mapsto \begin{bmatrix} e \\ u \end{bmatrix}$$

is a bounded operator from $\mathcal{L}_2(\mathbb{R})$ to $\mathcal{L}_2(\mathbb{R})$.

Now introduce the following uncertainty model for the plant, defined by an additive perturbation of the nominal plant:

$$\mathcal{P}_\gamma = \{P + \Delta W : \|\Delta\| < \gamma\}.$$

The fixed weighting system W is assumed to be FDLTI, causal, and stable; the variable perturbation Δ can be any bounded operator from $\mathcal{L}_2(\mathbb{R})$ to $\mathcal{L}_2(\mathbb{R})$, time-invariant or not. Typically, W models an uncertainty envelope in the magnitude of the plant, and Δ models uncertainty in the phase. Thus, $\mathcal{P}_\gamma$ is a weighted ball centered at P, with the size of the ball being characterized by the scalar γ.

The question we ask is, how large can γ be and yet for all plants in $\mathcal{P}_\gamma$ the mapping

$$\begin{bmatrix} r \\ d \end{bmatrix} \mapsto \begin{bmatrix} e \\ u \end{bmatrix}$$

is a bounded operator from $\mathcal{L}_2(\mathbb{R})$ to $\mathcal{L}_2(\mathbb{R})$? The maximum such γ is the *stability margin*. The answer involves a SD norm computation.

Replace P in Figure 11.13 by a representative $P + \Delta W$ from $\mathcal{P}_\gamma$ to get the block diagram shown in Figure 11.14. It is convenient to reconfigure this as shown in Figure 11.15. This figure highlights the input and output of interest, namely,

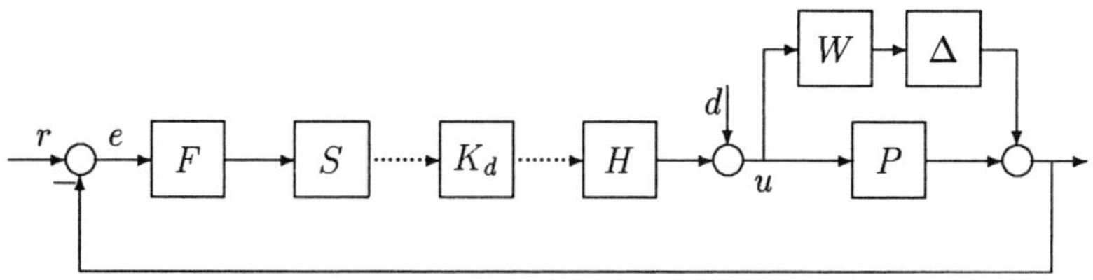

Figure 11.14: Additive plant perturbation.

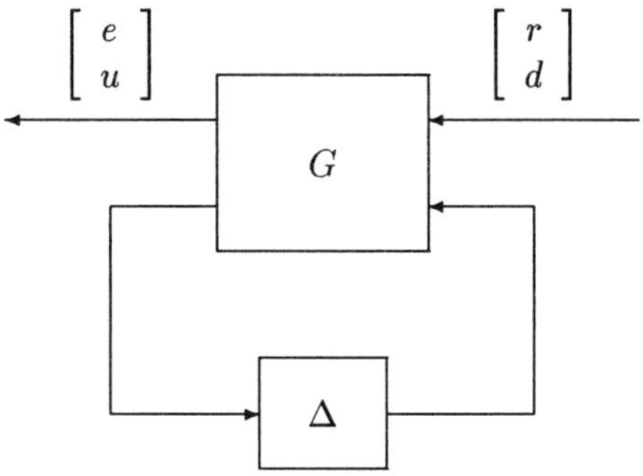

Figure 11.15: Perturbation of nominal system.

$$\begin{bmatrix} r \\ d \end{bmatrix} \text{ and } \begin{bmatrix} e \\ u \end{bmatrix},$$

and in addition isolates the variable perturbation Δ. The fixed system G is bounded as an operator from $\mathcal{L}_2(\mathbb{R})$ to $\mathcal{L}_2(\mathbb{R})$ (by stability for the nominal plant, P). It can be partitioned as a 2×2 matrix:

$$G = \begin{bmatrix} G_{11} & G_{12} \\ G_{21} & G_{22} \end{bmatrix}.$$

In particular, G_{22} equals the system in Figure 11.14 from the output of Δ around to the input of Δ, which can be computed to be

$$G_{22} = -WHK_dSF(I + PHK_dSF)^{-1}. \tag{11.4}$$

The mapping T in Figure 11.15 from

$$\begin{bmatrix} r \\ d \end{bmatrix} \text{ to } \begin{bmatrix} e \\ u \end{bmatrix},$$

equals

$$G_{11} + G_{12}(I - \Delta G_{22})^{-1}\Delta G_{21}.$$

As each G_{ij} and Δ are bounded, T will be a bounded operator provided the small-gain condition $||\Delta G_{22}|| < 1$ holds. A sufficient condition for this is $||\Delta|| \cdot ||G_{22}|| < 1$. Actually, this is necessary if Δ is time-varying. We conclude that if the uncertainty radius γ satisfies $\gamma < 1/||G_{22}||$, then robust stability will result. So, we can think of $1/||G_{22}||$ as an upper bound on the stability margin. It remains to see how to compute $||G_{22}||$.

From (11.4) we get

$$G_{22} = -WH(I + K_dSFPH)^{-1}K_dSF.$$

Defining

$$R_d := -(I + K_dSFPH)^{-1}K_d,$$

an LTI discrete-time system ($SFPH$ is the step-invariant transformation of FP), we have that

$$G_{22} = WHR_dSF.$$

In terms of lifted operators, we have that

$$||G_{22}|| = ||\underline{WH}\ R_d\ \underline{SF}||.$$

Bring in FDLTI discrete-time systems M and N such that

$$\underline{SF}\ \underline{SF}^* = MM^*, \quad \underline{WH}^*\ \underline{WH} = N^*N.$$

Explicit formulas for M and N in terms of F and W were derived in Sections 10.5 and 10.6. Then we compute as follows:

$$\begin{aligned}
\|G_{22}\| &= \|\underline{WH}R_d\underline{SF}\| \\
&= \|\underline{SF}^* R_d^*(\underline{WH}^*\ \underline{WH})R_d\underline{SF}\|^{1/2} \\
&= \|\underline{SF}^* R_d^*(N^*N)R_d\underline{SF}\|^{1/2} \\
&= \|NR_d\underline{SF}\| \\
&= \|NR_d(\underline{SF}\ \underline{SF}^*)R_d^*N^*\|^{1/2} \\
&= \|NR_d(MM^*)R_d^*N^*\|^{1/2} \\
&= \|NR_dM\|.
\end{aligned}$$

The three systems N, R_d, M are all FDLTI. We conclude that $\|G_{22}\|$ can be computed via a simple $\mathcal{H}_\infty(\mathbb{D})$-norm computation:

$$\|G_{22}\| = \|\hat{n}\hat{r}_d\hat{m}\|_\infty.$$

Let us summarize the computations.

Step 1 Start with F, P, W, and a controller K_d that stabilizes P.

Step 2 Discretize FP and compute

$$R_d := -(I + K_dSFPH)^{-1}K_d.$$

Step 3 As in Section 10.5, compute M such that

$$\underline{SF}\ \underline{SF}^* = MM^*.$$

Step 4 As in Section 10.6, compute N such that

$$\underline{WH}^*\ \underline{WH} = N^*N.$$

Step 5 An upper bound on the stability margin is $1/\|\hat{n}\hat{r}_d\hat{m}\|_\infty$.

Example 11.3.1 Let us reconsider Example 11.2.1, which had

$$\hat{f}(s) = \frac{1}{(0.5/\pi)s + 1}, \quad \hat{k}_d(\lambda) = 1, \quad \hat{p}(s) = \frac{1}{s}e^{-\tau s}.$$

The sampling period was $h = 0.5$. We computed the time-delay margin, $\tau_{\max}$ (the minimum value of τ such that the SD feedback system is unstable), to be exactly 1.2029 s. Let us apply the techniques of this section to this problem.

The plant can be modelled as a perturbation of the nominal plant $1/s$:

$$\frac{1}{s}e^{-\tau s} = \frac{1}{s} + \frac{1}{s}\left(e^{-\tau s} - 1\right).$$

Following the notation of this section, let P now denote the nominal plant, with transfer function $1/s$. The perturbation, denoted P_Δ with

$$\hat{p}_\Delta(s) = \frac{1}{s}\left(e^{-\tau s} - 1\right),$$

is covered by the weighted ball

$$\{\Delta W : \|\Delta\| < 1\}$$

provided the weighting function W is chosen to cover the frequency response of P_Δ, that is,

$$|\hat{p}_\Delta(j\omega)| < |\hat{w}(j\omega)|, \quad \forall\omega.$$

A suitable function that works for all τ (chosen by modifying a Padé approximation) is

$$\hat{w}(s) = 2\left[\left(\frac{\tau^3}{120}\right)^{1/2} s + \left(\frac{\tau}{2}\right)^{1/2}\right]^2 \Bigg/ \left[\frac{\tau^3}{120}s^3 + \frac{\tau^2}{10}s^2 + \frac{\tau}{2}s + 1\right].$$

Figure 11.16 shows graphs of the two frequency responses for $\tau = 0.4$, for example.

The feedback system is stable for all perturbations satisfying $\|\Delta\| < 1$ provided $\|G_{22}\| \le 1$, or equivalently, $\|\hat{n}\hat{r}_d\hat{m}\|_\infty \le 1$. Notice that $\hat{w}(s)$ depends on τ and $\hat{n}(s)$ depends on $\hat{w}(s)$. In this way, $\|\hat{n}\hat{r}_d\hat{m}\|_\infty$ depends on τ. The value of τ where $\|\hat{n}\hat{r}_d\hat{m}\|_\infty = 1$ is a lower bound on $\tau_{\max}$ (since the condition $\|\hat{n}\hat{r}_d\hat{m}\|_\infty \le 1$ is only sufficient). This lower bound can be computed by bisection search to be 0.6095. Figure 11.17 shows Bode plots of $\hat{n}\hat{r}_d\hat{m}$ for $\tau = 0.4$ and $\tau = 0.6095$.

In summary, the exact time-delay margin for the SD feedback system equals 1.2029, while robust stability theory gives the lower bound 0.6095. This conservativeness is because covering the perturbation by a weighted ball is a coarse approximation.

11.4 Step Tracking

In this section we look at the multi-input, multi-output setup in Figure 11.18 and we would like, in addition to internal stability, that the system be *step-tracking*, that is, $e(t) \to 0$ as $t \to \infty$ for every step input $r(t)$. This section will focus on the following question:

> If we design K_d in discrete time to achieve step-tracking for the discretized system, will step-tracking be achieved in continuous time, or will there be residual intersample ripple?

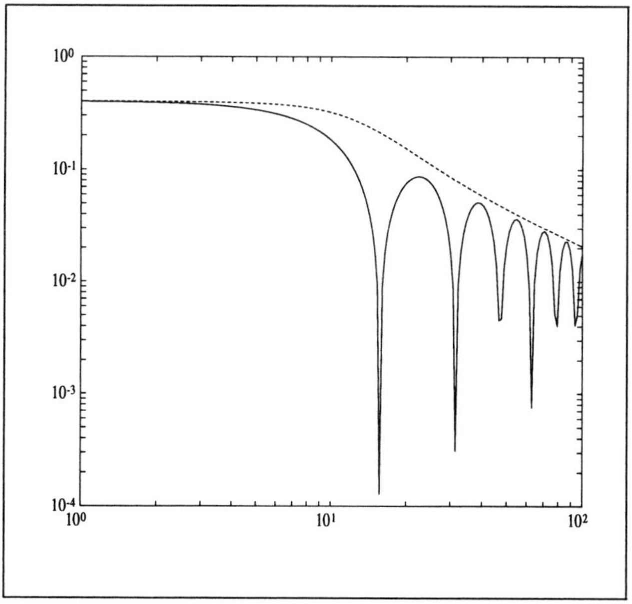

Figure 11.16: Magnitude Bode Plots of $\hat{p}_\Delta$ (solid) and $\hat{w}$ (dash) for $\tau = 0.4$.

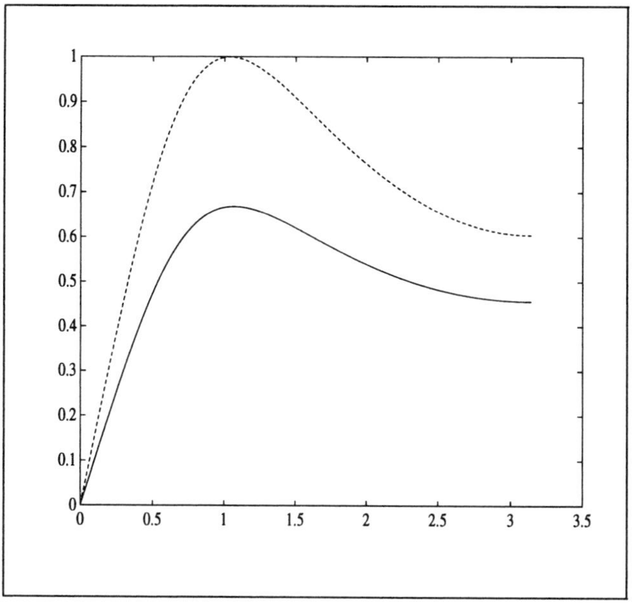

Figure 11.17: Magnitude Bode Plots of $\hat{n}\hat{r}_d\hat{m}$ for $\tau = 0.4$ (solid) and $\tau = 0.6095$ (dash).

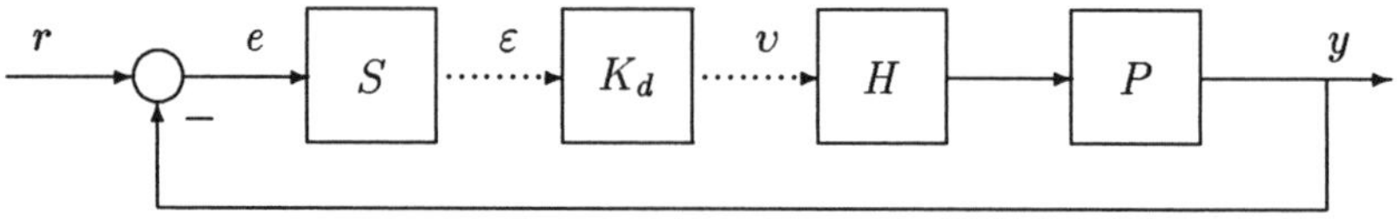

Figure 11.18: Unity-feedback digital control loop.

Before we answer the question, let us introduce a handy function. For a positive δ, define the function

$$f_\delta(s) = \int_0^\delta e^{\tau s}\, d\tau = \frac{e^{\delta s} - 1}{s}.$$

This is useful for the following reasons. Let (A, B) be continuous-time state-space data. Then

$$f_h(A)A = e^{hA} - I,$$

so if $(A_d, B_d) = c2d(A, B, h)$, then

$$B_d = f_h(A)B, \tag{11.5}$$

$$A_d - I = f_h(A)A. \tag{11.6}$$

Lemma 11.4.1 *The matrix $f_h(A)$ is invertible if 1 is not an uncontrollable eigenvalue of (A_d, B_d) (a fortiori, if (A_d, B_d) is stabilizable).*

Proof We have from (11.5) and (11.6)

$$\begin{aligned} \begin{bmatrix} A_d - I & B_d \end{bmatrix} &= \begin{bmatrix} f_h(A)A & f_h(A)B \end{bmatrix} \\ &= f_h(A) \begin{bmatrix} A & B \end{bmatrix}, \end{aligned}$$

so

$$\operatorname{rank} \begin{bmatrix} A_d - I & B_d \end{bmatrix} \le \operatorname{rank} f_h(A).$$

Thus, if 1 is not an uncontrollable eigenvalue of (A_d, B_d), then

$$\operatorname{rank} \begin{bmatrix} A_d - I & B_d \end{bmatrix} = n,$$

so $\operatorname{rank} f_h(A) = n$. ■

Now we return to our tracking problem. We assume in Figure 11.18 that P is strictly causal and K_d is causal. Start with stabilizable and detectable state models for P and K_d :

$$\hat{p}(s) = \left[\begin{array}{c|c} A & B \\ \hline C & 0 \end{array}\right], \quad \hat{k}_d(\lambda) = \left[\begin{array}{c|c} A_K & B_K \\ \hline C_K & D_K \end{array}\right].$$

Here the corresponding state vectors for P and K_d are x_P and ξ_K respectively. The discretized system is shown in Figure 11.19, where P_d, the step-invariant discretization of P, has the associated state model

$$\hat{p}_d(\lambda) = \left[\begin{array}{c|c} A_d & B_d \\ \hline C & 0 \end{array}\right]$$

with the state vector ξ_P defined by $\xi_P(k) = x_P(kh)$.

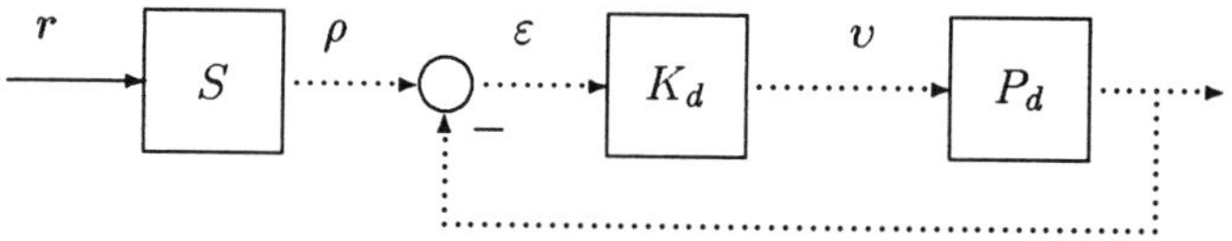

Figure 11.19: Discretized unity-feedback system.

The following theorem answers the question we asked earlier in the section.

Theorem 11.4.1 *Assume that the SD system in Figure 11.18 is internally stable and the input r is a step signal, say, $r(t) = r_0 1(t)$, where r_0 is an arbitrary but fixed vector. Then the continuous-time tracking error $e(t)$ is convergent as $t \to \infty$ and $e(\infty) = \varepsilon(\infty)$.*

Under the assumption of the theorem, the discretized system in Figure 11.19 is also internally stable (Theorem 11.1.1). Thus the discrete-time steady-state tracking error $\varepsilon(\infty)$ is finite. Theorem 11.4.1 says that no steady-state intersample ripple exists if the signal to be tracked is a step.

Proof of Theorem 11.4.1 For a step input, internal stability implies that all discrete signals are convergent, that is, $\xi_P(\infty), \xi_K(\infty), v(\infty)$, etc., all exist. As $k \to \infty$, the state equation for P_d becomes

$$\xi_P(\infty) = A_d \xi_P(\infty) + B_d v(\infty),$$

or

$$(A_d - I)\xi_P(\infty) + B_d v(\infty) = 0.$$

From (11.5) and (11.6), this is the same as

$$f_h(A) A \xi_P(\infty) + f_h(A) B v(\infty) = 0. \tag{11.7}$$

Now as in Section 11.1, internal stability of the discretized system implies that (A_d, B_d) is stabilizable and hence that $f_h(A)$ is invertible by Lemma 11.4.1. Thus from (11.7) we have

$$A\xi_P(\infty) + Bv(\infty) = 0. \tag{11.8}$$

In view of Figure 11.18, it suffices to show that $y(t)$ is convergent as $t \to \infty$, since $r(\infty)$ is finite. To do this, we look at the lifted model for $PH : v \mapsto y$, namely, $LPH : v \mapsto \underline{y}$, which has the following state model

$$\begin{aligned} \dot{\xi}_P(k) &= A_d\xi_P(k) + B_d v(k) \\ \underline{y}_k &= \underline{C}\xi_P(k) + \underline{D}v(k), \end{aligned} \tag{11.9}$$

where the operators $\underline{C}$ and $\underline{D}$ are as follows:

$$\begin{aligned} &\underline{C}: \ \mathcal{E} \to \mathcal{K}, \quad (\underline{C}\xi)(t) = C_1 e^{tA}\xi \\ &\underline{D}: \ \mathcal{E} \to \mathcal{K}, \quad (\underline{D}v)(t) = C\int_0^t e^{\tau A}\, d\tau\, Bv. \end{aligned}$$

Letting $k \to \infty$ in (11.9) gives

$$\underline{y}_\infty = \underline{C}\xi_P(\infty) + \underline{D}v(\infty).$$

After lifting, $\underline{y}_\infty$ is a function on $[0, h)$. Substituting the definitions of $\underline{C}$ and $\underline{D}$ into (11.9) gives

$$\begin{aligned} \underline{y}_\infty(t) &= Ce^{tA}\xi_P(\infty) + C\int_0^t e^{\tau A}\, d\tau\, Bv(\infty) \\ &= C[I + f_t(A)A]\xi_P(\infty) + Cf_t(A)Bv(\infty) \\ &= C\xi_P(\infty) + Cf_t(A)[A\xi_P(\infty) + Bv(\infty)] \\ &= C\xi_P(\infty) \quad \text{from (11.8)}. \end{aligned}$$

Since $C\xi_P(\infty)$ is constant, so is $\underline{y}_\infty$. Therefore, $y(t)$ converges. ■

The answer to the question at the start of this section follows readily from the theorem by forcing $\varepsilon(\infty) = 0$ for every step input $r(t)$:

Corollary 11.4.1 *Assume the SD system in Figure 11.18 is internally stable. Then the SD system is step-tracking iff the discretized system is step-tracking.*

Recall that in Examples 6.6.1 and 8.4.2 the controllers were designed for discrete-time (optimal) step-tracking and our simulations showed that SD step-tracking was also achieved. This is one way for digital design, which is guaranteed to work for step-tracking problems. Another way is to design an analog controller and then do a digital implementation, the subject of the next section.

11.5 Digital Implementation and Step Tracking

In this section we are interested in the following type of question:

> If we design an analog controller to achieve step-tracking and then discretize the controller, will step-tracking be achieved in the SD system?

We shall look at two ways to discretize an analog controller: step-invariant and bilinear transformation. First, consider digital implementation using step-invariant transformation.

We begin with the internally stabilized analog system in Figure 11.20 and

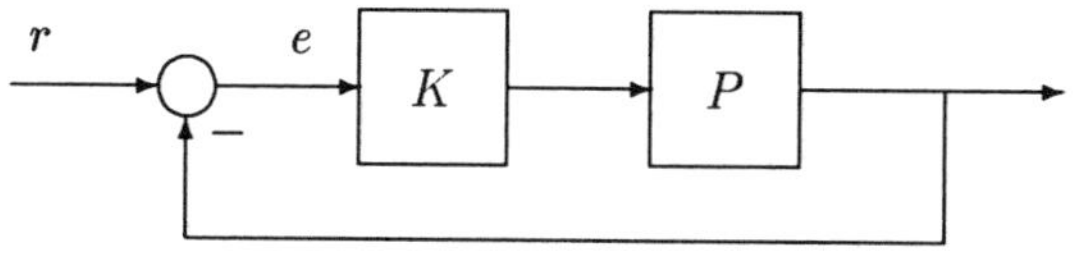

Figure 11.20: Unity-feedback analog system.

the following assumptions:

- $\hat{p}(s)$ is strictly proper
- $\hat{k}(s)$ is proper.

The analog controller is implemented by a SD controller as in Figure 11.21 with $K_d = SKH$. Bring S around to get Figure 11.22.

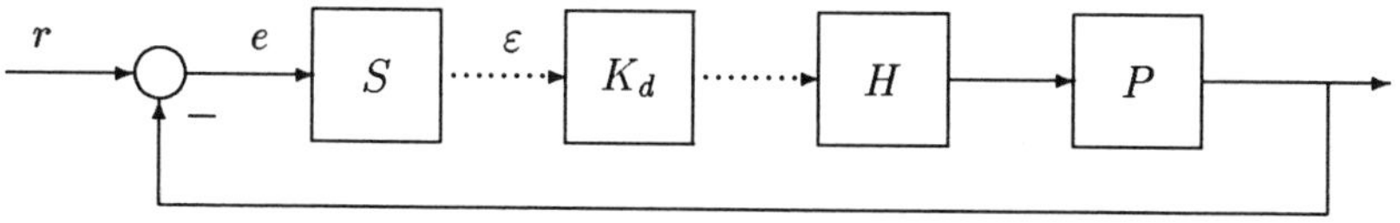

Figure 11.21: SD system: Digital implementation.

Intuitively, since P_d is the step-invariant transformation of P, the DC gains of P and P_d, if they exist, are equal (Exercise 11.7). Similarly about K_d and K. Thus one can hope that the two closed-loop DC gains, $r \mapsto e$ in Figure 11.20 and $\rho \mapsto \varepsilon$ in Figure 11.22, are equal as well. So the two systems have the same steady-state error for the same input r. Even though open-loop DC gains may not exist, the closed-loop DC gains always exist under the assumption of internal stability. In what follows we shall give a formal derivation that the two closed-loop DC gains are equal.

Bring in stabilizable and detectable realizations of P and K:

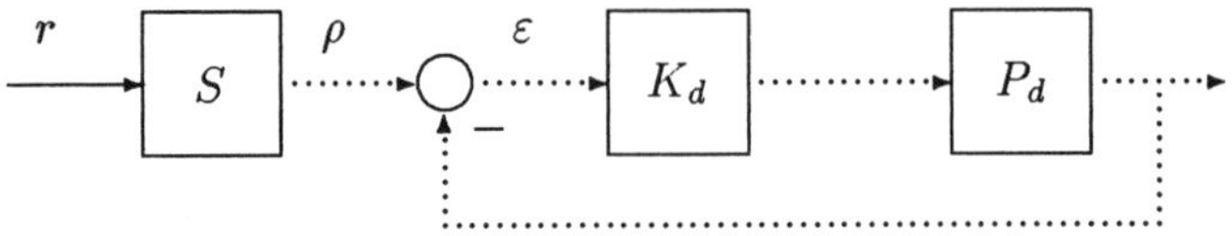

Figure 11.22: Discretized system.

$$\hat{p}(s) = \left[\begin{array}{c|c} A_P & B_P \\ \hline C_P & 0 \end{array}\right], \quad \hat{k}(s) = \left[\begin{array}{c|c} A_K & B_K \\ \hline C_K & D_K \end{array}\right].$$

They induce a realization from r to e in Figure 11.20:

$$\left[\begin{array}{c|c} A & B \\ \hline C & D \end{array}\right] := \left[\begin{array}{cc|c} A_P - B_P D_K C_P & B_P C_K & B_P D_K \\ -B_K C_P & A_K & B_K \\ \hline -C_P & 0 & I \end{array}\right]. \tag{11.10}$$

Therefore

$$\begin{aligned} \text{DC gain from } r \text{ to } e &= D + C(sI - A)^{-1} B|_{s=0} \\ &= D - CA^{-1}B. \end{aligned}$$

On the other hand,

$$\hat{p_d}(\lambda) = \left[\begin{array}{c|c} A_{P_d} & B_{P_d} \\ \hline C_P & 0 \end{array}\right], \quad \hat{k_d}(\lambda) = \left[\begin{array}{c|c} A_{K_d} & B_{K_d} \\ \hline C_P & 0 \end{array}\right].$$

The realization from ρ to ε in Figure 11.22 is

$$\left[\begin{array}{c|c} \underline{A} & \underline{B} \\ \hline \underline{C} & \underline{D} \end{array}\right] = \left[\begin{array}{cc|c} A_{P_d} - B_{P_d} D_K C_P & B_{P_d} C_K & B_{P_d} D_K \\ -B_{K_d} C_P & A_{K_d} & B_{K_d} \\ \hline -C_P & 0 & I \end{array}\right]. \tag{11.11}$$

Therefore

$$\begin{aligned} \text{DC gain from } \rho \text{ to } \varepsilon &= \underline{D} + \lambda \underline{C}(I - \lambda \underline{A})^{-1} \underline{B}|_{\lambda=1} \\ &= \underline{D} + \underline{C}(I - \underline{A})^{-1} \underline{B}. \end{aligned}$$

Note that the two DC gains exist if we assume that the two associated systems are internally stable, i.e., the matrices A and $\underline{A}$ are stable in continuous and discrete time respectively. To prove that they are in fact equal, let us see the relation between the continuous-time closed-loop data (A, B, C, D) and the discrete-time closed-loop data ($\underline{A}$, $\underline{B}$, $\underline{C}$, $\underline{D}$). Define

$$T := \begin{bmatrix} f_h(A_P) & 0 \\ 0 & f_h(A_K) \end{bmatrix}.$$

Comparing (11.10) and (11.11) and using (11.5) and (11.6), we get

$$
\begin{aligned}
\underline{A} - I &= \begin{bmatrix} A_{P_d} - B_{P_d} D_K C_P - I & B_{P_d} C_K \\ -B_{K_d} C_P & A_{K_d} - I \end{bmatrix} \\
&= \begin{bmatrix} f_h(A_P)A_P - f_h(A_P)B_P D_K C_P & f_h(A_P) B_P C_K \\ -f_h(A_K) B_K C_P & f_h(A_K) A_K \end{bmatrix} \\
&= TA \qquad (11.12) \\
\underline{B} &= TB \\
\underline{C} &= C \\
\underline{D} &= D.
\end{aligned}
$$

If $\underline{A}$ is stable (all eigenvalues inside $\mathbb{D}$), then T is invertible [from (11.12)] and the discrete-time DC gain equals the continuous-time DC gain:

$$
\begin{aligned}
\underline{D} + \underline{C}(I - \underline{A})^{-1}\underline{B} &= D - C(TA)^{-1}TB \\
&= D - CA^{-1}B.
\end{aligned}
$$

This means that for the same step input r in Figures 11.20 and 11.22 (K_d the step-invariant transformation of K), we have $e(\infty) = \varepsilon(\infty)$, if the two systems are internally stable. Using Theorem 11.4.1 we can relate the tracking error to that of the SD system in Figure 11.21.

Theorem 11.5.1 *Assume that the analog system in Figure 11.20 and the SD system in Figure 11.21 with $K_d = SKH$ are internally stable. Then for the same step input r, the steady-state tracking error in the analog system equals that in the SD system.*

In general, due to approximation in digital implementation, the SD performance will be worse than the analog one. However, this theorem says that what is deteriorating in step-input systems is not the steady-state performance but, possibly, the transient performance and this is so for all sampling periods for which the SD systems are internally stable.

Specializing the theorem to the step-tracking case, we get an answer to the question asked at the start of this section:

Corollary 11.5.1 *Under the same assumptions as in Theorem 11.5.1, the SD system is step-tracking iff the analog system is step-tracking.*

Next we turn to the second way to do digital implementation, namely, via bilinear transformation. The setup is the same as in Figures 11.20 through 11.22, but now the discrete-time controller K_d is obtained via bilinear transformation of the analog K. So given the realization for $\hat{k}(s)$ as before, a corresponding realization for $\hat{k}_d(\lambda)$ now is (Section 3.4)

$$
\hat{k}_d(\lambda) = \left[\begin{array}{c|c} A_{K_{bt}} & B_{K_{bt}} \\ \hline C_{K_{bt}} & D_{K_{bt}} \end{array} \right],
$$

where

$$\begin{aligned}
A_{K_{bt}} &= \left(I - \frac{h}{2}A_K\right)^{-1}\left(I + \frac{h}{2}A_K\right) \\
B_{K_{bt}} &= \frac{h}{2}\left(I - \frac{h}{2}A_K\right)^{-1} B_K \\
C_{K_{bt}} &= C_K(I + A_{K_{bt}}) \\
D_{K_{bt}} &= D_K + C_K B_{K_{bt}}.
\end{aligned}$$

(Assume $2/h$ is not an eigenvalue of A_K.) Thus the state matrices for the closed-loop map $\rho \mapsto \varepsilon$ in Figure 11.22 must be modified too:

$$\left[\begin{array}{c|c} \underline{A} & \underline{B} \\ \hline \underline{C} & \underline{D} \end{array}\right] = \left[\begin{array}{cc|c} A_{P_d} - B_{P_d} D_{K_{bt}} C_P & B_{P_d} C_{K_{bt}} & B_{P_d} D_{K_{bt}} \\ -B_{K_{bt}} C_P & A_{K_{bt}} & B_{K_{bt}} \\ \hline -C_P & 0 & I \end{array}\right].$$

The other matrices stay the same.

Recall that the bilinear transformation is

$$s = \frac{2}{h}\frac{1-\lambda}{1+\lambda}.$$

It takes $s = 0$ to $\lambda = 1$ and so preserves the DC gain: $\hat{k}(s)|_{s=0} = \hat{k}_d(\lambda)|_{\lambda=1}$. Therefore as before, we should expect intuitively the two closed-loop DC gains to be the same. A proof of this fact is summarized as follows: Assume internal stability in Figures 11.20 and 11.22. Let

$$M := \frac{h}{2}\left(I - \frac{h}{2}A_K\right)^{-1}$$

and define the matrices

$$T_1 = \begin{bmatrix} f_h(A_P) & f_h(A_P)B_P C_K M \\ 0 & M \end{bmatrix}, \quad T_2 = \begin{bmatrix} I & 0 \\ 0 & 2I \end{bmatrix}.$$

Then it can be verified (Exercise 11.9) that

$$\begin{aligned}
\underline{A} - I &= T_1 A T_2 \\
\underline{B} &= T_1 B \\
\underline{C} &= C T_2 \\
\underline{D} &= D,
\end{aligned}$$

and hence

$$\underline{D} + \underline{C}(I - \underline{A})^{-1}\underline{B} = D - CA^{-1}B.$$

This allows us to state the following results on digital implementation via bilinear transformation.

Theorem 11.5.2 *Assume the analog system in Figure 11.20 and the SD system in Figure 11.21 with*

$$\hat{k}_d(\lambda) = \hat{k}\left(\frac{2}{h}\frac{1-\lambda}{1+\lambda}\right)$$

are internally stable. Then for the same step input r, the steady-state tracking error in the analog system equals that in the SD system.

Corollary 11.5.2 *Under the same assumptions as in Theorem 11.5.2, the SD system is step-tracking iff the analog system is step-tracking.*

In conclusion, another way to design a step-tracking SD system is as follows: Design an analog step-tracking controller and do a digital implementation via step-invariant or bilinear transformation.

Example 11.5.1 A certain flexible beam is modeled by the transfer function (from torque input to tip deflection output)

$$\hat{p}(s) = \frac{1.6188s^2 - 0.1575s - 43.9425}{s^4 + 0.1736s^3 + 27.9001s^2 + 0.0186s}.$$

The following analog controller has been designed for step-tracking:

$$\hat{k}(s) = -\frac{0.0460s^5 + 1.5402s^4 + 1.5498s^3 + 42.75s^2 + 0.0285s + 0.000158s}{s^6 + 3.766s^5 + 34.9509s^4 + 106.2s^3 + 179.2s^2 + 166.43s + 0.0033}.$$

We now implement the analog controller digitally via step-invariant and bilinear transformations and compare the responses of the two implementations.

Note that the analog plant has a pole at 0, so for step-tracking in the digital implementations, by the preceding corollaries we need internal stability only. Let us take $h = 0.5$, for which the two SD systems are internally stable. Next we simulate the unit step responses for the two SD systems as in Figure 11.23; shown also is the unit step response of the analog system for comparison. All three responses approach 1, indicating step-tracking in all three cases. Clearly, the response via bilinear transformation is better than that via step-invariant transformation: It responds faster and has less overshoot. Both are inferior to the analog response due to the approximation introduced in digital implementation.

11.6 Tracking Other Signals

In Section 11.4 we saw that if the discretized system is step-tracking, the SD system is also step-tracking. What about tracking other signals? Can we generalize the step-tracking result to, say, the ramp-tracking case? These

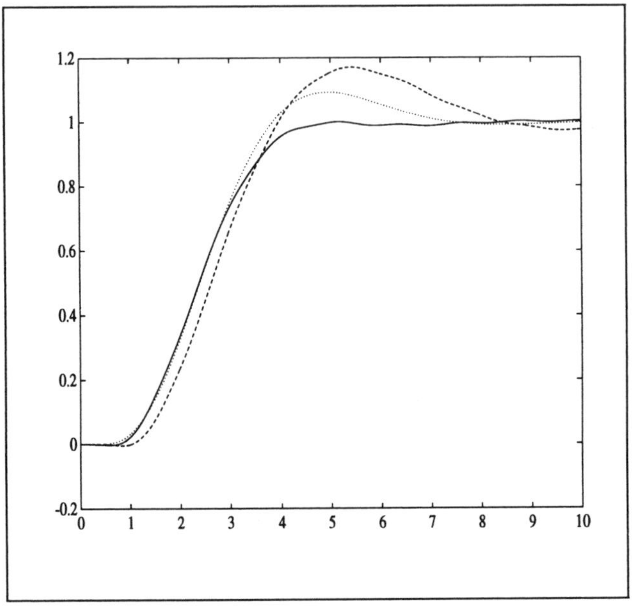

Figure 11.23: Beam example: step responses for analog controller (solid) and its discretization via step-invariant transformation (dash) and bilinear transformation (dot).

questions will be addressed briefly in this section. For simplicity, we look at the SISO setup only.

Consider the SD tracking setup in Figure 11.18, where now the reference input r is the unit ramp signal, $r(t) = t, t \geq 0$. Discretize the system as before and define $\rho = Sr$ (a discrete ramp) to get Figure 11.19. It is apparent that for the SD system to track r, it is necessary that the discretized system track ρ. Thus suppose we design K_d in discrete time so that the discretized system is stable and tracks ρ. The question is: Is this sufficient for intersample tracking? The answer is in general no. This is illustrated by the following example.

Example 11.6.1 In Figure 11.18, take

$$\hat{p}(s) = \frac{1}{s+1}, \qquad h = 1.$$

First, let us design a ramp-tracking controller for the discretized system. Discretize P via step-invariant transformation:

$$\hat{p}_d(\lambda) = \frac{0.6321\lambda}{1 - 0.3679\lambda}.$$

For asymptotic tracking in discrete time, $\hat{k}_d$ must have a double pole at 1; so $\hat{k}_d$ is of the form

$$\hat{k}_d(\lambda) = \frac{1}{(\lambda-1)^2}\hat{k}_{d1}(\lambda).$$

Now absorb the double pole into the plant and define

$$\hat{p}_{d1}(\lambda) = \frac{1}{(\lambda-1)^2}\hat{p}_d(\lambda).$$

Then we need to design K_{d1} to stabilize P_{d1} in the setup:

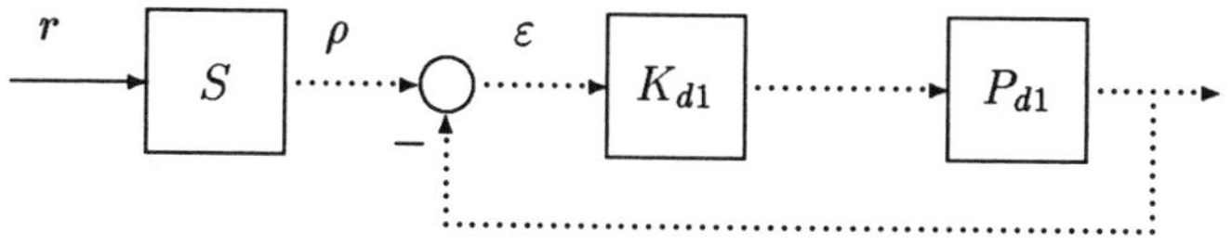

This can be done via the observer-based controller design of Section 5.2:

Step 1 Obtain a minimal realization for $\hat{p}_{d1}$:

$$\hat{p}_{d1}(\lambda) = \left[\begin{array}{c|c} A & B \\ \hline C & 0 \end{array}\right].$$

Step 2 Compute F and H such that the two matrices $A+BF$ and $A+HC$ are stable; we assign the eigenvalues of the two matrices all to be at 0. This will guarantee that ε settles to 0 in a finite time (deadbeat control).

Step 3 Then the controller is

$$\hat{k}_{d1}(\lambda) = \left[\begin{array}{c|c} A+BF+HC & H \\ \hline F & 0 \end{array}\right].$$

The computation gives $\hat{k}_{d1}(\lambda)$ and hence $\hat{k}_d(\lambda)$:

$$\hat{k}_d(\lambda) = \frac{0.5820\lambda(\lambda-1.7358)(\lambda-2.5601)}{(\lambda-1)^2(\lambda+0.4223)}.$$

The continuous-time tracking error e is simulated in Figure 11.24. It is clear that the discretized system tracks; in fact, ε settles to 0 in 4 sampling periods (4 seconds), reflecting the deadbeat response. But observe that there is a steady-state intersample ripple; so the SD system does not track.

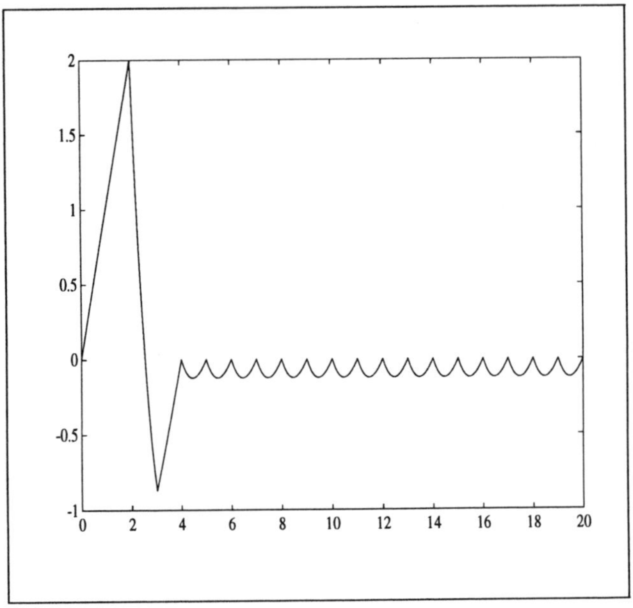

Figure 11.24: Tracking error.

This example indicated that the theorems in the preceding two sections do not generalize naively to the ramp-tracking case. Some modification is necessary.

To obtain SD ramp tracking, or to eliminate the steady-state intersample ripple, two methods can be considered. The first is to use a first-order hold instead of a zero-order hold; then we expect a similar statement as in Theorem 11.4.1 to hold. However, we shall not pursue this direction. The second method is to introduce analog pre-compensation so that the compensated plant includes an integrator. This again is illustrated by the example.

Example 11.6.2 For the same plant as in Example 11.6.1, pre-compensate to get:

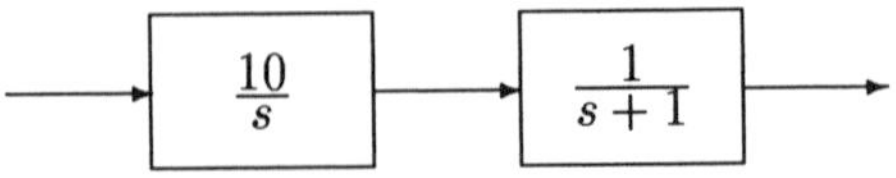

Take this as P, discretize to get

$$\hat{p}_d(\lambda) = \frac{7.1828(\lambda + 1.3922)}{(\lambda - 1)(\lambda - 2.7183)}.$$

Since $\hat{p}_d$ has one pole at 1, to get discrete tracking, we need just one additional pole at 1 in $\hat{k}_d$. So $\hat{k}_d$ is of the form

$$\hat{k}_d(\lambda) = \frac{1}{\lambda - 1}\hat{k}_{d1}(\lambda).$$

Again absorb this pole into $\hat{p}_d(\lambda)$ to get

$$\hat{p}_{d1}(\lambda) = \frac{1}{\lambda - 1}\hat{p}_d(\lambda).$$

As in Example 11.6.1, design $\hat{k}_{d1}$ for $\hat{p}_{d1}$ as an observer-based controller so that the closed-loop poles are all at 0; this gives $\hat{k}_{d1}$ and thus $\hat{k}_d$:

$$\hat{k}_d(\lambda) = -\frac{0.1392\lambda(\lambda - 2.6567)(\lambda - 1.5511)}{(\lambda - 1)(\lambda + 0.6327)(\lambda + 1.2699)}.$$

The tracking error is simulated in Figure 11.25. This confirms that there is no intersample ripple and in fact both ε and e are deadbeat in 5 periods.

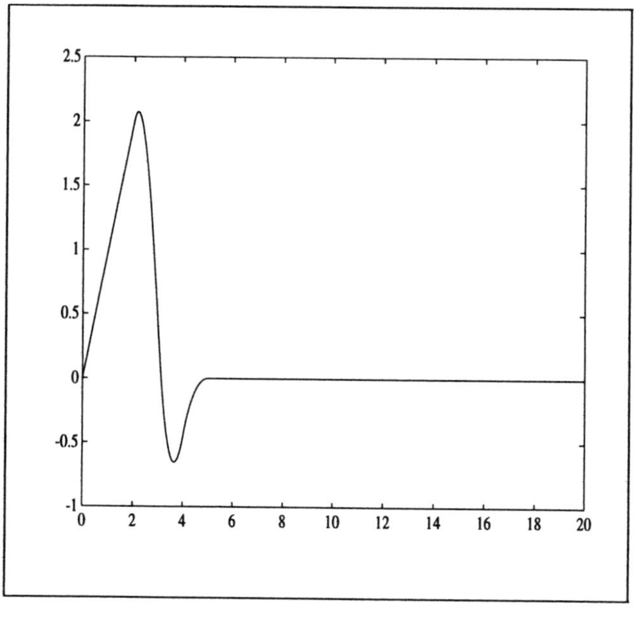

Figure 11.25: Tracking error.

In summary, if the analog plant $\hat{p}(s)$ does not have a pole at 0, the SD setup (with zero-order hold) is inherently incapable of tracking a ramp signal. This can be resolved by using either the first-order hold or analog precompensation to introduce a pole at 0; then a discrete-time tracking design also yields SD tracking.

This observation can be generalized to tracking other types of signals such as $e^{\alpha t}$ ($\alpha > 0$) and $\sin\omega t$. If zero-order hold is to be used, then to get SD tracking, one has to pre-compensate the analog plant so that the compensated plant contains an internal model of the signal; for example, if $\sin\omega t$ is to be tracked, the compensated plant must have a pair of poles at $s = \pm j\omega$.

Exercises

11.1 This problem concerns control of the double integrator

$$\hat{y}(s) = \frac{1}{s^2}\hat{u}(s).$$

The usual analog controller for this is a PD (proportional-derivative) controller:

$$u = K_p(r - y) - K_v\dot{y}.$$

Then the feedback system is internally stable for all positive K_p, K_v. For digital implementation of this controller, a common approximation of the derivative is the backward difference

$$\dot{y}(kh) \approx \frac{1}{h}\left\{y(kh) - y[(k-1)h]\right\}.$$

This leads to the SD system

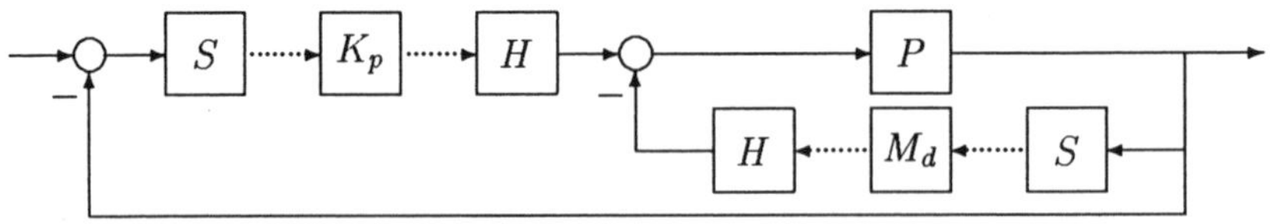

where

$$\hat{p}(s) = \frac{1}{s^2}, \quad \hat{m}_d(\lambda) = \frac{K_v}{h}(1 - \lambda).$$

Take $K_p = 2$, $K_v = 1$ and compute the range of h for which the SD system is stable.

11.2 Consider the flexible beam in Example 11.5.1.

1. Design an analog controller $\hat{k}(s)$ to achieve the step-response specs of less than 10% overshoot and less than 8 s settling time.

2. Discretize the controller using the step-invariant transformation. Compute h_{st}, the maximum sampling period for which the feedback system is stable.

3. Simulate the sampled-data system for a step input and plot overshoot and settling time versus h over the range $0 < h < h_{st}$. What is the maximum sampling period for which the specs are met?

11.3 Prove Theorem 11.2.1.

11.4 This exercise relates internal stability to input-output stability. For simplicity, consider the discrete-time setup:

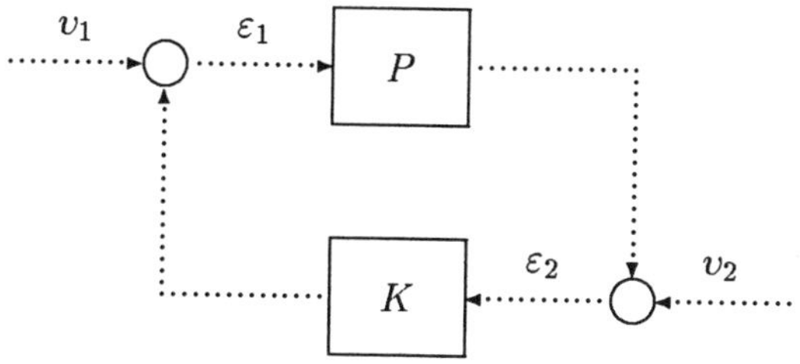

Here P and K both have stabilizable and detectable realizations:

$$\hat{p}(\lambda) = \left[\begin{array}{c|c} A & B \\ \hline C & 0 \end{array}\right], \quad \hat{k}(\lambda) = \left[\begin{array}{c|c} A_K & B_K \\ \hline C_K & D_K \end{array}\right].$$

Show that internal stability is equivalent to input-output stability: The closed-loop A-matrix

$$\underline{A} = \begin{bmatrix} A + BD_KC & BC_K \\ B_KC & A_K \end{bmatrix}$$

is stable iff the transfer matrix

$$\begin{bmatrix} v_1 \\ v_2 \end{bmatrix} \mapsto \begin{bmatrix} \varepsilon_1 \\ \varepsilon_2 \end{bmatrix}$$

is stable.

11.5 Consider Figure 11.14 with

$$h = 0.1, \quad \hat{f}(s) = \frac{1}{(0.1/\pi)s + 1}, \quad \hat{p}(s) = \frac{s-1}{s^2(s+1)}, \quad \hat{w}(s) = 0.2\frac{s+1}{s+10}.$$

Design an analog controller K to stabilize the feedback system consisting of P, K, and F; then implement $K_d = SKH$ to get Figure 11.14. Compute an upper bound on the stability margin.

11.6 This problem concerns deadbeat regulation. Look at the sampled-data state feedback setup:

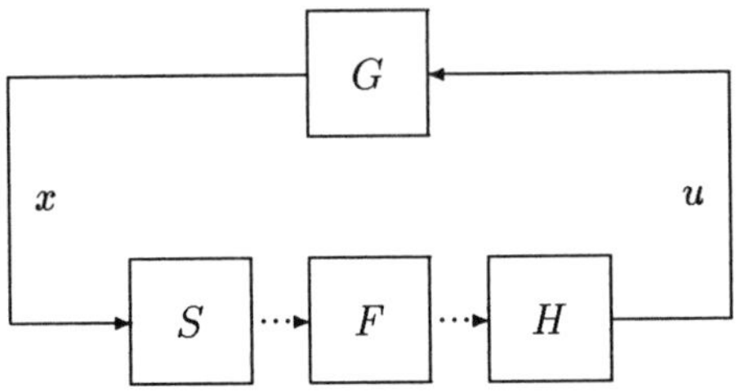

Here G is described by a state model with an initial condition

$$\dot{x}(t) = Ax(t) + Bu(t), \quad x(0) = x_0,$$

and F is a constant matrix. Discretize via *c2d* to obtain the usual matrices A_d and B_d. Let $\xi(k) := x(kh)$ and let the dimension of x be n. Suppose F is designed such that $A_d + B_d F$ has all eigenvalues at 0.

1. Prove that $\forall x_0,\ \xi(k) = 0$ for $k \geq n$.

2. Prove that $\forall x_0,\ x(t) = 0$ for $t \geq nh$.

Thus both ξ and x are deadbeat in n sampling periods.

11.7 Consider

$$\hat{p}(s) = \left[\begin{array}{c|c} A & B \\ \hline C & D \end{array}\right], \quad \hat{p}_d(\lambda) = \left[\begin{array}{c|c} A_d & B_d \\ \hline C & D \end{array}\right],$$

with A stable. Show that the DC gains of the two systems are equal: $\hat{p}(s)|_{s=0} = \hat{p}_d(\lambda)|_{\lambda=1}$.

11.8 For the same system as in Example 6.6.1, design a deadbeat controller K_d for step-tracking. How many steps does it take for the continuous-time error to settle to 0?

11.9 Assume the systems in Figure 11.20 and Figure 11.22 are internally stable and K_d is obtained via bilinear transformation of K. Provide the details in the proof in Section 11.5 that the two closed-loop maps $r \mapsto e$ and $\rho \mapsto \varepsilon$ have the same DC gain.

11.10 Another way to get $\hat{k}_d(\lambda)$ from $\hat{k}(s)$ in digital implementation is the *backward rule*: $\hat{k}_d(\lambda) = \hat{k}\left(\frac{1-\lambda}{h}\right)$. Suppose the analog system is internally stable and step-tracking. With the implementation of the controller via the backward rule, is the SD system step-tracking?

11.11 In Figure 11.18, take

$$\hat{p}(s) = \frac{1}{s}, \quad r(t) = t\ (t \geq 0), \quad h = 1.$$

Design a K_d so that the discretized system tracks $\rho := Sr$. Simulate the continuous-time error e. Is SD tracking achieved? With your design, show analytically that $e(t) \to 0$ as $t \to \infty$.

11.12 In Figure 11.18, take

$$\hat{p}(s) = \frac{1}{s+1}, \quad r(t) = \mathrm{e}^t \ (t \geq 0), \quad h = 1.$$

1. Design K_d for $P_d := SPH$ to achieve tracking at the sampling instants. Simulate e. Is SD tracking achieved?

2. Pre-compensate P to get

$$\hat{p}_1(s) = \frac{1}{(s-1)(s+1)}.$$

   Repeat the first part for $P_{d1} := SP_{d1}H$.

Notes and References

Sections 11.1 and 11.2 are based on [52], which treats boundedness on $\mathcal{L}_\infty$, and [28] and [23], which treat boundedness on $\mathcal{L}_p$ in general. For stability of SD systems with time-varying components, see [76]. A detailed discussion of the flexible beam example is in Section 10.3 of [39].

The robust stability result in Section 11.3 is from [30]. The same result is in [69]. It is an interesting fact that the stability bound obtained by the $\mathcal{L}_2$-induced norm is conservative when the perturbation is LTI, but nonconservative when the perturbation is linear, time-varying; see [44] and [125]. Other work on robust stability of sampled-data systems can be found in [88].

Tracking steps in purely continuous time is a special case of algebraic regulator theory, for example, [150]. The sampled-data case was treated by Dullerud [41]; Corollary 11.4.1 is from [41]. For more on ripple-free SD tracking systems, see [37], [53], [143], [151], and [71].

Chapter 12

$\mathcal{H}_2$-Optimal SD Control

In Chapter 6 we looked at how to design $\mathcal{H}_2$-optimal controllers for discrete-time FDLTI systems via the state-space approach; there in Example 6.6.1 we also saw that performance specs based on discretized systems could result in a severe intersample ripple. Our goal in this chapter is to incorporate intersample behaviour into design; specifically, we consider how to pose and solve $\mathcal{H}_2$-optimal control problems for SD systems.

12.1 A Simple $\mathcal{H}_2$ SD Problem

There are several ways to pose a SD $\mathcal{H}_2$ problem. As an introduction, we first look at a simple formulation which admits a solution using continuous lifting.

We begin with the continuous-time $\mathcal{H}_2$-optimal control problem for the standard setup in Figure 12.1, where the generalized plant G and the con-

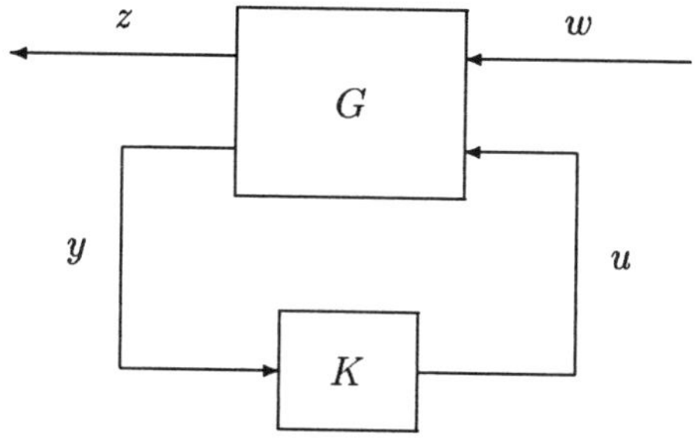

Figure 12.1: The continuous system.

troller K are both assumed to be LTI and causal. Let T_{zw} denote the closed-loop system from w to z. We can contemplate minimizing the $\mathcal{H}_2$ norm of the (matrix-valued) transfer function $\hat{t}_{zw}$, that is,

$$\|\hat{t}_{zw}\|_2 := \left\{ \frac{1}{2\pi} \int_{-\infty}^{\infty} \text{trace } \left[\hat{t}_{zw}(j\omega)^* \hat{t}_{zw}(j\omega)\right] d\omega \right\}^{1/2},$$

over all internally stabilizing K. For example, if w is standard white noise, then $\|\hat{t}_{zw}\|_2$ equals the root-mean-square value of z.

Another way to think of the same criterion is as follows. Let m denote the dimension of w and denote by $\{e_i\}_{i=1,\cdots,m}$ the standard basis in $\mathbb{R}^m$. An impulse at the ith component of the exogenous signal is achieved by setting $w(t) = \delta(t)e_i$, the resulting output z being $T_{zw}\delta e_i$. Then it is easy to derive (Theorem 2.1.1) that

$$\|\hat{t}_{zw}\|_2 = \left(\sum_i \|T_{zw}\delta e_i\|_2^2 \right)^{1/2}, \tag{12.1}$$

the right-hand norm being the usual one on $\mathcal{L}_2(\mathbb{R}_+)$.

In this chapter, we are concerned instead with SD controllers, so the appropriate setup is in Figure 12.2. For a chosen sampling period h, the

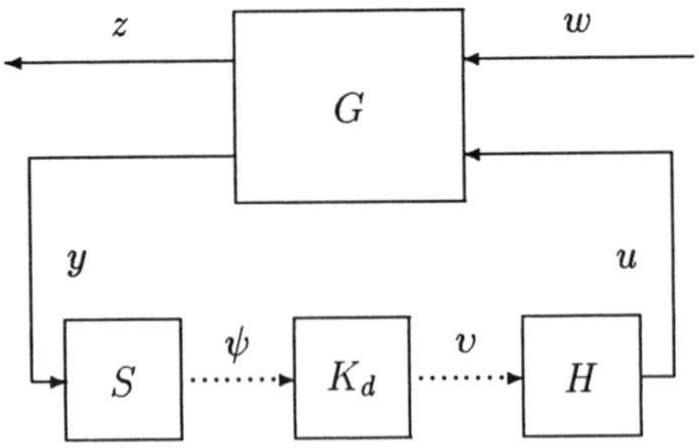

Figure 12.2: The SD system.

designable element is now the discrete LTI controller K_d. Of course, T_{zw} is now time-varying, or more precisely, periodic with period h, so there is no transfer function in the normal sense whose $\mathcal{H}_2$ norm could be minimized. However, the right-hand side of (12.1) still makes sense. The problem of this section is to minimize the quantity

$$J_0 = \left(\sum_i \|T_{zw}\delta e_i\|_2^2 \right)^{1/2}$$

over all FDLTI, causal K_ds which provide internal stability. [By contrast, in Example 8.4.1 we minimized $\left(\sum_i \|S_f T_{zw}\delta e_i\|_2^2\right)^{1/2}$ for the telerobot.]

Note that the performance measure is defined in terms of the *continuous* signals for impulsive inputs; thus intersample behaviour of signals of interest is taken into account in the design. This should be the natural way of looking at SD systems since they operate in a continuous-time environment.

We remark that this problem formulation relates closely to the continuous-time linear quadratic regulation (LQR) problem using SD control. Consider a continuous-time system P, driven by its initial condition and controlled by a SD controller as in Figure 12.3, where P is described by the state model:

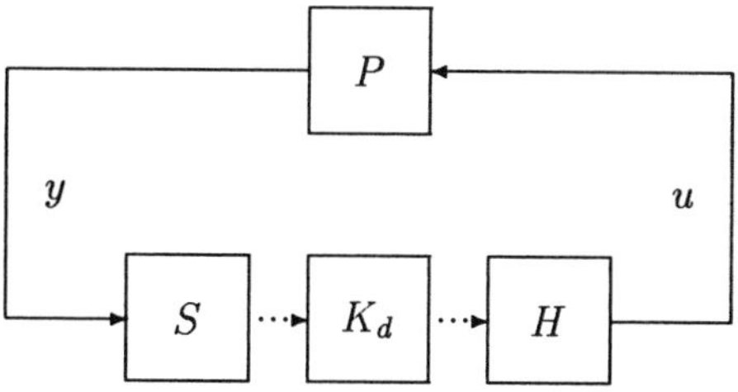

Figure 12.3: A SD system for the LQR problem.

$$\begin{aligned} \dot{x}(t) &= Ax(t) + B_2u(t), \quad x(0) = x_0, \\ y(t) &= C_2x(t). \end{aligned}$$

The LQR problem is to design K_d to minimize

$$J_{LQR} := \int_0^\infty [x(t)'Qx(t) + u(t)'Ru(t)]\, dt,$$

where Q and R are both positive semi-definite. Perform a Cholesky factorization

$$\begin{bmatrix} C_1 & D_{12} \end{bmatrix}' \begin{bmatrix} C_1 & D_{12} \end{bmatrix} = \begin{bmatrix} Q & 0 \\ 0 & R \end{bmatrix}$$

and define

$$z(t) = C_1x(t) + D_{12}u(t)$$

to get that J_{LQR} equals the square of the $\mathcal{L}_2(\mathbb{R}_+)$-norm of z. The problem can be put into the standard framework in Figure 12.2 if we introduce an exogenous input w and set $w(t) = x_0\delta(t)$, the corresponding G being

$$\begin{aligned} \dot{x}(t) &= Ax(t) + w(t) + B_2u(t), \quad x(0) = 0, \\ z(t) &= C_1x(t) + D_{12}u(t), \\ y(t) &= C_2x(t). \end{aligned}$$

Thus the problem reduces to minimizing

$$\|T_{zw}x_0\delta\|_2,$$

which is quite like the proposed optimization problem.

Returning to Figure 12.2, assume G is FDLTI and causal. Bring in a state model

$$\hat{g}(s) = \left[\begin{array}{c|cc} A & B_1 & B_2 \\ \hline C_1 & 0 & D_{12} \\ C_2 & 0 & 0 \end{array}\right].$$

Note that we have taken D_{11} and D_{21} to be zero; this is due to the facts that J_0 must be finite and that S is not defined on impulsive functions. For simplicity, we have also assumed that $D_{22} = 0$.

Now bring in the associated discrete-time LTI system as in Figure 12.4. The state matrices in the plant are defined as follows: First, as usual

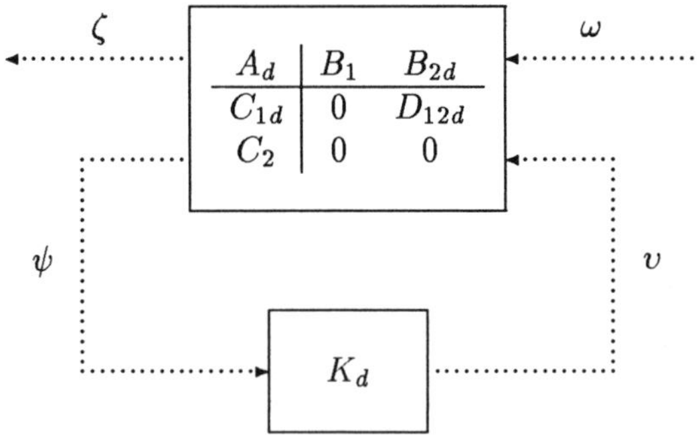

Figure 12.4: The associated discrete system.

$$A_d := e^{hA}, \quad B_{2d} := \int_0^h e^{tA} dt B_2.$$

Next, define the square matrix

$$\underline{A} := \begin{bmatrix} A & B_2 \\ 0 & 0 \end{bmatrix}$$

and then C_{1d} and D_{12d} by the equation

$$\begin{bmatrix} C_{1d} & D_{12d} \end{bmatrix}' \begin{bmatrix} C_{1d} & D_{12d} \end{bmatrix}$$

$$= \int_0^h e^{t\underline{A}'} \begin{bmatrix} C_1 & D_{12} \end{bmatrix}' \begin{bmatrix} C_1 & D_{12} \end{bmatrix} e^{t\underline{A}} dt. \tag{12.2}$$

[The matrix C_{1d} may have more rows than C_1 due to the fact that the rank of the matrix on the right-hand side of (12.2) may be greater than the number of rows in C_1.] Note that the (2,2)-block in the discrete plant is exactly the same as the discretized G_{22}. Thus the SD system is internally stable iff the discrete system in Figure 12.4 is internally stable (the two systems share the same K_d). Recall from Section 11.1 that a necessary and sufficient condition for internal stability to be achievable is

$$(A_d, B_{2d}) \text{ is stabilizable and } (C_2, A_d) \text{ is detectable.}$$

This condition will be assumed hereafter; it would be sufficient to assume non-pathological sampling, (A, B_2) stabilizable, and (C_2, A) detectable.

Let $T_{\zeta w}$ denote the closed-loop system from ω to ζ in Figure 12.4. We are set to state the main result of this section.

Theorem 12.1.1 *The SD spec J_0 for Figure 12.2 equals the $\mathcal{H}_2(\mathbb{D})$-norm of the transfer function $\hat{t}_{\zeta\omega}$ in Figure 12.4.*

This theorem provides a way to solve the SD problem of minimizing J_0: It is equivalent to the discrete-time $\mathcal{H}_2$ problem of minimizing $\|\hat{t}_{\zeta\omega}\|_2$ over all the internally stabilizing K_ds in Figure 12.4, that is,

$$\min_{K_d} J_0 = \min_{K_d} \|\hat{t}_{\zeta\omega}\|_2.$$

The latter problem was studied in Chapter 6.

Proof of Theorem 12.1.1 The proof is a nice application of continuous lifting. Fix a basis vector e_i, apply the input $w(t) = \delta(t)e_i$, and lift the output z to get Figure 12.5. Bringing L, S, and H into G gives the system in

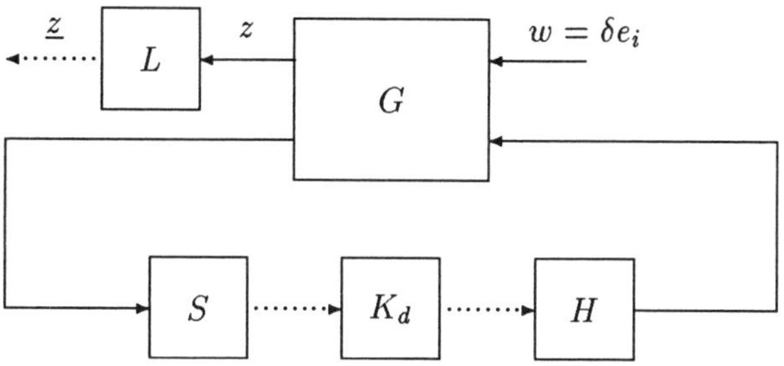

Figure 12.5: Lifting z.

Figure 12.6. We would like to convert the system in Figure 12.6 to one with a discrete-time input instead of δe_i. We could convert the input to $\delta_d e_i$, but it turns out to simplify things if the input is $U^* \delta_d e_i$ instead: Recall that U^*

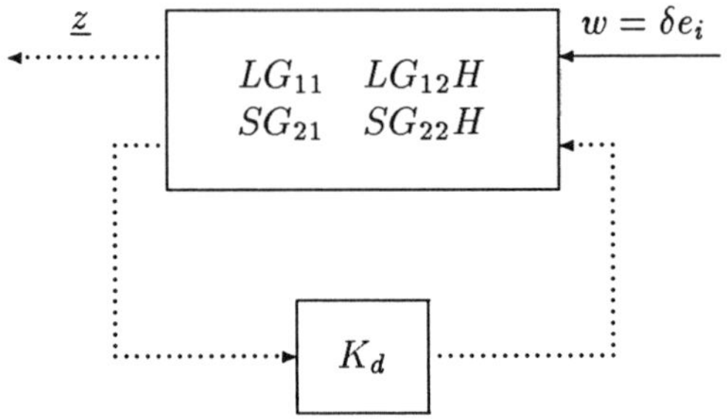

Figure 12.6: Lifted system with continuous-time input w.

is time advance, so $U^*\delta_d$ is the impulse applied at time $k = -1$. To do the conversion, consider first the (1,1)-block: Set $\underline{v} = LG_{11}\delta e_i$. Recalling from Section 10.2 the definition

$$\underline{C}_1 : \ \mathcal{E} \to \mathcal{K}, \quad (\underline{C}_1\xi)(t) = C_1 e^{tA}\xi,$$

we have

$$\begin{aligned} \underline{v} &= \{v_0, v_1, \ldots\} \\ v_0(t) &= (\underline{C}_1 B_1 e_i)(t) \\ v_1(t) &= (\underline{C}_1 A_d B_1 e_i)(t) \\ &\text{etc.} \end{aligned}$$

Thus $\underline{v}$ equals the response of the system

$$\left[\begin{array}{c|c} A_d & B_1 \\ \hline \underline{C}_1 & 0 \end{array}\right]$$

to the input $U^*\delta_d e_i$. Proceeding in a similar fashion with the other three blocks, we get that the output in Figure 12.6 equals the output in Figure 12.7. Here, the definition of $\underline{D}_{12,res}$ (the restriction of $\underline{D}_{12}$) is

$$\underline{D}_{12,res} : \ \mathcal{E} \to \mathcal{K}, \quad (\underline{D}_{12,res}v)(t) = \left[D_{12} + \int_0^t C_1 e^{\tau A} d\tau B_2\right] v.$$

Two steps remain before we get Figure 12.4. The output equation in Figure 12.7 is

$$\underline{z}_k = \underline{C}_1\xi(k) + \underline{D}_{12,res}v(k). \tag{12.3}$$

From Lemma 10.5.1,

$$e^{t\underline{A}} = \begin{bmatrix} e^{tA} & \int_0^t e^{\tau A} d\tau B_2 \\ 0 & I \end{bmatrix},$$

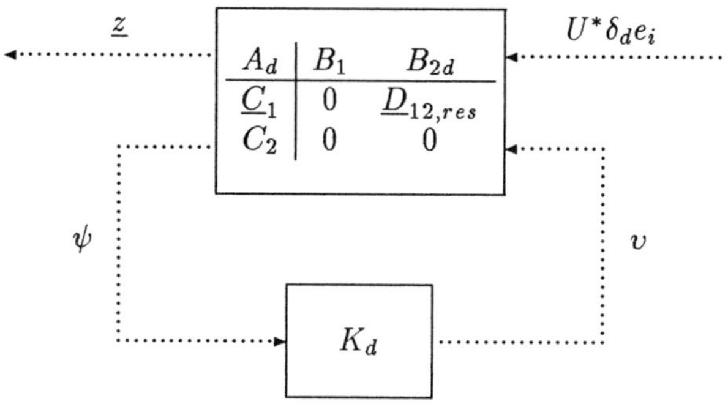

Figure 12.7: Equivalent discrete-time system.

which implies from (12.3) that

$$\underline{z}_k(t) = \begin{bmatrix} C_1 & D_{12} \end{bmatrix} e^{t\underline{A}} \begin{bmatrix} \xi(k) \\ v(k) \end{bmatrix}. \tag{12.4}$$

Defining

$$\zeta(k) = C_{1d}\xi(k) + D_{12d}v(k),$$

we get from (12.2) and (12.4) that the $\mathcal{K}$-norm of $\underline{z}_k$ equals the $\mathcal{E}$-norm of $\zeta(k)$. Thus the $\ell_2(\mathbb{Z}_+, \mathcal{K})$-norm of the output $\underline{z}$ in Figure 12.7 equals the $\ell_2(\mathbb{Z}_+, \mathcal{E})$-norm of the output ζ in Figure 12.8.

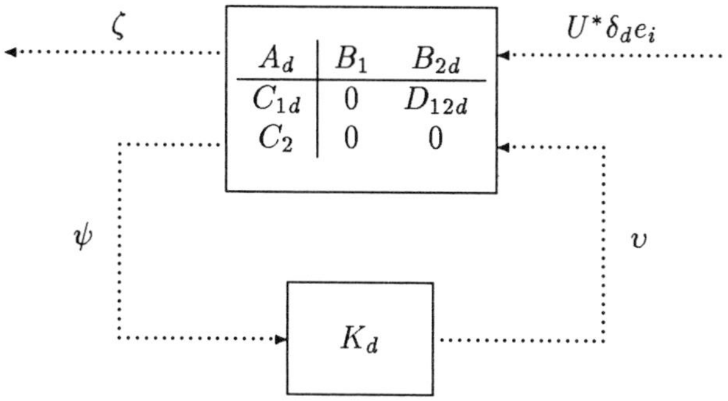

Figure 12.8: Equivalent discrete-time system.

Finally, since $T_{\zeta w}$ is time-invariant,

$$\|T_{\zeta w}U^*\delta_d e_i\|_2 = \|T_{\zeta w}\delta_d e_i\|_2.$$

The right-hand side pertains to Figure 12.4. ■

The matrix integral involved in computing C_{1d} and D_{12d} can be easily determined using a matrix exponential function; the formulas were given at the end of Section 10.6.

Note that the D_{21}-term in the discrete-time plant is zero; thus the discrete $\mathcal{H}_2$ problem is inherently singular and the formulas in Section 6.5 do not apply directly. Often this can be fixed by introducing some time advance in the exogenous input channel (see Exercise 6.12), or by applying the frequency-domain solution of Section 6.6. This will be further illustrated by the following example.

Example 12.1.1 Consider again the SD step-tracking system studied in Examples 6.6.1 and 8.4.2; it is redrawn in Figure 12.9. Here we have introduced

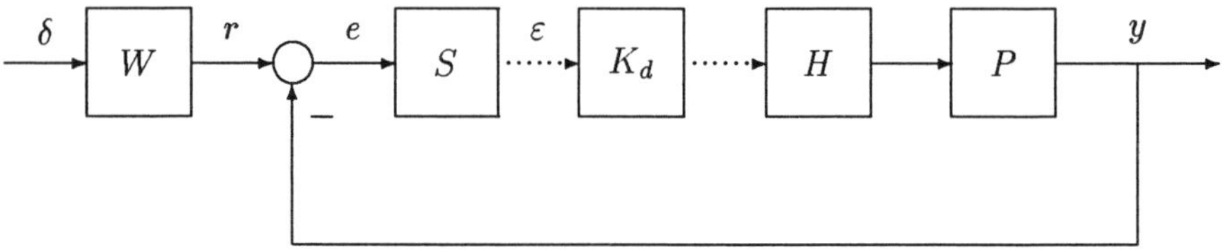

Figure 12.9: The SD tracking system.

a reference model W with $\hat{w}(s) = 1/s$. The fictitious input to W is the unit impulse δ. The plant transfer function and the sampling period are as before:

$$\hat{p}(s) = \frac{1}{(10s+1)(25s+1)}, \qquad h = 1.$$

We shall design K_d to provide internal stability and minimize the $\mathcal{L}_2(\mathbb{R}_+)$-norm of the tracking error e. This can be put into the framework in Theorem 12.1.1 by defining $w = \delta, z = y = e$; the corresponding standard setup (Figure 12.2) has the generalized plant

$$G = \begin{bmatrix} W & -P \\ W & -P \end{bmatrix}.$$

A realization is

$$\hat{g}(s) = \left[\begin{array}{c|cc} A & B_1 & B_2 \\ \hline C_1 & 0 & 0 \\ C_2 & 0 & 0 \end{array}\right],$$

where the matrices are

$$A = \begin{bmatrix} 0 & 0 & 0 \\ 0 & 0 & 1 \\ 0 & -0.004 & -0.14 \end{bmatrix}, \quad B_1 = \begin{bmatrix} 1 \\ 0 \\ 0 \end{bmatrix}, \quad B_2 = \begin{bmatrix} 0 \\ 0 \\ 1 \end{bmatrix}$$

$$C_1 = C_2 = \begin{bmatrix} 1 & -0.004 & 0 \end{bmatrix}.$$

Note that since W is unstable, (A, B_2) is not stabilizable (though there exists K_d to internally stabilize P). Nevertheless, Theorem 12.1.1 is still applicable to reduce the SD problem to a discrete problem. So the equivalent discrete $\mathcal{H}_2$ problem pertains to the setup in Figure 12.10, where

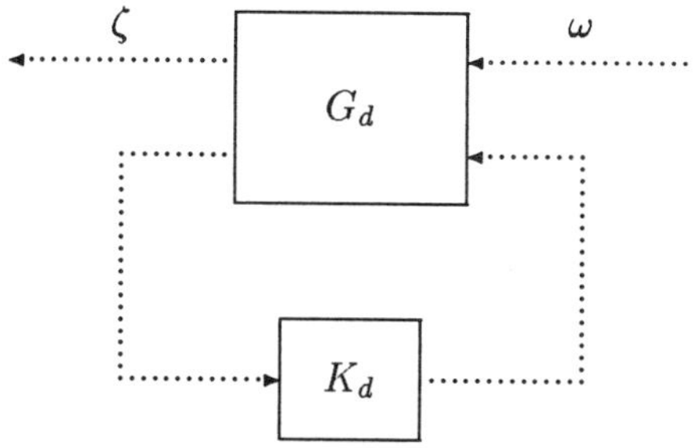

Figure 12.10: The equivalent discrete system.

$$\hat{g}_d(\lambda) = \begin{bmatrix} \hat{g}_{d_{11}} & \hat{g}_{d_{12}} \\ \hat{g}_{d_{21}} & \hat{g}_{d_{22}} \end{bmatrix}(\lambda) = \left[\begin{array}{c|cc} A_d & B_1 & B_{2d} \\ \hline C_{1d} & 0 & D_{12d} \\ C_2 & 0 & 0 \end{array}\right].$$

The matrices A_d and B_{2d} can be obtained from (A, B_2) via *c2d*. Now let us look closely at computing C_{1d} and D_{12d}. Define the square matrices

$$\begin{aligned} \underline{A} &= \begin{bmatrix} A & B_2 \\ 0 & 0 \end{bmatrix} \\ N &= \begin{bmatrix} C_1 & 0 \end{bmatrix}' \begin{bmatrix} C_1 & 0 \end{bmatrix} \\ M &= \int_0^h e^{t\underline{A}'} N e^{t\underline{A}} dt. \end{aligned}$$

The integral M can be computed via the matrix exponential discussed in Section 10.6:

$$\begin{aligned} \begin{bmatrix} P_{11} & P_{12} \\ 0 & P_{22} \end{bmatrix} &= \exp\left\{ h \begin{bmatrix} -\underline{A}' & N \\ 0 & \underline{A} \end{bmatrix} \right\} \\ M &= P_{22}' P_{12}. \end{aligned}$$

The matrix M computed has three nonzero eigenvalues: 1.0000, 1.4676×10^{-6}, and 1.7561×10^{-8}; regarding the latter two eigenvalues as zero, we get an approximate Cholesky factorization,

$$M \approx \begin{bmatrix} C_{1d} & D_{12d} \end{bmatrix}' \begin{bmatrix} C_{1d} & D_{12d} \end{bmatrix},$$

with

$$C_{1d} = \begin{bmatrix} 1 & -0.004 & -0.001901 \end{bmatrix}, \quad D_{12d} = -6.4384 \times 10^{-4}.$$

Because (A_d, B_{2d}) is not stabilizable, it is hard to proceed further in the time domain. Turning to the frequency domain, we note first that $G_{22d} = -SPH$, which is stable since P is. So the set of stabilizing controllers for P can be parametrized by

$$\hat{k}_d = \frac{\hat{q}}{1 + \hat{g}_{d_{22}}\hat{q}}, \quad \hat{q} \in \mathcal{RH}_\infty(\mathbb{D}).$$

Thus the closed-loop transfer function in Figure 12.10 is

$$\begin{aligned} \hat{t}_{\zeta w}(\lambda) &= \hat{g}_{d_{11}}(\lambda) + \hat{g}_{d_{12}}(\lambda)\hat{q}(\lambda)\hat{g}_{d_{21}}(\lambda) \\ &= [1 + \hat{g}_{d_{12}}(\lambda)\hat{q}(\lambda)]\frac{\lambda}{1-\lambda}, \end{aligned}$$

since $\hat{g}_{d_{11}}(\lambda) = \hat{g}_{d_{21}}(\lambda) = \lambda/(1-\lambda)$. Note that the function

$$\hat{g}_{d_{12}}(\lambda) = \frac{-6.9052 \times 10^{-4}(\lambda + 3.8657)(\lambda + 0.2774)}{(\lambda - 1.1052)(\lambda - 1.0408)}$$

is stable and $\hat{g}_{d_{12}}(1) = -1$. Thus for $\|\hat{t}_{\zeta w}\|_2 < \infty$, we must have $\hat{q}(1) = 1$. The $\hat{q}$s in $\mathcal{RH}_\infty(\mathbb{D})$ satisfying this condition are parametrized by

$$\hat{q}(\lambda) = 1 + (1 - \lambda)\hat{q}_1(\lambda), \quad \hat{q}_1(\lambda) \in \mathcal{RH}_\infty(\mathbb{D}).$$

Therefore

$$\hat{t}_{\zeta w} = \hat{t}_1 + \hat{t}_2\hat{q}_1,$$

where $\hat{t}_2(\lambda) = \lambda\hat{g}_{d_{12}}(\lambda)$ and $\hat{t}_1$ is in $\mathcal{RH}_\infty(\mathbb{D})$:

$$\hat{t}_1(\lambda) = \frac{\lambda[1 + \hat{g}_{d_{12}}(\lambda)]}{1 - \lambda} = \frac{-0.9993\lambda(\lambda - 1.1503)}{(\lambda - 1.1052)(\lambda - 1.0408)}.$$

In this way we arrive at the $\mathcal{H}_2$ model-matching problem:

$$\min_{\hat{q}_1 \in \mathcal{RH}_\infty(\mathbb{D})} \|\hat{t}_1 + \hat{t}_2\hat{q}_1\|_2.$$

This latter problem can be solved as in Section 6.7:

$$\hat{q}_1(\lambda) = \frac{-5218.8(\lambda - 1.1478)}{(\lambda + 3.8657)(\lambda + 3.6044)}.$$

Now substitute back to get $\hat{q}$ and then $\hat{k}_d$:

$$\hat{k}_d(\lambda) = \frac{-525.0970(\lambda - 1.1052)(\lambda - 1.0408)}{(\lambda - 1)(\lambda + 1.4017)}.$$

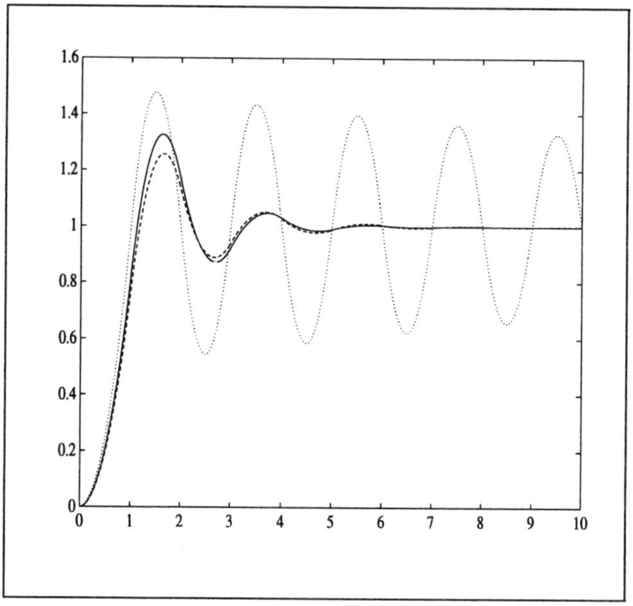

Figure 12.11: Step-response of example: SD design (solid), design by fast discretization (dash), design by slow discretization (dot).

The step response y (solid) with this controller was simulated and is displayed in Figure 12.11. Shown also in the figure, for comparison, are the step response (dashed) in Example 8.4.2 where the controller was designed via fast discretization and that of Example 6.6.1 via slow discretization. For this example, the fast-discretization method is nearly optimal.

12.2 Generalized $\mathcal{H}_2$ Measure for Periodic Systems

Generalized $\mathcal{H}_2$ Measure

Let us start by reconsidering the performance spec used in Section 12.1, namely,

$$J_0 = \left(\sum_i \|T_{zw} \delta e_i\|_2^2 \right)^{1/2}.$$

The impulsive functions are applied at $t = 0$. But since T_{zw} is time-varying, applying an input only at $t = 0$ may seem inappropriate. Letting $\delta_\tau(t) = \delta(t - \tau)$, we arrive at an alternative spec,

$$J_\tau = \left(\sum_i \|T_{zw}\delta_\tau e_i\|_2^2 \right)^{1/2}.$$

The fact that T_{zw} is a periodic system with period h implies that J_τ is an h-periodic function, that is, $J_{\tau+h} = J_\tau, \forall\tau$. Thus it suffices to consider J_τ for $0 \le \tau < h$. The *generalized* $\mathcal{H}_2$ *measure* is defined to be

$$J = \left(\int_0^h J_\tau^2 d\tau \right)^{1/2}.$$

Notice that $J/\sqrt{h}$ can be interpreted as the root-mean-square of J_τ, where "mean" means "time average." If T_{zw} is LTI, then $J_\tau = J_0, \forall\tau$, and $J/\sqrt{h}$ reduces to the $\mathcal{H}_2$ norm of the transfer function $\hat{t}_{zw}$.

The quantity J can be expressed in terms of impulse responses. For this we turn to a more general discussion.

Hilbert-Schmidt Operators

Let F be a continuous-time linear transformation mapping $\mathcal{L}_2(\mathbb{R}_+, \mathbb{R}^m)$ to $\mathcal{L}_2(\mathbb{R}_+, \mathbb{R}^p)$ and described by the integral equation

$$(Fu)(t) = \int_0^\infty f(t,\tau)u(\tau)d\tau. \tag{12.5}$$

Here f, a $p \times m$ matrix function defined on $\mathbb{R}_+ \times \mathbb{R}_+$, is the *impulse response* of F, for if $u(\tau) = \delta(\sigma - \tau)e_i$, an impulse at time $\sigma \ge 0$, then $(Fu)(t) = f(t,\sigma)e_i$. We shall assume that every element of the matrix f is square-integrable on $\mathbb{R}_+ \times \mathbb{R}_+$, or equivalently

$$\int_0^\infty \int_0^\infty \text{trace } [f(t,\tau)'f(t,\tau)]dtd\tau < \infty. \tag{12.6}$$

It turns out that such an F is a bounded operator from $\mathcal{L}_2(\mathbb{R}_+, \mathbb{R}^m)$ to $\mathcal{L}_2(\mathbb{R}_+, \mathbb{R}^p)$; moreover, the induced norm of F satisfies

$$\|F\| \le \left\{ \int_0^\infty \int_0^\infty \text{trace } [f(t,\tau)'f(t,\tau)]dtd\tau \right\}^{1/2}$$

This latter quantity defines a different norm on the operator F, the *Hilbert-Schmidt norm*, denoted $\|F\|_{\mathbf{HS}}$:

$$\|F\|_{\mathbf{HS}} := \left\{ \int_0^\infty \int_0^\infty \text{trace } [f(t,\tau)'f(t,\tau)]dtd\tau \right\}^{1/2}.$$

An operator F of the form (12.5) satisfying condition (12.6) is called a *Hilbert-Schmidt operator.* The class of all Hilbert-Schmidt operators, denoted by $\mathbf{HS}$, forms a Hilbert space with the inner product

$$\langle F, G\rangle := \int_0^\infty \int_0^\infty \text{trace } [f(t,\tau)'g(t,\tau)]dtd\tau.$$

Example 12.2.1 The usual state-space system

$$\begin{aligned} \dot{x} &= Ax + Bu, \quad x(0) = 0 \\ y &= Cx + Du, \end{aligned}$$

with A stable, defines a linear transformation $y = Fu$ with impulse response

$$f(t,\tau) = D\delta(t-\tau) + Ce^{(t-\tau)A}B1(t-\tau).$$

Observe that $F \in \mathsf{HS}$ iff $D = 0$.

The above Hilbert-Schmidt operator is in terms of inputs and outputs defined on all of $\mathbb{R}_+$, but the concept applies in a similar way to signals defined on intervals of $\mathbb{R}_+$. For example, we would say that a continuous-time linear transformation F mapping $\mathcal{L}_2((a,b),\mathbb{R}^m)$ to $\mathcal{L}_2((c,d),\mathbb{R}^p)$ and described by the integral equation

$$(Fu)(t) = \int_a^b f(t,\tau)u(\tau)d\tau, \quad c < t < d$$

is a Hilbert-Schmidt operator if

$$\int_a^b \int_c^d \text{trace } [f(t,\tau)'f(t,\tau)]dtd\tau < \infty.$$

Example 12.2.2 Continuing Example 12.2.1, consider the compression F to $\mathcal{K}$:

$$\underline{D}: \mathcal{K} \to \mathcal{K}, \ (\underline{D}u)(t) = Du(t) + \int_0^t Ce^{(t-\tau)A}Bu(\tau)d\tau.$$

This arises when we lift the system; see Section 10.2. Here the time interval for input and output is $[0,h)$. It is clear that $\underline{D}$ is Hilbert-Schmidt iff $D = 0$. If this is true, the impulse response for $\underline{D}$ is

$$\underline{d}(t,\tau) = Ce^{(t-\tau)A}B1(t-\tau)$$

and the Hilbert-Schmidt norm for $\underline{D}$ can be found via

$$\begin{aligned} \|\underline{D}\|_{\mathsf{HS}}^2 &= \int_0^h \int_0^h \text{trace } \left[B'e^{(t-\tau)A'}C'Ce^{(t-\tau)A}B1(t-\tau)\right] d\tau\, dt \\ &= \int_0^h \int_0^t \text{trace } \left[B'e^{(t-\tau)A'}C'Ce^{(t-\tau)A}B\right] d\tau\, dt \\ &= \text{trace } \left(B' \int_0^h \int_0^t e^{\tau A'}C'Ce^{\tau A}d\tau dt B\right). \end{aligned}$$

Since the time interval is finite, the matrix A need not be stable for $\underline{D}$ to be well-defined.

Let us return to F in (12.5). It is causal iff $f(t,\tau) = 0$ whenever $t < \tau$ and is h-periodic iff $f(t+h, \tau+h) = f(t,\tau)$ (Exercise 12.3). Let us assume that F is causal and h-periodic. Then

$$(Fu)(t) = \int_0^t f(t,\tau)u(\tau)d\tau$$

and F is completely characterized by $f(t,\tau)$ for $0 \leq t < \infty$ and $0 \leq \tau < h$. It is not hard to check that the generalized $\mathcal{H}_2$ measure for F, denoted $J(F)$, equals

$$J(F) = \left\{ \int_0^h \int_0^\infty \text{trace}\ [f(t,\tau)' f(t,\tau)] dt d\tau \right\}^{1/2} \qquad (12.7)$$

(Exercise 12.4). The right-hand side is the Hilbert-Schmidt norm when F is considered as mapping $\mathcal{L}_2([0,h), \mathbb{R}^m)$ to $\mathcal{L}_2(\mathbb{R}_+, \mathbb{R}^p)$.

Lifting Periodic Operators

Let F be a linear, h-periodic, causal system mapping $\mathcal{L}_2(\mathbb{R}_+)$ to $\mathcal{L}_2(\mathbb{R}_+)$:

$$y = Fu \Longleftrightarrow y(t) = \int_0^t f(t,\tau)u(\tau)d\tau, \quad t \geq 0.$$

Assume that

$$\int_0^h \int_0^\infty \text{trace}\ [f(t,\tau)' f(t,\tau)] dt d\tau < \infty. \qquad (12.8)$$

Lift the input and output: $\underline{u} = Lu, \underline{y} = Ly$. Then the lifted operator $\underline{F} := LFL^{-1}$ maps $\ell_2(\mathbb{Z}_+, \mathcal{K})$ to $\ell_2(\mathbb{Z}_+, \mathcal{K})$. It is not difficult to derive as in Section 10.2 that the matrix for $\underline{F}$ is

$$[\underline{F}] = \begin{bmatrix} \underline{f}_0 & 0 & 0 & \cdots \\ \underline{f}_1 & \underline{f}_0 & 0 & \cdots \\ \vdots & \vdots & \vdots & \end{bmatrix},$$

where $\underline{f}_k : \mathcal{K} \to \mathcal{K}$ $(k \geq 0)$ is given by

$$(\underline{f}_k u)(t) = \int_0^h f(t+kh, \tau)u(\tau)d\tau.$$

Thus we have the convolution equation

$$\underline{y}_k = \sum_{l=0}^{k} \underline{f}_{k-l}\underline{u}_l, \quad k \geq 0.$$

It follows from (12.8) that every $\underline{f}_k \in \mathbf{HS}$, that is,

$$\int_0^h \int_0^h \text{trace } [f(t+kh,\tau)'f(t+kh,\tau)]dtd\tau < \infty.$$

Thus $\underline{F}$ can be represented by the Hilbert-Schmidt operator sequence $\underline{f} := \{\underline{f}_0, \underline{f}_1, \cdots\}$. The norm of $\underline{f}$, denoted $\|\underline{f}\|_2$, is defined as

$$\|\underline{f}\|_2 := \left(\sum_{k=0}^{\infty} \|\underline{f}_k\|_{\mathbf{HS}}^2\right)^{1/2}.$$

To recap, associated with F is a sequence $\underline{f}$ of Hilbert-Schmidt operators. This association is important because of the next result.

Lemma 12.2.1 $J(F) = \|\underline{f}\|_2$.

Proof

$$\begin{aligned}
\|\underline{f}\|_2^2 &= \sum_{k=0}^{\infty} \|\underline{f}_k\|_{\mathbf{HS}}^2 \\
&= \sum_{k=0}^{\infty} \int_0^h \int_0^h \text{trace } [f(t+kh,\tau)'f(t+kh,\tau)]dtd\tau \\
&= \int_0^h \left[\sum_{k=0}^{\infty} \int_0^h \text{trace } [f(t+kh,\tau)'f(t+kh,\tau)]dt\right] d\tau \\
&= \int_0^h \int_0^{\infty} \text{trace } [f(t,\tau)'f(t,\tau)]dtd\tau \\
&= J(F)^2
\end{aligned}$$

■

It is a fact that the set of all Hilbert-Schmidt operator sequences $\underline{f}$ with $\|\underline{f}\|_2 < \infty$ forms a Hilbert space, denoted $\ell_2(\mathbb{Z}_+, \mathbf{HS})$, with the inner product

$$\langle \underline{f}, \underline{g} \rangle = \sum_{k=0}^{\infty} \langle \underline{f}_k, \underline{g}_k \rangle,$$

the right-hand inner product being on $\mathbf{HS}$.

Operator-Valued Transfer Functions

If F is h-periodic, the lifted $\underline{F}$ is apparently LTI in discrete time. Thus we can associate an operator-valued transfer function with $\underline{F}$:

$$\hat{\underline{f}}(\lambda) := \sum_{k=0}^{\infty} \underline{f}_k \lambda^k.$$

Defining the λ-transform for the input

$$\hat{\underline{u}}(\lambda) = \sum_{k=0}^{\infty} \underline{u}_k \lambda^k$$

and similarly for the output $\hat{\underline{y}}(\lambda)$, we get

$$\hat{\underline{y}}(\lambda) = \hat{\underline{f}}(\lambda)\hat{\underline{u}}(\lambda).$$

Here, for every λ in their respective regions of convergence, $\hat{\underline{u}}(\lambda) \in \mathcal{K}$, a function on $[0, h)$, and $\hat{\underline{f}}(\lambda) \in \mathbf{HS}$, a Hilbert-Schmidt operator.

The Hardy space $\mathcal{H}_2(\mathbf{D}, \mathbf{HS})$ consists of all operator-valued functions $\hat{\underline{f}}(\lambda)$ which are analytic in $\mathbf{D}$ (with power series expansion in the neighborhood of every $\lambda_0 \in \mathbf{D}$), have boundary functions on $\partial\mathbf{D}$, and satisfy

$$\left[\frac{1}{2\pi}\int_0^{2\pi} \|\hat{\underline{f}}(e^{j\theta})\|_{\mathbf{HS}}^2 d\theta\right]^{1/2} < \infty.$$

The left-hand side is defined to be the norm on $\mathcal{H}_2(\mathbf{D}, \mathbf{HS})$, denoted $\|\hat{\underline{f}}\|_2$. Moreover, $\mathcal{H}_2(\mathbf{D}, \mathbf{H})$ is a Hilbert space with the inner product

$$\langle \hat{\underline{f}}, \hat{\underline{g}} \rangle = \frac{1}{2\pi}\int_0^{2\pi} \langle \hat{\underline{f}}(e^{j\theta}), \hat{\underline{g}}(e^{j\theta}) \rangle d\theta.$$

The right-hand inner product is on $\mathbf{HS}$.

Recall in the matrix case that the $\mathcal{H}_2(\mathbf{D})$-norm is

$$\|\hat{f}\|_2 = \left\{\frac{1}{2\pi}\int_0^{2\pi} \text{trace } [\hat{f}(e^{j\theta})^* \hat{f}(e^{j\theta})] d\theta\right\}^{1/2}.$$

The integrand is the square of the trace norm for matrices, which can be denoted by $\|\hat{f}(e^{j\theta})\|_{\mathbf{HS}}^2$. Thus we get

$$\|\hat{f}\|_2 = \left[\frac{1}{2\pi}\int_0^{2\pi} \|\hat{f}(e^{j\theta})\|_{\mathbf{HS}}^2 d\theta\right]^{1/2}.$$

Therefore the norm on $\mathcal{H}_2(\mathbf{D}, \mathbf{HS})$ is a generalization of the norm on $\mathcal{H}_2(\mathbf{D})$ by replacing the trace norm by the Hilbert-Schmidt norm.

As one may expect, the λ-transformation should preserve the norm from $\ell_2(\mathbb{Z}_+, \mathbb{HS})$ to $\mathcal{H}_2(\mathbb{D}, \mathbb{H})$. This can be verified as follows. For $\hat{\underline{f}} \in \mathcal{H}_2(\mathbb{D}, \mathbb{HS})$,

$$\begin{aligned} \|\hat{\underline{f}}\|_2^2 &= \frac{1}{2\pi}\int_0^{2\pi} \langle \hat{\underline{f}}(e^{j\theta}), \hat{\underline{f}}(e^{j\theta})\rangle d\theta \\ &= \frac{1}{2\pi}\int_0^{2\pi} \langle \sum_{k=0}^{\infty} \underline{f}_k e^{jk\theta}, \sum_{l=0}^{\infty} \underline{f}_l e^{jl\theta}\rangle d\theta \\ &= \frac{1}{2\pi}\sum_{k=0}^{\infty}\sum_{l=0}^{\infty} \langle \underline{f}_k, \underline{f}_l\rangle \int_0^{2\pi} e^{j(l-k)\theta} d\theta \\ &= \sum_{k=0}^{\infty} \langle \underline{f}_k, \underline{f}_k\rangle, \end{aligned}$$

the last equality following from the fact that

$$\frac{1}{2\pi}\int_0^{2\pi} e^{j(k-l)\theta} d\theta = \begin{cases} 1, & k = l \\ 0, & k \neq l. \end{cases}$$

So $\|\hat{\underline{f}}\|_2 = \|\underline{f}\|_2$.

A stronger result, which is quite similar to the matrix case in Section 4.5, can be stated:

Theorem 12.2.1 *The λ-transformation is an isomorphism from $\ell_2(\mathbb{Z}_+, \mathbb{HS})$ onto $\mathcal{H}_2(\mathbb{D}, \mathbb{HS})$.*

In summary, the following quantities are all equal:

$$\begin{aligned} J(F) &= \left(\int_0^h \sum_i \|F\delta_\tau e_i\|_2^2 d\tau\right)^{1/2} \\ \|F\|_2 &= \left\{\int_0^h \int_0^\infty \text{trace } [f(t,\tau)' f(t,\tau)] dt d\tau\right\}^{1/2} \\ \|\underline{f}\|_2 &= \left(\sum_{k=0}^{\infty} \|\underline{f}_k\|_{\mathbb{HS}}^2\right)^{1/2} \\ \|\hat{\underline{f}}\|_2 &= \left[\frac{1}{2\pi}\int_0^{2\pi} \|\hat{\underline{f}}(e^{j\theta})\|_{\mathbb{HS}}^2 d\theta\right]^{1/2}. \end{aligned}$$

12.3 Generalized $\mathcal{H}_2$ SD Problem

With reference to Figure 12.2, we pose the $\mathcal{H}_2$-optimal SD control problem: Design a K_d to provide internal stability and minimize the generalized $\mathcal{H}_2$ measure discussed in the preceding section.

Defining the lifted plant

$$\underline{G} = \begin{bmatrix} \underline{G}_{11} & \underline{G}_{12} \\ \underline{G}_{21} & \underline{G}_{22} \end{bmatrix} = \begin{bmatrix} LG_{11}L^{-1} & LG_{12}H \\ SG_{21}L^{-1} & G_{22d} \end{bmatrix},$$

we arrive at the lifted configuration in Figure 12.12. Note that the (2,2)

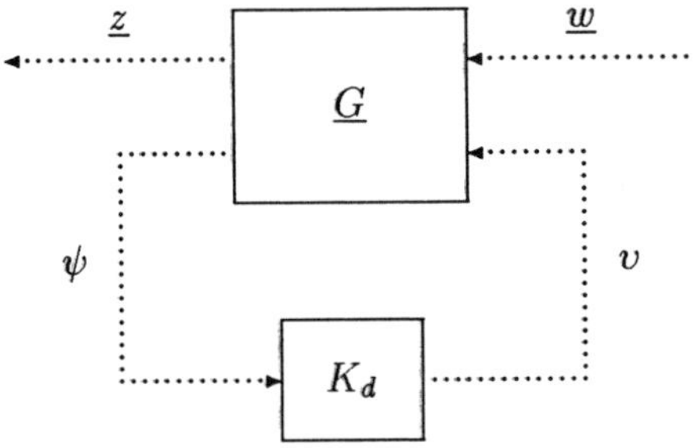

Figure 12.12: The lifted system.

block in $\underline{G}$ is exactly the discretized G_{22}. The lifted system is now LTI and therefore has an operator-valued transfer function. To simplify notation, let T be $T_{zw} : w \mapsto z$ in Figure 12.2. Then the closed-loop map $\underline{w} \mapsto \underline{z}$ in Figure 12.12 is LTL^{-1}, denoted $\underline{T}$.

In this way, we can recast the SD $\mathcal{H}_2$ problem in the lifted system of Figure 12.12: Design K_d to give internal stability and minimize the norm of $\hat{\underline{t}}$ in $\mathcal{H}_2(\mathbb{D}, \mathsf{HS})$.

This latter problem cannot be solved directly via our known techniques because it involves infinite-dimensional input and output spaces. Our goal in this section is to reduce it to the standard discrete-time problem whose solution was treated in Chapter 6.

Bring in a state model for G:

$$\hat{g}(s) = \left[\begin{array}{c|cc} A & B_1 & B_2 \\ \hline C_1 & 0 & D_{12} \\ C_2 & 0 & 0 \end{array}\right].$$

Based on this, a state model for $\underline{G}$ was given in Section 10.3. Using transfer functions, we can write

$$\hat{\underline{g}}(\lambda) = \left[\begin{array}{c|cc} A_d & \underline{B}_1 & B_{2d} \\ \hline \underline{C}_1 & \underline{D}_{11} & \underline{D}_{12} \\ C_2 & 0 & 0 \end{array}\right],$$

where A_d and B_{2d} are as usual; the operator-valued entries are:

$$\underline{B}_1 : \ \mathcal{K} \to \mathcal{E}, \quad \underline{B}_1 w = \int_0^h e^{(h-\tau)A} B_1 w(\tau) d\tau$$

$$\underline{C}_1 : \ \mathcal{E} \to \mathcal{K}, \quad (\underline{C}_1 x)(t) = C_1 e^{tA} x$$

$$\underline{D}_{11}: \ \mathcal{K} \to \mathcal{K}, \quad (\underline{D}_{11}w)(t) = C_1 \int_0^t e^{(t-\tau)A} B_1 w(\tau) d\tau$$

$$\underline{D}_{12}: \ \mathcal{E} \to \mathcal{K}, \quad (\underline{D}_{12}v)(t) = D_{12}v + C_1 \int_0^t e^{(t-\tau)A} d\tau B_2 v.$$

With this setup we associate the standard discrete-time system shown in Figure 12.13, where the matrices in the plant $G_{eq,d}$ are defined as follows: First, B_{1d} is obtained by the equation

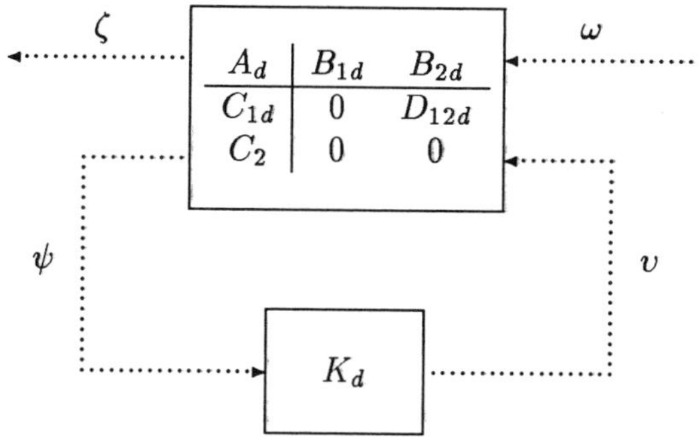

Figure 12.13: The associated discrete-time system.

$$B_{1d}B_{1d}' = \int_0^h e^{tA} B_1 B_1' e^{tA'} dt.$$

And second, C_{1d} and D_{12d} are found in the same way as in Section 12.1, that is,

$$\begin{bmatrix} C_{1d} & D_{12d} \end{bmatrix}' \begin{bmatrix} C_{1d} & D_{12d} \end{bmatrix} = \int_0^h e^{t\underline{A}'} \begin{bmatrix} C_1 & D_{12} \end{bmatrix}' \begin{bmatrix} C_1 & D_{12} \end{bmatrix} e^{t\underline{A}} dt,$$

where $\underline{A}$ is the square matrix

$$\underline{A} = \begin{bmatrix} A & B_2 \\ 0 & 0 \end{bmatrix}.$$

Note the following relations:

$$B_{1d}B_{1d}' = \underline{B}_1 \underline{B}_1^*$$

$$\begin{bmatrix} C_{1d} & D_{12d} \end{bmatrix}' \begin{bmatrix} C_{1d} & D_{12d} \end{bmatrix} = \begin{bmatrix} \underline{C}_1 & \underline{D}_{12} \end{bmatrix}^* \begin{bmatrix} \underline{C}_1 & \underline{D}_{12} \end{bmatrix}.$$

Letting $T_{\zeta\omega}$ be the closed-loop system $\omega \mapsto \zeta$ in Figure 12.13, we are ready to state the main result.

Theorem 12.3.1 *The $\mathcal{H}_2$-norms of the two systems in Figures 12.12 and 12.13 satisfy*

$$\|\hat{\underline{t}}\|_2^2 = \|\underline{D}_{11}\|_{\mathsf{HS}}^2 + \|\hat{t}_{\zeta w}\|_2^2.$$

Note that the plants $\underline{G}$ and G_d share the same (2,2) block; thus the two closed-loop systems share internal stability. The theorem implies that the lifted problem of minimizing the $\mathcal{H}_2(\mathbb{D}, \mathsf{HS})$-norm of $\hat{\underline{t}}$ over the class of stabilizing controllers is equivalent to the discrete problem of minimizing the $\mathcal{H}_2(\mathbb{D})$-norm of $\hat{t}_{\zeta w}$ over the same class of controllers, that is,

$$\min_{K_d} \|\hat{\underline{t}}\|_2^2 = \|\underline{D}_{11}\|_{\mathsf{HS}}^2 + \min_{K_d} \|\hat{t}_{\zeta w}\|_2^2.$$

Proof of Theorem 12.3.1 The closed-loop transfer function in Figure 12.12 is

$$\hat{\underline{t}} = \hat{\underline{g}}_{11} + \hat{\underline{g}}_{12}(I - \hat{k}_d \hat{g}_{d22})^{-1} \hat{k}_d \hat{\underline{g}}_{21},$$

where $\hat{\underline{g}}_{11}, \hat{\underline{g}}_{12}, \hat{\underline{g}}_{21}$ are all operator-valued:

$$\hat{\underline{g}}_{11}(\lambda) = \left[\begin{array}{c|c} A_d & \underline{B}_1 \\ \hline \underline{C}_1 & \underline{D}_{11} \end{array}\right], \quad \hat{\underline{g}}_{12}(\lambda) = \left[\begin{array}{c|c} A_d & B_{2d} \\ \hline \underline{C}_1 & \underline{D}_{12} \end{array}\right], \quad \hat{\underline{g}}_{21}(\lambda) = \left[\begin{array}{c|c} A_d & \underline{B}_1 \\ \hline C_2 & 0 \end{array}\right].$$

Define the three matrix-valued transfer functions

$$\hat{\bar{g}}_{11}(\lambda) = \left[\begin{array}{c|c} A_d & I \\ \hline I & 0 \end{array}\right], \quad \hat{\bar{g}}_{12}(\lambda) = \left[\begin{array}{c|c} A_d & B_{2d} \\ \hline I & 0 \\ 0 & I \end{array}\right], \quad \hat{\bar{g}}_{21}(\lambda) = \left[\begin{array}{c|c} A_d & I \\ \hline C_2 & 0 \end{array}\right]$$

to get

$$\hat{\underline{t}} = \underline{D}_{11} + \left[\begin{array}{cc} \underline{C}_1 & \underline{D}_{12} \end{array}\right] \left\{ \left[\begin{array}{c} \hat{\bar{g}}_{11} \\ 0 \end{array}\right] + \hat{\bar{g}}_{12}(I - \hat{k}_d \hat{g}_{22d})^{-1} \hat{k}_d \hat{\bar{g}}_{21} \right\} \underline{B}_1. \quad (12.9)$$

Let $\hat{q}$ be the quantity in the curly braces; $\hat{q}(\lambda)$ is a matrix-valued function. Since $\hat{q}(0) = 0$ ($\hat{\bar{g}}_{11}$ and $\hat{\bar{g}}_{21}$ have this property), it follows that the two functions on the right-hand side of (12.9) are orthogonal, so

$$\|\hat{\underline{t}}\|_2^2 = \|\underline{D}_{11}\|_{\mathsf{HS}}^2 + \|\left[\begin{array}{cc} \underline{C}_1 & \underline{D}_{12} \end{array}\right] \hat{q} \underline{B}_1\|_2^2.$$

The second norm on the right is an $\mathcal{H}_2(\mathbb{D}, \mathsf{HS})$-norm; by definition this is

$$\frac{1}{2\pi} \int_0^{2\pi} \|\left[\begin{array}{cc} \underline{C}_1 & \underline{D}_{21} \end{array}\right] \hat{q}(e^{j\theta}) \underline{B}_1\|_{\mathsf{HS}}^2 d\theta. \quad (12.10)$$

Fix θ. Then

$$F := \left[\begin{array}{cc} \underline{C}_1 & \underline{D}_{21} \end{array}\right] \hat{q}(e^{j\theta}) \underline{B}_1.$$

is a Hilbert-Schmidt operator; its impulse response is given by

$$f(t,\tau) = \begin{bmatrix} C_1 & D_{12} \end{bmatrix} e^{t\underline{A}} \hat{q}(e^{j\theta}) e^{(h-\tau)A} B_1 \tag{12.11}$$

(Exercise 12.6). Thus

$$\begin{aligned}
\|F\|_{\mathbf{HS}}^2 &= \text{trace} \int_0^h \int_0^h f(t,\tau)^* f(t,\tau)\, dt d\tau \\
&= \text{trace} \int_0^h B_1' e^{(h-\tau)A'} \hat{q}(e^{j\theta})^* \begin{bmatrix} C_{1d}' \\ D_{12d}' \end{bmatrix} \times \\
&\qquad \begin{bmatrix} C_{1d} & D_{12d} \end{bmatrix} \hat{q}(e^{j\theta}) e^{(h-\tau)A} B_1 d\tau \\
&= \text{trace} \left(\int_0^h e^{(h-\tau)A} B_1 B_1' e^{(h-\tau)A'} d\tau \right) \hat{q}(e^{j\theta})^* \times \\
&\qquad \begin{bmatrix} C_{1d}' \\ D_{12d}' \end{bmatrix} \begin{bmatrix} C_{1d} & D_{12d} \end{bmatrix} \hat{q}(e^{j\theta}) \\
&= \text{trace}\, B_{1d} B_{1d}' \hat{q}(e^{j\theta})^* \begin{bmatrix} C_{1d}' \\ D_{12d}' \end{bmatrix} \begin{bmatrix} C_{1d} & D_{12d} \end{bmatrix} \hat{q}(e^{j\theta}) \\
&= \text{trace}\, B_{1d}' \hat{q}(e^{j\theta})^* \begin{bmatrix} C_{1d}' \\ D_{12d}' \end{bmatrix} \begin{bmatrix} C_{1d} & D_{12d} \end{bmatrix} \hat{q}(e^{j\theta}) B_{1d}.
\end{aligned}$$

So the quantity in (12.10) is the $\mathcal{H}_2(\mathbb{D})$-norm of the function

$$\begin{bmatrix} C_{1d} & D_{12d} \end{bmatrix} \hat{q}(e^{j\theta}) B_{1d},$$

which simplifies to $\hat{t}_{\zeta\omega}$. ■

Comparing the equivalent discrete plant $G_{eq,d}$ with that in Section 12.1, we see that the matrices are the same except for one: In Section 12.1 we had B_1, but here we have B_{1d}. Thus the two problems are strongly related.

Let us now recap and summarize the design steps:

Step 1 Start with a state model for G:

$$\hat{g}(s) = \left[\begin{array}{c|cc} A & B_1 & B_2 \\ \hline C_1 & 0 & D_{12} \\ C_2 & 0 & 0 \end{array}\right].$$

Step 2 Compute A_d and B_{2d} via

$$(A_d, B_{2d}) = c2d(A, B_2, h).$$

Step 3 Compute

$$\begin{bmatrix} P_{11} & P_{12} \\ 0 & P_{22} \end{bmatrix} = \exp\left\{ h \begin{bmatrix} -A & B_1 B_1' \\ 0 & A' \end{bmatrix} \right\}.$$

Then compute B_{1d} (via Cholesky factorization) satisfying

$$B_{1d}B_{1d}' = P_{22}'P_{12}.$$

Step 4 Define the square matrices

$$\underline{A} = \begin{bmatrix} A & B_2 \\ 0 & 0 \end{bmatrix}, \quad Q = [\ C_1 \quad D_{12}\ ]'[\ C_1 \quad D_{12}\ ]$$

and compute

$$\begin{bmatrix} M_{11} & M_{12} \\ 0 & M_{22} \end{bmatrix} = \exp\left\{h \begin{bmatrix} -\underline{A}' & Q \\ 0 & \underline{A} \end{bmatrix}\right\}.$$

Then compute C_{1d} and D_{12d} (via Cholesky factorization) satisfying

$$[\ C_{1d} \quad D_{12d}\ ]'[\ C_{1d} \quad D_{12d}\ ] = M_{22}'M_{12}.$$

Step 5 Form the equivalent discrete-time system $G_{eq,d}$:

$$\hat{g}_{eq,d}(\lambda) = \left[\begin{array}{c|cc} A_d & B_{1d} & B_{2d} \\ \hline C_{1d} & 0 & D_{12d} \\ C_2 & 0 & 0 \end{array}\right].$$

and compute for $G_{eq,d}$ the $\mathcal{H}_2$-optimal controller $K_{d,opt}$ and the optimal performance $\min \|\hat{t}_{\zeta\omega}\|_2$.

Step 6 The $\mathcal{H}_2$-optimal K_d for the SD system is $K_{d,opt}$ and the optimal performance is

$$\min \|\underline{\hat{t}}\|_2 = \left(\|\underline{D}_{11}\|_{\mathbf{HS}}^2 + \min \|\hat{t}_{\zeta\omega}\|_2^2\right)^{1/2},$$

where the Hilbert-Schmidt norm of $\underline{D}_{11}$ is (Example 12.2.2)

$$\|\underline{D}_{11}\|_{\mathbf{HS}}^2 = \operatorname{trace}\left(B_1' \int_0^h \int_0^t e^{\tau A'} C_1' C_1 e^{\tau A} d\tau dt B_1\right).$$

12.4 Examples

This section presents a few examples to compare the two techniques presented in this chapter.

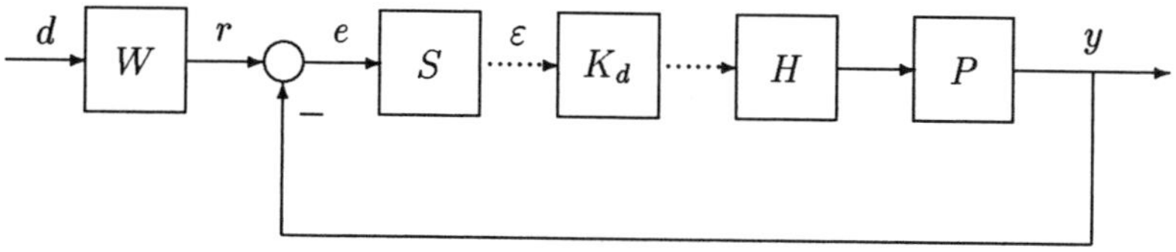

Figure 12.14: The SD tracking system.

Example 12.4.1 Consider again the SD step-tracking system studied in Example 12.1.1; it is redrawn in Figure 12.14. Here we assume no *a priori* knowledge of when the step input is applied, that is, the step input r is not assumed to be synchronized with the sampling operation. This means that the fictitious input d to the reference model W $[\hat{w}(s) = 1/s]$ is a shifted impulsive function $\delta_\tau(t) := \delta(t-\tau)$, where $\tau \in [0, h)$ is unknown. Let T_{ed} be the linear system $d \mapsto e$ in Figure 12.14. It makes sense now to minimize the time average of the quantity $\|T_{ed}\delta_\tau\|_2^2$, namely,

$$\int_0^h \|T_{ed}\delta_\tau\|_2^2 d\tau.$$

(Recall that in Example 12.1.1 we minimized $\|T_{ed}\delta_0\|_2^2$.) This is the generalized $\mathcal{H}_2$ measure introduced in Section 12.2. Putting things in the standard framework as in Example 12.1.1, we get the generalized plant

$$\hat{g}(s) = \begin{bmatrix} \hat{w}(s) & -\hat{p}(s) \\ \hat{w}(s) & -\hat{p}(s) \end{bmatrix} = \left[\begin{array}{c|cc} A & B_1 & B_2 \\ \hline C_1 & 0 & 0 \\ C_2 & 0 & 0 \end{array}\right],$$

where the matrices are given in Example 12.1.1. By Theorem 12.3.1 the equivalent discrete-time system $G_{eq,d}$ is

$$\hat{g}_{eq,d} = \left[\begin{array}{c|cc} A_d & B_{1d} & B_{2d} \\ \hline C_{1d} & 0 & D_{12d} \\ C_2 & 0 & 0 \end{array}\right].$$

All matrices are the same as in Example 12.1.1 except B_{1d}, which we compute via Step 3 of the procedure in the preceding section:

$$B_{1d}B_{1d}' = \begin{bmatrix} 1 & 0 & 0 \\ 0 & 0 & 0 \\ 0 & 0 & 0 \end{bmatrix}.$$

Thus

$$B_{1d} = B_1 = \begin{pmatrix} 1 \\ 0 \\ 0 \end{pmatrix}.$$

This means that the equivalent discrete-time system here is exactly the same as that in Example 12.1.1 and hence the optimal controller is the same too. It turns out that this is the case in general for the SD step-tracking setup; in other words, this observation is true regardless of what P and h we have (Exercise 12.7). So in the step-tracking case, the generalized $\mathcal{H}_2$ measure is essentially the same as the simpler measure introduced in Section 12.1.

Example 12.4.2 Figure 12.15 shows half the telerobot of Example 2.3.1, namely, the master side (see Figure 2.3). A human provides a force command,

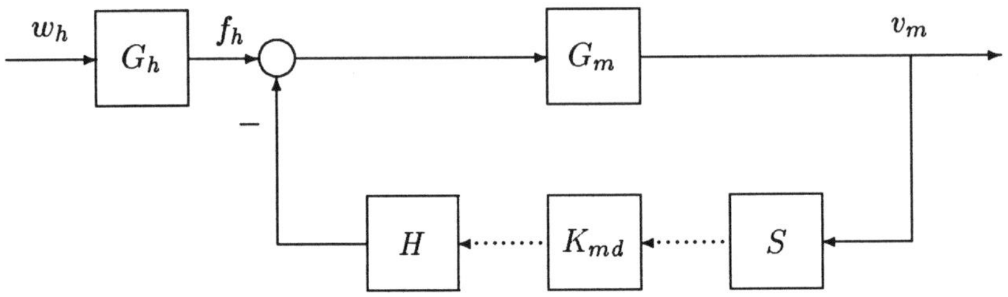

Figure 12.15: Master manipulator.

f_h, to a manipulator, G_m; the sampled-data controller, $HK_{md}S$, measures the velocity, v_m. Ideally, we want a desired compliance, say $v_m = f_h$ for simplicity. As in Example 6.5.1, the manipulator dynamics are taken as simple as possible, $\hat{g}_m(s) = 1/s$, and f_h is taken to be the triangular pulse

$$f_h(t) = \begin{cases} 2t, & 0 \le t \le 1 \\ -2t+4, & 1 \le t \le 2 \\ 0, & t > 2, \end{cases}$$

to mimic a ramp-up, ramp-down command. This can be approximated as the output of the model

$$\hat{g}_h(s) = 2 \Big/ \left(1 + \frac{s}{2} + \frac{s^2}{12}\right)^2$$

with input w_h the unit impulse.

Let us first design K_{md} by the simple method of Section 12.1, that is, K_{md} should minimize $\|f_h - v_m\|_2$ for $w_h = \delta$. The first step is to put the system into the standard form of Figure 12.2. Taking state models

$$\hat{g}_h(s) = \left[\begin{array}{c|c} A_h & B_h \\ \hline C_h & 0 \end{array}\right], \quad \hat{g}_m(s) = \left[\begin{array}{c|c} A_m & B_m \\ \hline C_m & 0 \end{array}\right],$$

we get the following state model for G:

$$\hat{g}(s) = \left[\begin{array}{c|cc} A & B_1 & B_2 \\ \hline C_1 & 0 & 0 \\ C_2 & 0 & 0 \end{array}\right] = \left[\begin{array}{cc|cc} A_h & 0 & B_h & 0 \\ B_m C_h & A_m & 0 & -B_m \\ \hline C_h & -C_m & 0 & 0 \\ 0 & C_m & 0 & 0 \end{array}\right].$$

Let us take $h = 1$, a value large enough to highlight the distinction between the two methods. The equivalent discrete-time generalized plant of Figure 12.4 is

$$\left[\begin{array}{c|cc} A_d & B_1 & B_{2d} \\ \hline C_{1d} & 0 & D_{12d} \\ C_2 & 0 & 0 \end{array}\right].$$

The corresponding discrete-time $\mathcal{H}_2$ problem is not regular because $D_{21} = 0$. It cannot be regularized by advancing the input ω by one time step because $C_2 B_1 = 0$ too. Instead, we shall regularize by perturbing to

$$\left[\begin{array}{c|ccc} A_d & B_1 & 0 & B_{2d} \\ \hline C_{1d} & 0 & 0 & D_{12d} \\ C_2 & 0 & \varepsilon & 0 \end{array}\right].$$

The design turns out to be relatively insensitive to small enough ε; we can take $\varepsilon = 0.01$. The problem can now be routinely solved by the formulas in Theorem 6.5.3.

Second, let us design K_{md} by the generalized method of Section 12.3; the criterion is $\|f_h - v_m\|_2^2$ averaged over τ for $w_h = \delta_\tau$. The equivalent discrete-time system is

$$\left[\begin{array}{c|cc} A_d & B_{1d} & B_{2d} \\ \hline C_{1d} & 0 & D_{12d} \\ C_2 & 0 & 0 \end{array}\right],$$

which can be regularized to

$$\left[\begin{array}{c|cc} A_d & A_d B_{1d} & B_{2d} \\ \hline C_{1d} & C_{1d} B_{1d} & D_{12d} \\ C_2 & C_2 B_{1d} & 0 \end{array}\right].$$

This satisfies all the required assumptions for Theorem 6.5.3.

Figure 12.16 shows simulated responses for the two designs: The solid line is $v_m(t)$ for the simple method, the dashed line is $v_m(t)$ for the generalized method, and the dotted line is $f_h(t)$ used in the simulation (this is the exact f_h, not the approximation used in the design). The two v_m-reponses are quite close. By contrast, Figure 12.17 shows simulated responses when the force input is applied some time during the first sampling period. The response of the generalized design is much better.

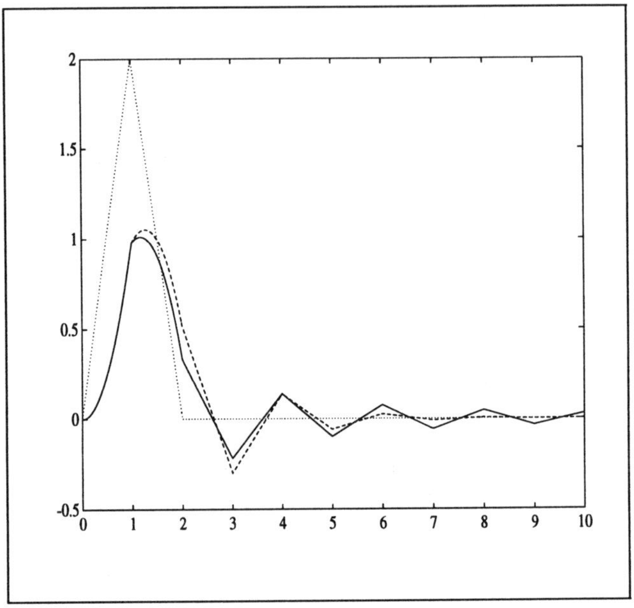

Figure 12.16: Manipulator velocity, v_m, for the simple design (solid) and the generalized design (dash); force input, f_h, (dot).

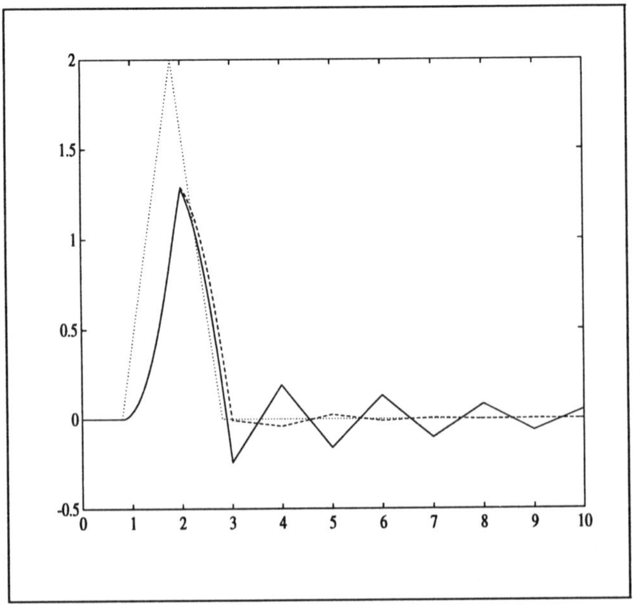

Figure 12.17: Manipulator velocity, v_m, for the simple design (solid) and the generalized design (dash); force input, f_h, (dot).

Exercises

12.1 Show that D_{12d} has full column-rank if the system with state matrices (A, B_2, C_1, D_{12}) is left-invertible, that is, only the trivial input produces the trivial output. A sufficient condition for this is that D_{12} has full column-rank.

12.2 Repeat the design in Example 12.1.1 for $h = 3$ and simulate the step response.

12.3 Let F be described by (12.5). Show that F is causal iff $f(t,\tau) = 0$ whenever $t < \tau$ and is h-periodic iff $f(t+h, \tau+h) = f(t,\tau)$.

12.4 Show (12.7).

12.5 Prove Lemma 12.2.1.

12.6 Derive the impulse response in (12.11) for the operator F.

12.7 For the step-tracking setup in Figure 12.14, let

$$\hat{g} = \begin{bmatrix} \hat{w} & -\hat{p} \\ \hat{w} & -\hat{p} \end{bmatrix} = \left[\begin{array}{c|cc} A & B_1 & B_2 \\ \hline C_1 & 0 & 0 \\ C_2 & 0 & 0 \end{array}\right].$$

Show that regardless of what P we have, the following relation is always true:

$$\int_0^h e^{tA} B_1 B_1' e^{tA'} dt = h B_1 B_1'.$$

Thus the problem of minimizing the generalized $\mathcal{H}_2$ measure of $d \mapsto e$ in Figure 12.14 is essentially the same as that of minimizing $\|e\|_2$ in Figure 12.9.

Notes and References

The idea of using continuous-time performance specs in SD design was first reflected in the work of Levis, Schlueter, and Athans [102] and Dorato and Levis [38] on LQR design with sampled state feedback; their technique of converting a SD problem into an equivalent discrete-time problem was later generalized in [27] to $\mathcal{H}_2$ design with SD dynamic feedback. Section 12.1 is based on [27]; however, the proof of Theorem 12.1.1 is rewritten using continuous lifting. Based on this performance criterion, it is interesting to note that linear time-varying control sometimes has advantage over LTI control [29].

The generalized $\mathcal{H}_2$ control problem was independently posed and solved by Khargonekar and Sivashankar [91] and Bamieh and Pearson [17]. The proof of Theorem 12.3.1 is adapted from [17]; an elementary proof, without resorting to the lifting technique, is given in [24]. There is also a sensible stochastic interpretation for the generalized $\mathcal{H}_2$ measure; for this see [17]. For more general setup to handle discrete-time exogenous inputs, see [91].

The materials on Hilbert-Schmidt operators in Section 12.2 can be found in many books on operator theory, e.g., [59]. Theorem 12.2.1 is from [135].

Chapter 13

$\mathcal{H}_\infty$-Optimal SD Control

In Chapter 7 we dealt with $\mathcal{H}_\infty$-optimal control for discrete-time FDLTI systems, where intersample behaviour was ignored in design. To continue our study of SD systems from a continuous-time viewpoint, we shall devote this chapter to $\mathcal{H}_\infty$-optimal control of SD systems.

The $\mathcal{H}_\infty$-norm of a continuous-time transfer function equals the $\mathcal{L}_2(\mathbb{R}_+)$-induced norm of the corresponding linear system. For SD systems, there exist no transfer functions in the normal sense in continuous time; thus we define the $\mathcal{H}_\infty$-norm of a SD system as the $\mathcal{L}_2(\mathbb{R}_+)$-induced norm. Using continuous signals, this definition captures the behaviour between samples.

Throughout the chapter we are concerned with the standard SD system in Figure 13.1. For simplicity, we write T for T_{zw}, the system from w to z in Figure 13.1. Two questions which are of primary interest to us are as follows:

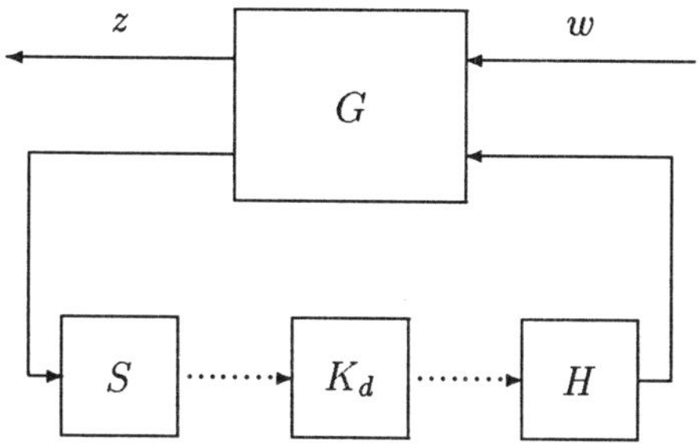

Figure 13.1: The standard SD system.

(1) Given G and K_d which provides internal stability, how to compute the

$\mathcal{L}_2(\mathbb{R}_+)$-induced norm of the system T?

(2) Given G, how to design K_d to minimize the $\mathcal{L}_2(\mathbb{R}_+)$-induced norm of T?

The first question is for $\mathcal{H}_\infty$ analysis and the second for synthesis. The questions are complicated because the systems in consideration are not LTI; however, by periodicity, they are still tractable using the continuous lifting technique.

The solutions to the analysis and synthesis problems are based on a process called $\mathcal{H}_\infty$ discretization: For $\gamma > 0$, construct an LTI discrete-time system $G_{eq,d}$ connected to K_d as in Figure 13.2; the two systems T and

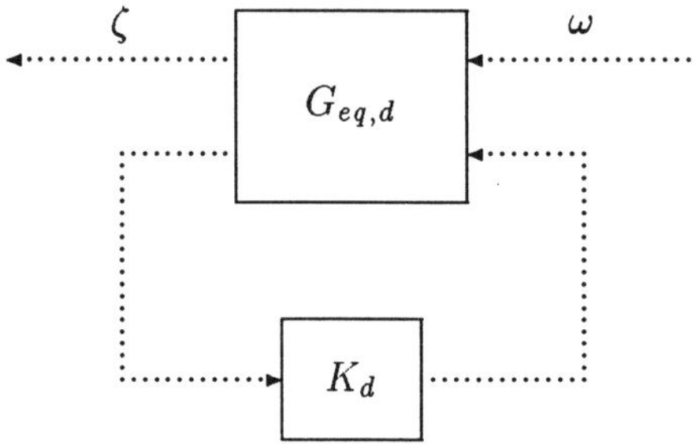

Figure 13.2: The equivalent discrete-time system.

$T_{eq,d} : \omega \mapsto \zeta$ in Figure 13.2, are *equivalent* in that $\|T\| < \gamma$ iff $\|T_{eq,d}\| < \gamma$, where the latter norm is $\ell_2(\mathbb{Z}_+)$-induced and, since $T_{eq,d}$ is LTI in discrete time, it equals the $\mathcal{H}_\infty(\mathbb{D})$-norm of the corresponding transfer function $\hat{t}_{eq,d}$. Thus the techniques in Chapter 7 are immediately applicable.

As will be seen, the $\mathcal{H}_\infty$ discretization process is not quite *exact* in the sense that $G_{eq,d}$ depends on γ. Recall that in Chapter 12 we got an exact equivalence between the SD $\mathcal{H}_2$ problem and the associated discrete-time $\mathcal{H}_2$ problem. The difference can perhaps be explained by the fact that $\mathcal{H}_\infty$ problems are in general considerably harder—even the simplest problem of computing the $\mathcal{H}_\infty(\mathbb{D})$-norm of a discrete-time transfer function requires a search on γ—see Corollary 7.1.1.

13.1 Frequency Response

We begin this chapter by extending the notion of frequency response to SD systems; it will then turn out that the natural notion for the $\mathcal{H}_\infty$-norm of a SD system is the maximum magnitude (norm) of the frequency response.

To motivate the SD case, consider a SISO, continuous-time, stable, LTI system F, with transfer function $\hat{f}(s)$. The frequency response, $\hat{f}(j\omega)$, can be generated experimentally as shown in the following diagram:

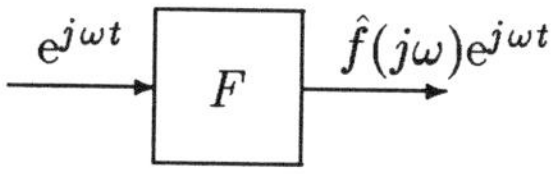

That is, the sinusoidal input $e^{j\omega t}$ (applied over $-\infty < t < \infty$) produces the sinusoidal output $\hat{f}(j\omega)e^{j\omega t}$. More generally, if we define the vector space of all sinusoids of frequency ω,

$$\mathcal{V}_\omega = \left\{v : v(t) = e^{j\omega t}u, u \in \mathbb{C}\right\},$$

then this space is invariant under F (that is, $v \in \mathcal{V}_\omega$ implies $Fv \in \mathcal{V}_\omega$) and moreover the following eigenvalue equation holds:

$$Fv = \hat{f}(j\omega)v, \quad v \in \mathcal{V}_\omega.$$

Let us extend the preceding to the MIMO case. Suppose $\hat{f}(j\omega)$ is $p \times m$. If $u \in \mathbb{C}^m$, then the sinusoidal input $v(t) = e^{j\omega t}u$ produces the sinusoidal output $y(t) = \hat{f}(j\omega)e^{j\omega t}u$. Defining

$$\mathcal{V}_\omega = \left\{v : v(t) = e^{j\omega t}u, u \text{ a complex vector}\right\},$$

we get again that $\mathcal{V}_\omega$ is invariant under F and that

$$Fv = \hat{f}(j\omega)v, \quad v \in \mathcal{V}_\omega.$$

What happens if we put a sinusoid into a SD system?

Example 13.1.1 Consider the simplest SD system, namely, sample-and-hold HS. Let $u(t) = e^{j\omega t}$ and $y = HSu$. As in (3.6), define

$$r(t) = \begin{cases} 1/h, & 0 \le t < h \\ 0, & \text{elsewhere.} \end{cases}$$

Then

$$\begin{aligned} y(t) &= h\sum_k e^{j\omega kh} r(t-kh) \\ &= e^{j\omega t} h \sum_k e^{-j\omega(t-kh)} r(t-kh). \end{aligned}$$

The function

$$w(t) := h\sum_k e^{-j\omega(t-kh)} r(t-kh)$$

is periodic, of period h, and $y(t) = e^{j\omega t}w(t)$. Thus the response of HS to a sinusoid is the product of the sinusoid times a periodic signal of period h. Perhaps a more concrete way to represent $y(t)$ is in terms of the Fourier series of $w(t)$:

$$w(t) = \sum_k e^{jk\omega_s t} a_k.$$

Then $y(t)$ has the representation

$$y(t) = \sum_k e^{j(\omega + k\omega_s)t} a_k;$$

that is, $y(t)$ contains harmonics at all frequencies $\omega + k\omega_s$, $k = 0, \pm 1, \pm 2, \ldots$.

The extension to the general case goes like this: Suppose F is a continuous-time, h-periodic, stable system—such as a SD system. Let $u_0 \in \mathcal{K} = \mathcal{L}_2[0, h)$ and let u denote the periodic extension of u_0; that is, $u_0(t)$ is defined on $[0, h)$, $u(t)$ is defined on all of $(-\infty, \infty)$, $u(t)$ is periodic of period h, and $u(t) = u_0(t)$ for $t \in [0, h)$. Let $\mathcal{P}$ denote the class of all such signals u; $\mathcal{P}$ is the periodic extension of $\mathcal{K}$. Conversely, from u in $\mathcal{P}$ we can recover u_0 in $\mathcal{K}$ as the *projection* of u onto $\mathcal{K}$, simply the restriction of u to the time interval $[0, h)$. Define the linear space

$$\mathcal{V}_\omega = \left\{ v : v(t) = e^{j\omega t} u(t), u \in \mathcal{P} \right\}.$$

In the lemma to follow, $\underline{\hat{f}}(\lambda)$ denotes the transfer function of the lifted system $\underline{F} = LFL^{-1}$.

Lemma 13.1.1 *Assume F is a continuous-time, h-periodic, stable system. Then $\mathcal{V}_\omega$ is invariant under F. Moreover, if $v \in \mathcal{V}_\omega$, $y = Fv$, and v_0 and y_0 denote the projections of v and y onto $\mathcal{K}$, then*

$$y_0 = \underline{\hat{f}}\left(e^{-j\omega h}\right) v_0.$$

Proof Let $u \in \mathcal{P}$, $v(t) = e^{j\omega t}u(t)$, and $y = Fv$. Let us lift v:

$$\underline{v} = \begin{bmatrix} \vdots \\ v_{-1} \\ \hline v_0 \\ v_1 \\ v_2 \\ \vdots \end{bmatrix}.$$

The zeroth component is

$$v_0(t) = v(t) = e^{j\omega t}u(t), \quad t \in [0,h).$$

Since u is h-periodic, the next component is

$$\begin{aligned} v_1(t) &= v(t+h), \quad t \in [0,h) \\ &= e^{j\omega(t+h)}u(t+h), \quad t \in [0,h) \\ &= e^{j\omega h}e^{j\omega t}u(t), \quad t \in [0,h) \\ &= e^{j\omega h}v_0(t), \quad t \in [0,h). \end{aligned}$$

And so on. Thus $v_k = e^{kj\omega h}v_0$. The equation $\underline{y} = \underline{Fv}$ can be written out in full as

$$\begin{bmatrix} \vdots \\ y_{-1} \\ \hline y_0 \\ y_1 \\ y_2 \\ \vdots \end{bmatrix} = \left[\begin{array}{c|cccc} & \vdots & \vdots & \vdots & \vdots \\ \cdots & \underline{f}(0) & 0 & 0 & 0 & \cdots \\ \hline \cdots & \underline{f}(1) & \underline{f}(0) & 0 & 0 & \cdots \\ \cdots & \underline{f}(2) & \underline{f}(1) & \underline{f}(0) & 0 & \cdots \\ & \vdots & \vdots & \vdots & \vdots \end{array}\right] \begin{bmatrix} \vdots \\ e^{-j\omega h}v_0 \\ \hline v_0 \\ e^{j\omega h}v_0 \\ e^{2j\omega h}v_0 \\ \vdots \end{bmatrix},$$

from which it follows that $\underline{y}$ has the form

$$\underline{y} = \begin{bmatrix} \vdots \\ e^{-j\omega h}y_0 \\ \hline y_0 \\ e^{j\omega h}y_0 \\ e^{2j\omega h}y_0 \\ \vdots \end{bmatrix}, \tag{13.1}$$

where

$$\begin{aligned} y_0 &= \sum_{k=0}^{\infty} \underline{f}(k)\left[e^{-kj\omega h}v_0\right] \\ &= \hat{\underline{f}}\left(e^{-j\omega h}\right)v_0. \end{aligned}$$

Equation 13.1 implies that $y \in \mathcal{V}_\omega$, thus showing that $\mathcal{V}_\omega$ is invariant under F. ■

The lemma justifies the definition that $\hat{\underline{f}}\left(e^{-j\omega h}\right)$ is the *frequency response* of F; it is an operator on $\mathcal{K}$. Since $\hat{\underline{f}}\left(e^{-j\omega h}\right)$ is a periodic function of ω of period ω_s, it makes sense to define it on the interval $[-\omega_N, \omega_N]$.

13.2 $\mathcal{H}_\infty$-Norm in the Frequency Domain

Recall from Section 7.1 that computing the $\mathcal{H}_\infty$-norm of an LTI discrete-time system is related to the eigenproblem of a symplectic pair. The goal of this and the next sections is to derive a similar result for the $\mathcal{H}_\infty$-norm of SD systems. In this section we focus on an alternative, frequency-domain expression of the $\mathcal{H}_\infty$-norm of SD systems.

For the standard SD setup in Figure 13.1, assume the generalized plant G and the discrete controller K_d are both FDLTI with stabilizable and detectable realizations:

$$\hat{g}(s) = \left[\begin{array}{c|cc} A & B_1 & B_2 \\ \hline C_1 & 0 & D_{12} \\ C_2 & 0 & 0 \end{array}\right] \tag{13.2}$$

$$\hat{k}_d(\lambda) = \left[\begin{array}{c|c} A_K & B_K \\ \hline C_K & D_K \end{array}\right]. \tag{13.3}$$

Here, we have taken $D_{21} = 0$ because the sampler must be lowpass filtered for $\|T\|$ to be finite. We have also assumed that $D_{11} = 0$; this is for a technical simplification—in the lifted system the operator $\underline{D}_{11}$ is *compact* iff $D_{11} = 0$.

Using continuous lifting, we lift T to get $\underline{T} := LTL^{-1}$; this is exactly the map $\underline{w} \mapsto \underline{z}$ in the lifted system of Figure 13.3, where $\underline{G}$ has the state model

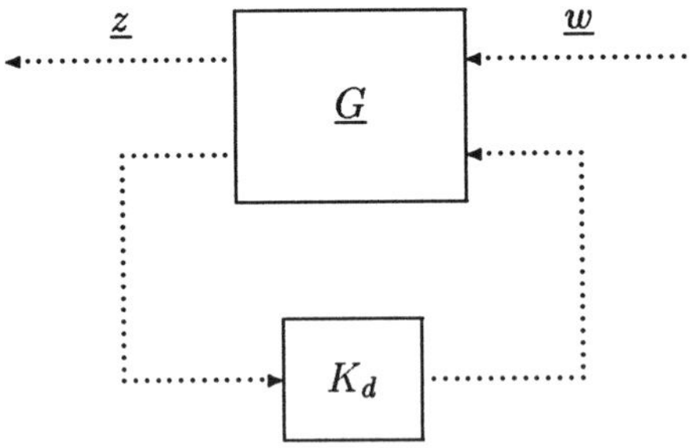

Figure 13.3: The lifted system.

$$\left[\begin{array}{c|cc} A_d & \underline{B}_1 & B_{2d} \\ \hline \underline{C}_1 & \underline{D}_{11} & \underline{D}_{12} \\ C_2 & 0 & 0 \end{array}\right]$$

and the operators are given in Section 10.3. The corresponding closed-loop state model for $\underline{T}$ is

$$\left[\begin{array}{c|c} A_{cl} & \underline{B}_{cl} \\ \hline \underline{C}_{cl} & \underline{D}_{11} \end{array}\right] := \left[\begin{array}{cc|c} A_d + B_{2d}D_KC_2 & B_{2d}C_K & \underline{B}_1 \\ B_KC_2 & A_K & 0 \\ \hline \underline{C}_1 + \underline{D}_{12}D_KC_2 & \underline{D}_{12}C_K & \underline{D}_{11} \end{array}\right]. \tag{13.4}$$

Note that A_{cl} is a matrix, while $\underline{B}_{cl}$, $\underline{C}_{cl}$, and $\underline{D}_{11}$ are operators on appropriate spaces.

Since L and L^{-1} are isometries, we have that $\|T\| = \|\underline{T}\|$, that is, the $\mathcal{L}_2(\mathbb{R}_+)$-induced norm of T equals the $\ell_2(\mathbb{Z}_+, \mathcal{K})$-induced norm of $\underline{T}$. The advantage of this lifting construction lies in that $\underline{T}$ is now LTI (albeit with infinite-dimensional input and output spaces) and thus, as in Section 12.2, we can associate an operator-valued transfer function with $\underline{T}$:

$$\hat{\underline{t}}(\lambda) = \underline{D}_{11} + \lambda \underline{C}_{cl}(I - \lambda A_{cl})^{-1}\underline{B}_{cl}.$$

If the SD system is internally stable, or equivalently, A_{cl} is stable, then for every $\lambda \in \mathbb{D}$, $\hat{\underline{t}}(\lambda)$ is a bounded operator from $\mathcal{K}$ to $\mathcal{K}$.

Let $\mathbb{B}$ denote the set of all bounded operators from $\mathcal{K}$ to $\mathcal{K}$ with the $\mathcal{K}$-induced norm. The Hardy space $\mathcal{H}_\infty(\mathbb{D}, \mathbb{B})$ consists of all operator-valued functions $\hat{\underline{f}} : \mathbb{D} \mapsto \mathbb{B}$ that are analytic in $\mathbb{D}$ (with power series expansion in the neighborhood of every $\lambda_0 \in \mathbb{D}$), have boundary functions on $\partial\mathbb{D}$, and satisfy

$$\sup_{0 \le \theta < 2\pi} \|\hat{\underline{f}}(e^{j\theta})\| < \infty.$$

The left-hand side is defined to be the norm on $\mathcal{H}_\infty(\mathbb{D}, \mathbb{B})$, denoted $\|\hat{\underline{f}}\|_\infty$. It is a fact that $\mathcal{H}_\infty(\mathbb{D}, \mathbb{B})$ is a Banach space. Note that since $\hat{\underline{f}}(e^{j\theta}) \in \mathbb{B}$, $\|\hat{\underline{f}}(e^{j\theta})\|$ is the $\mathcal{K}$-induced norm:

$$\|\hat{\underline{f}}(e^{j\theta})\| = \sup_{0 \ne w \in \mathcal{K}} \frac{\|\hat{\underline{f}}(e^{j\theta})w\|}{\|w\|}.$$

The space $\mathcal{H}_\infty(\mathbb{D}, \mathbb{B})$ can be regarded as a generalization of $\mathcal{H}_\infty(\mathbb{D})$ from matrix-valued functions to operator-valued functions. Recall that the norm on $\mathcal{H}_\infty(\mathbb{D})$ is defined by

$$\|\hat{f}\|_\infty = \sup_{0 \le \theta < 2\pi} \sigma_{max}\left[\hat{f}(e^{j\theta})\right].$$

The largest singular value of $\hat{f}(e^{j\theta})$ is the induced norm on Euclidean spaces; so if we regard $\hat{f}(e^{j\theta})$ as an operator on Euclidean spaces, we have

$$\|\hat{f}\|_\infty = \sup_{0 \le \theta < 2\pi} \|\hat{f}(e^{j\theta})\|,$$

just like the norm on $\mathcal{H}_\infty(\mathbb{D}, \mathbb{B})$.

Now let us return to the lifted SD system. It follows that if A_{cl} is stable, then the transfer function $\hat{\underline{t}}$ belongs to $\mathcal{H}_\infty(\mathbb{D}, \mathbb{B})$ and thus $\|\hat{\underline{t}}\|_\infty$ is well-defined. Does this frequency-domain norm relate to the time-domain norm of $\underline{T}$?

Theorem 13.2.1 *Assume the SD system is internally stable. Then* $\|\underline{T}\| = \|\hat{\underline{t}}\|_\infty$.

Thus the $\mathcal{L}_2(\mathbb{R}_+)$-induced norm of the SD system T equals the $\mathcal{H}_\infty(\mathbb{D},\mathbb{B})$-norm of the transfer function of the lifted system. This justifies our terminology of using $\mathcal{H}_\infty$ in the SD context.

Finally, we remark that SD systems form a subclass of periodic systems and the discussion in this section could be extended to the more general class of h-periodic systems.

13.3 $\mathcal{H}_\infty$-Norm Characterization

From the preceding section we saw that the $\mathcal{L}_2(\mathbb{R}_+)$-induced norm of T is equal to the $\mathcal{H}_\infty(\mathbb{D},\mathbb{B})$-norm of $\hat{\underline{t}}$:

$$\|\hat{\underline{t}}\|_\infty = \sup_{0\le\theta<2\pi} \|\hat{\underline{t}}(e^{j\theta})\|.$$

However, this does not immediately provide a way for computing the $\mathcal{H}_\infty$-norm since given a θ, it is not clear yet how to evaluate the norm of the operator $\hat{\underline{t}}(e^{j\theta})$ in $\mathbb{B}$. The goal of this section is to move one step forward, namely, to relate the $\mathcal{H}_\infty(\mathbb{D},\mathbb{B})$-norm of $\hat{\underline{t}}$ to some symplectic pair of matrices.

Fix $0 \le \theta < 2\pi$. The first question we address is the following: Is the norm of $\hat{\underline{t}}(e^{j\theta})$ equal to its largest singular value? The answer turns out to be yes. First, let us see that $\|\hat{\underline{t}}(e^{j\theta})\|$ is no less than the largest singular value of $\hat{\underline{t}}(e^{j\theta})$.

Let $\gamma > 0$ be a singular value of $\hat{\underline{t}}(e^{j\theta})$. Then γ^2 is an eigenvalue of $\hat{\underline{t}}(e^{j\theta})^*\hat{\underline{t}}(e^{j\theta})$, so there exists a nonzero u in $\mathcal{K}$ such that

$$\hat{\underline{t}}(e^{j\theta})^*\hat{\underline{t}}(e^{j\theta})u = \gamma^2 u.$$

It follows that

$$\begin{aligned}
\|\hat{\underline{t}}(e^{j\theta})u\|^2 &= \langle \hat{\underline{t}}(e^{j\theta})u, \hat{\underline{t}}(e^{j\theta})u\rangle \\
&= \langle \hat{\underline{t}}(e^{j\theta})^*\hat{\underline{t}}(e^{j\theta})u, u\rangle \\
&= \langle \gamma^2 u, u\rangle \\
&= \gamma^2\|u\|^2.
\end{aligned}$$

This implies that $\|\hat{\underline{t}}(e^{j\theta})\| \ge \gamma$.

Recall that

$$\hat{\underline{t}}(e^{j\theta}) = \underline{D}_{11} + e^{j\theta}\underline{C}_{cl}(I - e^{j\theta}A_{cl})^{-1}\underline{B}_{cl}.$$

Since $\underline{D}_{11}$ is compact ($D_{11} = 0$) and A_{cl} is a matrix, the operator $\hat{\underline{t}}(e^{j\theta})$ is compact too. It is a fact that for any compact operator, its norm equals its largest singular value. Thus the norm of $\hat{\underline{t}}(e^{j\theta})$ equals indeed its largest singular value. This fact gives a potential way to compute the norm via computing the singular values.

Notice that by the maximum modulus theorem,

$$\|\hat{\underline{t}}\|_\infty \geq \|\hat{\underline{t}}(0)\| = \|\underline{D}_{11}\|.$$

Thus to compute $\|\hat{\underline{t}}\|_\infty$, we just need to search over those singular values of $\hat{\underline{t}}(e^{j\theta})$ which are greater than $\|\underline{D}_{11}\|$.

For γ a positive number, introduce the symplectic pair (S_l, S_r)

$$S_l = \begin{bmatrix} A_{cl} + \underline{B}_{cl}\underline{D}_{11}(\gamma^2 - \underline{D}_{11}\underline{D}_{11}^*)^{-1}\underline{C}_{cl} & 0 \\ -\gamma\underline{C}_{cl}^*(\gamma^2 - \underline{D}_{11}\underline{D}_{11}^*)^{-1}\underline{C}_{cl} & I \end{bmatrix}$$

$$S_r = \begin{bmatrix} I & -\gamma\underline{B}_{cl}(\gamma^2 - \underline{D}_{11}^*\underline{D}_{11})^{-1}\underline{B}_{cl}^* \\ 0 & [A_{cl} + \underline{B}_{cl}\underline{D}_{11}(\gamma^2 - \underline{D}_{11}\underline{D}_{11}^*)^{-1}\underline{C}_{cl}]^* \end{bmatrix}.$$

The pair (S_l, S_r) is associated with the transfer function $\hat{\underline{t}}$ in (13.4). The operators $\gamma^2 - \underline{D}_{11}\underline{D}_{11}^*$ and $\gamma^2 - \underline{D}_{11}^*\underline{D}_{11}$ are invertible over $\mathbb{B}$ if $\gamma > \|\underline{D}_{11}\|$. Examining the domains and co-domains of the operator compositions, we see that S_l and S_r are in fact matrices, not operators.

Theorem 13.3.1 *Assume $\gamma > \|\underline{D}_{11}\|$ and $0 \leq \theta < 2\pi$. Then γ is a singular value of $\hat{\underline{t}}(e^{j\theta})$ iff $e^{-j\theta}$ is an eigenvalue of (S_l, S_r).*

The proof of the theorem is exactly the same as that of Theorem 7.1.1, with the simple modification of replacing matrix transposes by operator adjoints where appropriate.

Similar to Corollary 7.1.1, the following is a consequence of Theorem 13.3.1.

Corollary 13.3.1 *Let γ_{max} be the maximum γ such that (S_l, S_r) has an eigenvalue on the unit circle. Then*

$$\|\hat{\underline{t}}\|_\infty = \max\{\|\underline{D}_{11}\|, \gamma_{max}\}.$$

Starting from the computation of $\|\underline{D}_{11}\|$ (Section 13.5) and the matrices in (S_l, S_r), we could give a procedure to compute $\|\hat{\underline{t}}\|_\infty$ based on this corollary. However, we shall not pursue this direction because in the next section we shall associate to the SD system a discrete-time system with finite-dimensional input and output spaces from which we can compute $\|\hat{\underline{t}}\|_\infty$.

13.4 $\mathcal{H}_\infty$ Discretization of SD Systems

Continuing with our discussion and notation of the preceding sections, in this section we describe a process, called $\mathcal{H}_\infty$ discretization, which associates a usual discrete-time system to the SD system in Figure 13.1; both systems satisfy the same $\mathcal{H}_\infty$-norm bound.

Fix $\gamma > \|\underline{D}_{11}\|$. From Corollary 13.3.1, the $\mathcal{H}_\infty(\mathbb{D}, \mathbb{B})$-norm condition $\|\hat{\underline{t}}\|_\infty < \gamma$ can be characterized by the eigenvalues of the symplectic pair (S_l, S_r). Since (S_l, S_r) is a matrix pair, it is conceivable that there exists a matrix-valued discrete-time transfer function $\hat{t}_{eq,d}$ such that the condition $\|\hat{t}_{eq,d}\|_\infty < \gamma$ gives rise to the same symplectic pair (S_l, S_r). We will show later that the linear system $T_{eq,d}$ is the map $\omega \mapsto \zeta$ in Figure 13.4, where the FDLTI discrete-time system $G_{eq,d}$ is defined by

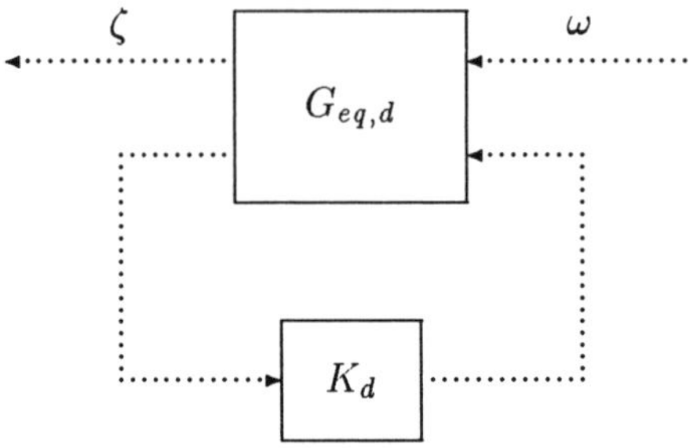

Figure 13.4: The $\mathcal{H}_\infty$-discretized system.

$$\hat{g}_{eq,d}(\lambda) = \left[\begin{array}{c|cc} A_{dd} & B_{1d} & B_{2dd} \\ \hline C_{1d} & 0 & D_{12d} \\ C_2 & 0 & 0 \end{array}\right],$$

with the matrices A_{dd} and B_{2dd} given by

$$\begin{aligned} A_{dd} &= A_d + \underline{B}_1 \underline{D}_{11}^*(\gamma^2 - \underline{D}_{11}\underline{D}_{11}^*)^{-1}\underline{C}_1 \\ B_{2dd} &= B_{2d} + \underline{B}_1 \underline{D}_{11}^*(\gamma^2 - \underline{D}_{11}\underline{D}_{11}^*)^{-1}\underline{D}_{12}, \end{aligned}$$

and B_{1d}, C_{1d}, and D_{12d} via

$$B_{1d}B_{1d}' = \gamma^2 \underline{B}_1(\gamma^2 - \underline{D}_{11}^*\underline{D}_{11})^{-1}\underline{B}_1^*$$

$$\begin{bmatrix} C_{1d}' \\ D_{12d}' \end{bmatrix} \begin{bmatrix} C_{1d} & D_{12d} \end{bmatrix} = \gamma^2 \begin{bmatrix} \underline{C}_1^* \\ \underline{D}_{12}^* \end{bmatrix} (\gamma^2 - \underline{D}_{11}\underline{D}_{11}^*)^{-1} \begin{bmatrix} \underline{C}_1 & \underline{D}_{12} \end{bmatrix}.$$

The realizations of $\hat{g}_{eq,d}$ and $\hat{k}_d$ in (13.3) induce a realization for the closed-loop transfer function $\hat{t}_{eq,d} : \omega \mapsto \zeta$ in Figure 13.4:

$$\left[\begin{array}{c|c} A_{cld} & B_{cld} \\ \hline C_{cld} & 0 \end{array}\right] = \left[\begin{array}{cc|c} A_{dd} + B_{2dd}D_K C_2 & B_{2dd}C_K & B_{1d} \\ B_K C_2 & A_K & 0 \\ \hline C_{1d} + D_{12d}D_K C_2 & D_{12d}C_K & 0 \end{array}\right]. \tag{13.5}$$

It is clear that the SD system in Figure 13.1 is internally stable iff the matrix A_{cl} defined in (13.4) is stable and the discrete-time system in Figure 13.4 is internally stable iff the matrix A_{cld} is stable.

Now we can state the main result of this chapter.

Theorem 13.4.1 *Assume $\gamma > \|\underline{D}_{11}\|$. Then the following statements are equivalent:*

1. *A_{cl} is stable and $\|\hat{\underline{t}}\|_\infty < \gamma$;*

2. *A_{cld} is stable and $\|\hat{t}_{eq,d}\|_\infty < \gamma$.*

Notice that by Theorem 13.2.1, $\|\underline{\hat{t}}\|_\infty = \|T\|$. Thus this theorem establishes an equivalence between the SD $\mathcal{H}_\infty$ problem, that K_d internally stabilize G and achieve $\|T\| < \gamma$, and the associated discrete $\mathcal{H}_\infty$ problem, that K_d internally stabilize $G_{eq,d}$ and achieve $\|\hat{t}_{eq,d}\|_\infty < \gamma$; the latter problem is solvable using techniques discussed in Chapter 7 once the matrices in $G_{eq,d}$ are computed.

Proof of Theorem 13.4.1 First, we show that the first statement implies stability of A_{cld}. Putting definitions (13.4) and (13.5) together, we have

$$\begin{aligned} A_{cld} &= A_{cl} + \left[\begin{array}{c} \underline{B}_1 \\ 0 \end{array}\right] \underline{D}_{11}^*(\gamma^2 - \underline{D}_{11}\underline{D}_{11}^*)^{-1} \cdot \\ &\qquad \left[\begin{array}{cc} \underline{C}_1 + \underline{D}_{12}D_K C_2 & \underline{D}_{12}C_K \end{array}\right] \\ &= A_{cl} + \underline{B}_{cl}\underline{D}_{11}^*(\gamma^2 - \underline{D}_{11}\underline{D}_{11}^*)^{-1}\underline{C}_{cl} \\ &= A_{cl} + \underline{B}_{cl}(\gamma^2 - \underline{D}_{11}^*\underline{D}_{11})^{-1}\underline{D}_{11}^*\underline{C}_{cl}. \end{aligned}$$

In addition, the following state model can be derived for the operator-valued transfer function $(\gamma^2 - \underline{D}_{11}^*\underline{\hat{t}})^{-1}$:

$$\left[\begin{array}{c|c} A_{cl} + \underline{B}_{cl}(\gamma^2 - \underline{D}_{11}^*\underline{D}_{11})^{-1}\underline{D}_{11}^*\underline{C}_{cl} & \underline{B}_{cl}(\gamma^2 - \underline{D}_{11}^*\underline{D}_{11})^{-1} \\ \hline (\gamma^2 - \underline{D}_{11}^*\underline{D}_{11})^{-1}\underline{D}_{11}^*\underline{C}_{cl} & (\gamma^2 - \underline{D}_{11}^*\underline{D}_{11})^{-1} \end{array}\right]. \qquad (13.6)$$

Thus A_{cld} equals the A-matrix of $(\gamma^2 - \underline{D}_{11}^*\underline{\hat{t}})^{-1}$. Since $\|\underline{\hat{t}}\|_\infty < \gamma$ and $\|\underline{D}_{11}\| < \gamma$, it follows that

$$(\gamma^2 - \underline{D}_{11}^*\underline{\hat{t}})^{-1} \in \mathcal{H}_\infty(\mathbb{D}, \mathbb{B}),$$

and hence that

$$\underline{C}_{cl}^*\underline{D}_{11}(\gamma^2 - \underline{D}_{11}^*\underline{D}_{11})(\gamma^2 - \underline{D}_{11}^*\underline{\hat{t}})^{-1}(\gamma^2 - \underline{D}_{11}^*\underline{D}_{11})\underline{B}_{cl}^* \in \mathcal{H}_\infty(\mathbb{D}).$$

The latter is a matrix-valued, not operator-valued, transfer function. From (13.6) a state model for the latter transfer function is

$$\left[\begin{array}{c|c} A_{cl} + \underline{B}_{cl}(\gamma^2 - \underline{D}_{11}^*\underline{D}_{11})^{-1}\underline{D}_{11}^*\underline{C}_{cl} & \underline{B}_{cl}\underline{B}_{cl}^* \\ \hline \underline{C}_{cl}^*\underline{D}_{11}\underline{D}_{11}^*\underline{C}_{cl} & \underline{C}_{cl}^*\underline{D}_{11}(\gamma^2 - \underline{D}_{11}^*\underline{D}_{11})\underline{B}_{cl}^* \end{array}\right].$$

These latter state parameters are matrices, not operators. To conclude that A_{cld} is stable, it remains to show that the pair

$$A_{cl} + \underline{B}_{cl}(\gamma^2 - \underline{D}_{11}^*\underline{D}_{11})^{-1}\underline{D}_{11}^*\underline{C}_{cl}, \quad \underline{B}_{cl}\underline{B}_{cl}^*$$

is stabilizable and the pair

$$\underline{C}_{cl}^*\underline{D}_{11}\underline{D}_{11}^*\underline{C}_{cl}, \quad A_{cl} + \underline{B}_{cl}(\gamma^2 - \underline{D}_{11}^*\underline{D}_{11})^{-1}\underline{D}_{11}^*\underline{C}_{cl}$$

is detectable. These follow from stability of A_{cl}.

Next, we need to show that the second statement implies stability of A_{cl}. This requires introducing an intermediate system and is left as an exercise (Exercise 13.3).

Finally, since $\hat{t}_{eq,d}$ is a usual discrete-time system with the realization in (13.5), we form the associated symplectic pair for $\hat{t}_{eq,d}$ (see Section 7.1):

$$(S_{ld}, S_{rd}) = \left(\begin{bmatrix} A_{cld} & 0 \\ -\frac{1}{\gamma} C'_{cld} C_{cld} & I \end{bmatrix}, \begin{bmatrix} I & -\frac{1}{\gamma} B_{cld} B'_{cld} \\ 0 & A'_{cld} \end{bmatrix} \right).$$

Then the proof is completed by observing that $(S_l, S_r) = (S_{ld}, S_{rd})$: For example, for the $(2,1)$ blocks in the left matrices, we need to show

$$-\gamma \underline{C}_{cl}^* (\gamma^2 - \underline{D}_{11} \underline{D}_{11}^*)^{-1} \underline{C}_{cl} = -\frac{1}{\gamma} C'_{cld} C_{cld}. \tag{13.7}$$

Indeed, since

$$\underline{C}_{cl} = \begin{bmatrix} \underline{C}_1 & \underline{D}_{12} \end{bmatrix} \begin{bmatrix} I & 0 \\ D_K C_2 & C_K \end{bmatrix},$$

we get that the left-hand side of (13.7) is

$$\begin{aligned} -\gamma & \begin{bmatrix} I & 0 \\ D_K C_2 & C_K \end{bmatrix}' \begin{bmatrix} \underline{C}_1^* \\ \underline{D}_{12}^* \end{bmatrix} (\gamma^2 - \underline{D}_{11} \underline{D}_{11}^*)^{-1} \times \\ & \qquad\qquad \begin{bmatrix} \underline{C}_1 & \underline{D}_{12} \end{bmatrix} \begin{bmatrix} I & 0 \\ D_K C_2 & C_K \end{bmatrix} \\ = & -\frac{1}{\gamma} \begin{bmatrix} I & 0 \\ D_K C_2 & C_K \end{bmatrix}' \begin{bmatrix} C'_{1d} \\ D'_{12d} \end{bmatrix} \begin{bmatrix} C_{1d} & D_{12d} \end{bmatrix} \begin{bmatrix} I & 0 \\ D_K C_2 & C_K \end{bmatrix} \\ = & -\frac{1}{\gamma} C'_{cld} C_{cld}. \end{aligned}$$

■

Notice that $G_{eq,d}$ depends on γ. Thus until a satisfactory γ is found, iteration on γ is necessary. Remember that γ has to be chosen such that $\gamma > \|\underline{D}_{11}\|$. Hence it is desirable to compute $\|\underline{D}_{11}\|$; this is a non-trivial task and will be treated in the next section. Once γ is chosen, the computation of $G_{eq,d}$ requires evaluating several operator compositions which are summarized below:

$$\begin{aligned} & \underline{B}_1 \underline{D}_{11}^* (\gamma^2 - \underline{D}_{11} \underline{D}_{11}^*)^{-1} \underline{C}_1 \\ & \underline{B}_1 \underline{D}_{11}^* (\gamma^2 - \underline{D}_{11} \underline{D}_{11}^*)^{-1} \underline{D}_{12} \\ & \gamma^2 \underline{B}_1 (\gamma^2 - \underline{D}_{11}^* \underline{D}_{11})^{-1} \underline{B}_1 \\ & \gamma^2 \begin{bmatrix} \underline{C}_1^* \\ \underline{D}_{12}^* \end{bmatrix} (\gamma^2 - \underline{D}_{11} \underline{D}_{11}^*)^{-1} \begin{bmatrix} \underline{C}_1 & \underline{D}_{12} \end{bmatrix}. \end{aligned}$$

These matrices are computed in Section 13.6 using matrix exponentials.

13.5 Computing the $\mathcal{L}_2[0,h)$-Induced Norm of a System

This section presents two methods for computing the $\mathcal{L}_2[0,h)$-induced norm of $\underline{D}_{11}$: The first is based on computing the singular values of $\underline{D}_{11}\underline{D}_{11}^*$, the second on fast discretization.

Recall that $\underline{D}_{11}$ is the compression of G_{11} to $\mathcal{K} = \mathcal{L}_2[0, h)$. For the state model of G in (13.2), G_{11} has the state model

$$\hat{g}_{11}(s) = \left[\begin{array}{c|c} A & B_1 \\ \hline C_1 & 0 \end{array}\right].$$

To simplify notation, let us drop the subscript 1 in this section. Thus, G is a continuous-time system with

$$\hat{g}(s) = \left[\begin{array}{c|c} A & B \\ \hline C & 0 \end{array}\right]$$

and the goal is to compute the norm of its compression $\underline{D}$ to $\mathcal{K}$.

The operator $\underline{D} : \mathcal{K} \to \mathcal{K}$ is defined via

$$y = \underline{D}u \Leftrightarrow y(t) = \int_0^t Ce^{(t-\tau)A} Bu(\tau)\, d\tau.$$

This can also be described via the state equations

$$\dot{x}_1 = Ax_1 + Bu, \quad x_1(0) = 0 \tag{13.8}$$

$$y = Cx_1, \quad 0 \leq t \leq h. \tag{13.9}$$

Since $\underline{D}$ is a compact operator (since $D = 0$), it follows that $\|\underline{D}\|$ equals the largest singular value of $\underline{D}$, or the square root of the largest eigenvalue of $\underline{D}\underline{D}^*$. Therefore $\|\underline{D}\|$ can be computed by characterizing the (nonzero) eigenvalues of $\underline{D}\underline{D}^*$.

It is left as an exercise (Exercise 13.5) to verify that $\underline{D}^* : u \mapsto y$ is described by the following state equations:

$$\dot{x}_2 = -A'x_2 - C'u, \quad x_2(h) = 0, \tag{13.10}$$

$$y = Bx_2, \quad 0 \leq t \leq h. \tag{13.11}$$

For $\gamma > 0$, define the matrix-valued function

$$Q(t) = \begin{bmatrix} Q_{11}(t) & Q_{12}(t) \\ Q_{21}(t) & Q_{22}(t) \end{bmatrix} := \exp\left\{t \begin{bmatrix} -A' & -C'C \\ \gamma^{-2}BB' & A \end{bmatrix}\right\}, \tag{13.12}$$

where the partitioning is conformable with that of the right-hand matrix.

Theorem 13.5.1 *Assume $\gamma > 0$. Then γ^2 is an eigenvalue of $\underline{D}\underline{D}^*$ iff*

$$\det[Q_{11}(h)] = 0.$$

Thus $\|\underline{D}\|$ can be computed as follows: Compute $Q_{11}(h)$ as a function of $\gamma > 0$; then $\|\underline{D}\|$ equals the largest γ such that $Q_{11}(h)$ has an eigenvalue at 0.

Proof of Theorem 13.5.1 We shall prove the theorem under the assumption that (A, B) is controllable and (C, A) is observable; then we shall discuss the proof in the general case.

($\Longrightarrow$) Let γ^2 be an eigenvalue of $\underline{D}\underline{D}^*$. Then there exists a nonzero $f \in \mathcal{K}$ such that

$$\underline{DD}^* f = \gamma^2 f.$$

Define $g = \gamma^{-2}\underline{D}^* f$ to get the pair of equations

$$\underline{D}g = f, \quad \underline{D}^* f = \gamma^2 g. \tag{13.13}$$

In terms of state equations we have

$$\begin{aligned} \dot{x}_1 &= Ax_1 + Bg, \quad x_1(0) = 0 \\ f &= Cx_1, \quad 0 \le t \le h \\ \dot{x}_2 &= -A'x_2 - C'f, \quad x_2(h) = 0 \\ g &= \gamma^{-2}B'x_2, \quad 0 \le t \le h. \end{aligned}$$

Eliminate f and g to get

$$\begin{bmatrix} \dot{x}_2 \\ \dot{x}_1 \end{bmatrix} = \begin{bmatrix} -A' & -C'C \\ \gamma^{-2}BB' & A \end{bmatrix} \begin{bmatrix} x_2 \\ x_1 \end{bmatrix}, \quad \begin{bmatrix} x_2(h) \\ x_1(0) \end{bmatrix} = 0.$$

Thus for $0 \le t \le h$,

$$\begin{bmatrix} x_2(t) \\ x_1(t) \end{bmatrix} = Q(t) \begin{bmatrix} x_2(0) \\ 0 \end{bmatrix} = \begin{bmatrix} Q_{11}(t)x_2(0) \\ Q_{21}(t)x_2(0) \end{bmatrix}.$$

To satisfy the boundary condition $x_2(h) = 0$, we must have $Q_{11}(h)x_2(0) = 0$. Since $f \ne 0$, it follows that $x_2(0) \ne 0$, and so $\det[Q_{11}(h)] = 0$.

($\Longleftarrow$) Assume $\det[Q_{11}(h)] = 0$. Then we can choose a nonzero x_{20} satisfying $Q_{11}(h)x_{20} = 0$. By reversing the steps above we see that f and g satisfying (13.13) are given by

$$\begin{aligned} \begin{bmatrix} \dot{x}_2 \\ \dot{x}_1 \end{bmatrix} &= \begin{bmatrix} -A' & -C'C \\ \gamma^{-2}BB' & A \end{bmatrix} \begin{bmatrix} x_2 \\ x_1 \end{bmatrix}, \quad \begin{bmatrix} x_2(0) \\ x_1(0) \end{bmatrix} = \begin{bmatrix} x_{20} \\ 0 \end{bmatrix} \\ \begin{bmatrix} f \\ g \end{bmatrix} &= \begin{bmatrix} 0 & C \\ \gamma^{-2}B' & 0 \end{bmatrix} \begin{bmatrix} x_2 \\ x_1 \end{bmatrix}. \end{aligned}$$

Thus the proof is completed if we can show $f \ne 0$, or equivalently, $\begin{bmatrix} f \\ g \end{bmatrix} \ne 0$. Since $x_{20} \ne 0$, this follows from observability of the pair

$$\left(\begin{bmatrix} 0 & C \\ \gamma^{-2}B' & 0 \end{bmatrix}, \begin{bmatrix} -A' & -C'C \\ \gamma^{-2}BB' & A \end{bmatrix} \right), \tag{13.14}$$

which is a consequence of the minimality assumption of $(A, B, C, 0)$ (Exercise 13.6). ■

Now let us look at the proof without the controllability and observability assumptions. We look at the case that (C, A) is observable but (A, B) is not controllable; the general case follows similarly. By a suitable similarity transformation, we can take

$$\begin{aligned} A &= \begin{bmatrix} A_m & A_{12} \\ 0 & A_{22} \end{bmatrix}, \quad B = \begin{bmatrix} B_m \\ 0 \end{bmatrix} \\ C &= \begin{bmatrix} C_m & C_2 \end{bmatrix}, \end{aligned}$$

where A_m contains all the controllable modes and A_{22} all the uncontrollable modes. Thus a minimal realization for $\hat{g}$ is

$$\hat{g}(s) = \left[\begin{array}{c|c} A_m & B_m \\ \hline C_m & 0 \end{array}\right].$$

For $\gamma > 0$, define

$$Q_m(t) = \begin{bmatrix} Q_{m11}(t) & Q_{m12}(t) \\ Q_{m21}(t) & Q_{m22}(t) \end{bmatrix} = \exp\left\{t \begin{bmatrix} -A_m' & -C_m'C_m \\ \gamma^{-2}B_mB_m' & A_m \end{bmatrix}\right\}.$$

Thus by the above proof of Theorem 13.5.1, γ^2 is an eigenvalue of $\underline{DD^*}$ iff $\det[Q_{m11}(h)] = 0$. We shall show that $\det[Q_{11}(h)] = 0$ iff $\det[Q_{m11}(h)] = 0$.

Define

$$\begin{aligned} E &:= \begin{bmatrix} -A' & -C'C \\ \gamma^{-2}BB' & A \end{bmatrix} \\ &= \begin{bmatrix} -A_m' & 0 & -C_m'C_m & -C_m'C_2 \\ -A_{12}' & -A_{22}' & -C_2'C_m & -C_2'C_2 \\ \gamma^{-2}B_mB_m' & 0 & A_m & A_{12} \\ 0 & 0 & 0 & A_{22} \end{bmatrix} \end{aligned}$$

and

$$J_{23} := \begin{bmatrix} I & 0 & 0 & 0 \\ 0 & 0 & I & 0 \\ 0 & I & 0 & 0 \\ 0 & 0 & 0 & I \end{bmatrix}$$

conformably. It can be verified that $J_{23}^{-1} = J_{23}$ and $J_{23}^{-1}EJ_{23}$ amounts to switching the second and third block rows and the second and third block columns in E:

$$J_{23}^{-1}EJ_{23} = \begin{bmatrix} -A_m' & -C_m'C_m & 0 & -C_mC_2 \\ \gamma^{-2}B_mB_m' & A_m & 0 & A_{12} \\ -A_{12}' & -C_2'C_m & -A_{22}' & -C_2'C_2 \\ 0 & 0 & 0 & A_{22} \end{bmatrix}.$$

The special structure of the right-hand matrix yields

$$J_{23}^{-1}Q(h)J_{23} = \begin{bmatrix} Q_{m11}(h) & Q_{m12}(h) & 0 & ? \\ Q_{m21}(h) & Q_{m22}(h) & 0 & ? \\ ? & ? & e^{-hA_{22}'} & ? \\ 0 & 0 & 0 & e^{hA_{22}} \end{bmatrix},$$

where ? denotes an irrelevant block, and so

$$Q_{11}(h) = \begin{bmatrix} Q_{m11}(h) & 0 \\ ? & e^{-hA'_{22}} \end{bmatrix}.$$

Since $e^{-hA'_{22}}$ is nonsingular, this proves that $\det[Q_{11}(h)] = 0$ iff

$$\det[Q_{m11}(h)] = 0.$$

The second method to compute $\|\underline{D}\|$ is via fast discretization. Recall from Section 8.2 that the fast-rate discretization of G is $S_f G H_f$, where the fast sampler S_f and hold H_f have period h/n. The transfer function for $S_f G H_f$ is

$$\left[\begin{array}{c|c} A_f & B_f \\ \hline C & 0 \end{array}\right],$$

where

$$(A_f, B_f) = c2d(A, B, h/n).$$

Now we lift the system in discrete time so that the lifted system corresponds to the base period h. A realization of the lifted system is given in Section 8.2:

$$\left[\begin{array}{c|c} A_n & B_n \\ \hline C_n & D_n \end{array}\right].$$

Here we need only D_n:

$$D_n = \begin{bmatrix} 0 & 0 & 0 & \cdots & 0 \\ CB_f & 0 & 0 & \cdots & 0 \\ CA_f B_f & CB_f & 0 & \cdots & 0 \\ \vdots & \vdots & \vdots & & \vdots \\ CA_f^{n-2}B_f & CA_f^{n-3}B_f & CA_f^{n-4}B_f & \cdots & 0 \end{bmatrix}.$$

This D_n captures the behaviour of $S_f G H_f$ in the first sampling interval, $[0, h)$. Since we would expect $S_f G H_f$ to emulate G for sufficiently large n, so also we would expect D_n to emulate $\underline{D}$, the compression of G to the first sampling interval. It can be shown that this is indeed so and

$$\|\underline{D}\| = \lim_{n\to\infty} \sigma_{\max}(D_n).$$

Example 13.5.1 In Section 13.8 we shall perform a SD $\mathcal{H}_\infty$ design for the system in Example 7.2.1. Here, let us compute the norm of the associated $\underline{D}_{11}$. Note that $h = 0.5$ and

$$G_{11} : \begin{bmatrix} w_1 \\ w_2 \end{bmatrix} \mapsto \begin{bmatrix} z_1 \\ z_2 \end{bmatrix}$$

is given by

$$\hat{g}_{11}(s) = \begin{bmatrix} \hat{w}(s) & 0 \\ 0 & 0 \end{bmatrix}, \quad \hat{w}(s) = \frac{1}{[(2.5/\pi)s+1]^2}.$$

We shall compute $\|\underline{D}_{11}\|$ by the two methods presented in this section.

First, we plot in Figure 13.5 the minimum magnitude of the eigenvalues of $Q_{11}(h)$ versus γ. The value $\det[Q_{11}(h)] = 0$ occurs when $\gamma = 0.016, 0.0775$. These two values of γ are singular values of $\underline{D}_{11}$ by Theorem 13.5.1 and therefore $\|\underline{D}_{11}\|$ equals the maximum singular value, namely, $\|\underline{D}_{11}\| = 0.0775$. This equals the norm of the compression of G_{11} to $\mathcal{K}$. [In contrast, the norm

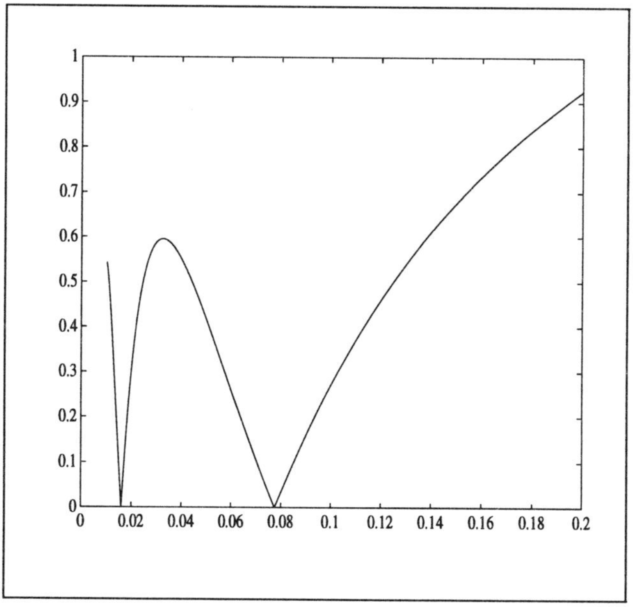

Figure 13.5: Minimum magnitude of the eigenvalues of $Q_{11}(h)$ versus γ.

of G_{11} on all of $\mathcal{L}_2(\mathbb{R})$ equals $\|\hat{w}\|_\infty = 1$.]

Next, via fast discretization we form the matrix D_{11n} (the D-matrix of $S_f G_{11} H_f$ lifted) and compute $\sigma_{\max}(D_{11n})$ for n from 1 to 50; this is shown in Figure 13.6. We see that $\sigma_{\max}(D_{11n})$ converges quite slowly. For $n = 50$, $\sigma_{\max}(D_{11n}) = 0.0762$, which is not yet very close to $\|\underline{D}_{11}\|$. An additional disadvantage of this method is that D_n grows in size as n increases. For these two reasons the first method seems better.

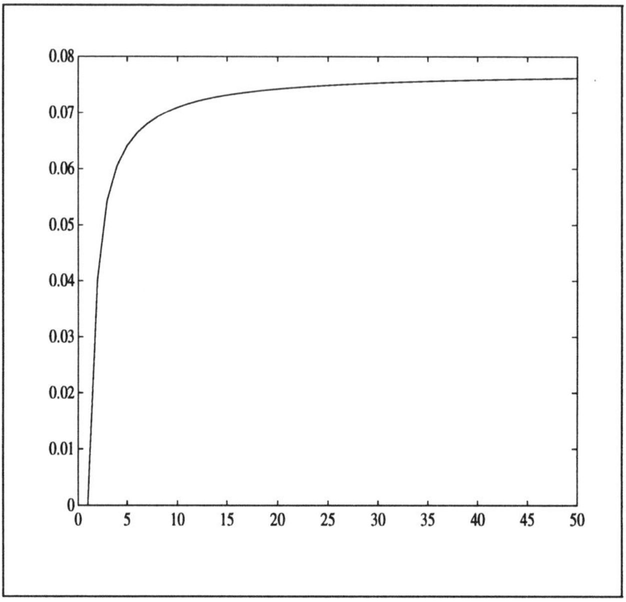

Figure 13.6: $\sigma_{\max}(D_{11n})$ versus n.

13.6 Computing the Matrices in $G_{eq,d}$

Under the assumption $\gamma > \|\underline{D}_{11}\|$, Theorem 13.4.1 says that the SD $\mathcal{H}_\infty$ control problem, $\|T\| < \gamma$, is equivalent to the discrete-time $\mathcal{H}_\infty$ control problem, $\|\hat{t}_{eq,d}\|_\infty < \gamma$, where $T_{eq,d}$ is the closed-loop system composed of a discrete-time generalized plant $G_{eq,d}$ and the same discrete controller K_d. For a realization of the analog plant

$$\hat{g}(s) = \left[\begin{array}{c|cc} A & B_1 & B_2 \\ \hline C_1 & 0 & D_{12} \\ C_2 & 0 & 0 \end{array}\right],$$

the $\mathcal{H}_\infty$ discretization $G_{eq,d}$ is given by

$$\hat{g}_{eq,d}(\lambda) = \left[\begin{array}{c|cc} A_{dd} & B_{1d} & B_{2dd} \\ \hline C_{1d} & 0 & D_{12d} \\ C_2 & 0 & 0 \end{array}\right],$$

with

$$\begin{aligned} A_{dd} &= A_d + \underline{B}_1\underline{D}_{11}^*(\gamma^2 - \underline{D}_{11}\underline{D}_{11}^*)^{-1}\underline{C}_1 \\ B_{2dd} &= B_{2d} + \underline{B}_1\underline{D}_{11}^*(\gamma^2 - \underline{D}_{11}\underline{D}_{11}^*)^{-1}\underline{D}_{12}, \end{aligned}$$

and B_{1d}, C_{1d}, D_{12d} obtained via the factorizations

$$B_{1d}B_{1d}' = \gamma^2\underline{B}_1(\gamma^2 - \underline{D}_{11}^*\underline{D}_{11})^{-1}\underline{B}_1^*$$

$$\begin{bmatrix} C_{1d}' \\ D_{12d}' \end{bmatrix} \begin{bmatrix} C_{1d} & D_{12d} \end{bmatrix} = \gamma^2 \begin{bmatrix} \underline{C}_1^* \\ \underline{D}_{12}^* \end{bmatrix} (\gamma^2 - \underline{D}_{11}\underline{D}_{11}^*)^{-1} \begin{bmatrix} \underline{C}_1 & \underline{D}_{12} \end{bmatrix}.$$

The goal in this section is to compute the matrices in the realization of $\hat{g}_{eq,d}$ using matrix exponentials. The standing assumption is $\gamma > \|\underline{D}_{11}\|$, which is easily checked since computing $\|\underline{D}_{11}\|$ was already treated in the preceding section.

Computing C_{1d} and D_{12d}

We start with computing the matrix

$$J := \begin{bmatrix} \underline{C}_1^* \\ \underline{D}_{12}^* \end{bmatrix} \left(I - \gamma^{-2}\underline{D}_{11}\underline{D}_{11}^*\right)^{-1} \begin{bmatrix} \underline{C}_1 & \underline{D}_{12} \end{bmatrix}.$$

Once J is computed, C_{1d} and D_{12d} are obtained by performing a Cholesky factorization of J, or simply by computing the square root of J.

Recall from Section 10.3 that $\underline{C}_1$ and $\underline{D}_{12}$ are defined by

$$\underline{C}_1 : \ \mathcal{E} \to \mathcal{K}, \quad (\underline{C}_1 x)(t) = C_1 e^{tA} x$$

$$\underline{D}_{12} : \ \mathcal{E} \to \mathcal{K}, \quad (\underline{D}_{12} v)(t) = D_{12} v + C_1 \int_0^t e^{(t-\tau)A}\, d\tau\, B_2 v.$$

Introduce the square matrix

$$\underline{A} = \begin{bmatrix} A & B_2 \\ 0 & 0 \end{bmatrix}.$$

Then it follows from Lemma 10.5.1 that

$$e^{t\underline{A}} = \begin{bmatrix} e^{tA} & \int_0^t e^{(t-\tau)A} B_2\, d\tau \\ 0 & I \end{bmatrix}.$$

Thus the matrix operator $\begin{bmatrix} \underline{C}_1 & \underline{D}_{12} \end{bmatrix} : \mathcal{E} \to \mathcal{K}$ is

$$\left(\begin{bmatrix} \underline{C}_1 & \underline{D}_{12} \end{bmatrix} \begin{bmatrix} x \\ v \end{bmatrix} \right)(t) = \begin{bmatrix} C_1 & D_{12} \end{bmatrix} e^{t\underline{A}} \begin{bmatrix} x \\ v \end{bmatrix}. \tag{13.15}$$

Based on this, it is easily derived that the adjoint operator mapping $\mathcal{K}$ to $\mathcal{E}$ is

$$\begin{bmatrix} \underline{C}_1^* \\ \underline{D}_{12}^* \end{bmatrix} u = \int_0^h e^{t\underline{A}'} \begin{bmatrix} C_1' \\ D_{12}' \end{bmatrix} u(t)\, dt. \tag{13.16}$$

Now in order to evaluate the matrix J, we need to compute the action of the operator $(I - \gamma^{-2}\underline{D}_{11}\underline{D}_{11}^*)^{-1} : \mathcal{K} \to \mathcal{K}$, which is well-defined by the assumption $\gamma > \|\underline{D}_{11}\|$. Let $u \in \mathcal{K}$ and define

$$y = (I - \gamma^{-2}\underline{D}_{11}\underline{D}_{11}^*)^{-1} u.$$

Then

$$u = y - \gamma^{-2}\underline{D}_{11}\underline{D}_{11}^* y,$$

or

$$u = y - \gamma^{-1}\underline{D}_{11} w, \quad w = \gamma^{-1}\underline{D}_{11}^* y.$$

Using the state equations for $\underline{D}_{11}$ and $\underline{D}_{11}^*$ in (13.8-13.11), we have that

$$\begin{aligned} \dot{x}_1 &= Ax_1 + \gamma^{-1}B_1 w, \quad x_1(0) = 0 \\ u &= y - C_1 x_1, \quad 0 \le t \le h \end{aligned}$$

and

$$\begin{aligned} \dot{x}_2 &= -A'x_2 - C_1'y, \quad x_2(h) = 0 \\ w &= \gamma^{-1}B_1'x_2, \quad 0 \le t \le h. \end{aligned}$$

Thus

$$\begin{aligned} \begin{bmatrix} \dot{x}_2 \\ \dot{x}_1 \end{bmatrix} &= \begin{bmatrix} -A' & 0 \\ \gamma^{-2}B_1B_1' & A \end{bmatrix} \begin{bmatrix} x_2 \\ x_1 \end{bmatrix} + \begin{bmatrix} -C_1' \\ 0 \end{bmatrix} y, \quad \begin{bmatrix} x_2(h) \\ x_1(0) \end{bmatrix} = 0 \\ u &= \begin{bmatrix} 0 & -C_1 \end{bmatrix} \begin{bmatrix} x_2 \\ x_1 \end{bmatrix} + y, \quad 0 \le t \le h. \end{aligned}$$

Rewrite the equations from u to y:

$$\begin{bmatrix} \dot{x}_2 \\ \dot{x}_1 \end{bmatrix} = \begin{bmatrix} -A' & -C_1'C_1 \\ \gamma^{-2}B_1B_1' & A \end{bmatrix} \begin{bmatrix} x_2 \\ x_1 \end{bmatrix} + \begin{bmatrix} -C_1' \\ 0 \end{bmatrix} u, \tag{13.17}$$

$$\begin{bmatrix} x_2(h) \\ x_1(0) \end{bmatrix} = 0$$

$$y = \begin{bmatrix} 0 & C_1 \end{bmatrix} \begin{bmatrix} x_2 \\ x_1 \end{bmatrix} + u, \quad 0 \le t \le h. \tag{13.18}$$

This system has two boundary conditions $x_1(0) = 0$ and $x_2(h) = 0$ and is therefore a *two-point boundary value problem*. Since $(I - \gamma^{-2}\underline{D}_{11}\underline{D}_{11}^*)^{-1}$ exists, we expect to be able to solve the equations for y in terms of u.

Define $Q(t)$ as in Theorem 13.5.1, that is,

$$Q(t) := \mathrm{e}^{tE}, \quad E := \begin{bmatrix} -A' & -C_1'C_1 \\ \gamma^{-2}B_1B_1' & A \end{bmatrix},$$

and partition $Q(t)$ conformably. Integrate equation (13.17) from 0 to t, $0 \le t \le h$:

$$\begin{bmatrix} x_2(t) \\ x_1(t) \end{bmatrix} = Q(t) \begin{bmatrix} x_2(0) \\ x_1(0) \end{bmatrix} + \int_0^t Q(t-\tau) \begin{bmatrix} -C_1' \\ 0 \end{bmatrix} u(\tau)\, d\tau.$$

Since $x_1(0) = 0$, we get

$$x_2(h) = Q_{11}(h)x_2(0) + \begin{bmatrix} I & 0 \end{bmatrix} \int_0^h Q(h-\tau) \begin{bmatrix} -C_1' \\ 0 \end{bmatrix} u(\tau)\, d\tau.$$

Since $\gamma > \|\underline{D}_{11}\|$, by Theorem 13.5.1, $Q_{11}(h)$ is nonsingular. Hence the condition $x_2(h) = 0$ yields

$$x_2(0) = -\begin{bmatrix} Q_{11}(h)^{-1} & 0 \end{bmatrix} \int_0^h Q(h-\tau) \begin{bmatrix} -C_1' \\ 0 \end{bmatrix} u(\tau)\, d\tau.$$

Thus

$$\begin{aligned} \begin{bmatrix} x_2(t) \\ x_1(t) \end{bmatrix} &= Q(t) \begin{bmatrix} Q_{11}(h)^{-1} & 0 \\ 0 & 0 \end{bmatrix} \int_0^h Q(h-\tau) \begin{bmatrix} C_1' \\ 0 \end{bmatrix} u(\tau)\, d\tau \\ &\quad - \int_0^t Q(t-\tau) \begin{bmatrix} C_1' \\ 0 \end{bmatrix} u(\tau)\, d\tau. \end{aligned}$$

We conclude by (13.18) that

$$\begin{aligned} y(t) &= \begin{bmatrix} 0 & C_1 \end{bmatrix} \left\{ Q(t) \begin{bmatrix} Q_{11}(h)^{-1} & 0 \\ 0 & 0 \end{bmatrix} \int_0^h Q(h-\tau) \begin{bmatrix} C_1' \\ 0 \end{bmatrix} u(\tau)\, d\tau \right. \\ &\quad \left. - \int_0^t Q(t-\tau) \begin{bmatrix} C_1 \\ 0 \end{bmatrix} u(\tau)\, d\tau \right\} + u(t). \qquad (13.19) \end{aligned}$$

This determines the action of $(I - \gamma^{-2}\underline{D}_{11}\underline{D}_{11}^*)^{-1} : u \mapsto y$. With this, we can proceed to computing J.

Define a new operator $\underline{F} : \mathcal{E} \to \mathcal{K}$ by

$$\underline{F} = \left(I - \frac{1}{\gamma^2}\underline{D}_{11}\underline{D}_{11}^* \right)^{-1} \begin{bmatrix} \underline{C}_1 & \underline{D}_{12} \end{bmatrix}.$$

It follows from (13.15) and (13.19) that $(\underline{F}w)(t) = f(t)w$, where $f(t)$ is the matrix

$$\begin{aligned} f(t) &= \begin{bmatrix} 0 & C_1 \end{bmatrix} \left\{ Q(t) \begin{bmatrix} Q_{11}(h)^{-1} & 0 \\ 0 & 0 \end{bmatrix} \int_0^h Q(h-\tau) \begin{bmatrix} C_1' \\ 0 \end{bmatrix} \times \right. \\ &\quad \left. \begin{bmatrix} C_1 & D_{12} \end{bmatrix} e^{\tau \underline{A}}\, d\tau - \int_0^t Q(t-\tau) \begin{bmatrix} C_1' \\ 0 \end{bmatrix} \begin{bmatrix} C_1 & D_{12} \end{bmatrix} e^{\tau \underline{A}}\, d\tau \right\} + \\ &\quad \begin{bmatrix} C_1 & D_{12} \end{bmatrix} e^{t \underline{A}}. \end{aligned}$$

To evaluate the two integrals in $f(t)$, define

$$H := \begin{bmatrix} E & \begin{bmatrix} C_1' \\ 0 \end{bmatrix} \begin{bmatrix} C_1 & D_{12} \end{bmatrix} \\ 0 & \underline{A} \end{bmatrix}.$$

Then by Lemma 10.5.1

$$e^{tH} = \begin{bmatrix} Q(t) & N(t) \\ 0 & e^{t\underline{A}} \end{bmatrix},$$

where

$$N(t) := \int_0^t Q(t-\tau) \begin{bmatrix} C_1' \\ 0 \end{bmatrix} \begin{bmatrix} C_1 & D_{12} \end{bmatrix} e^{\tau \underline{A}} \, d\tau.$$

Thus

$$\begin{aligned} f(t) &= \begin{bmatrix} 0 & C_1 \end{bmatrix} \left\{ Q(t) \begin{bmatrix} Q_{11}(h)^{-1} & 0 \\ 0 & 0 \end{bmatrix} N(h) - N(t) \right\} + \\ &\quad \begin{bmatrix} C_1 & D_{12} \end{bmatrix} e^{t\underline{A}}. \end{aligned}$$

With $f(t)$ computed, we can now compute J using (13.16):

$$\begin{aligned} J &= \int_0^h e^{t\underline{A}'} \begin{bmatrix} C_1' \\ D_{12}' \end{bmatrix} f(t)\, dt \\ &= \int_0^h e^{t\underline{A}'} \begin{bmatrix} C_1' \\ D_{12}' \end{bmatrix} \begin{bmatrix} 0 & C_1 \end{bmatrix} Q(t)\, dt \begin{bmatrix} Q_{11}(h)^{-1} & 0 \\ 0 & 0 \end{bmatrix} N(h) \\ &\quad - \int_0^h e^{t\underline{A}'} \begin{bmatrix} C_1' \\ D_{12}' \end{bmatrix} \begin{bmatrix} 0 & C_1 \end{bmatrix} N(t)\, dt \\ &\quad + \int_0^h e^{t\underline{A}'} \begin{bmatrix} C_1' \\ D_{12}' \end{bmatrix} \begin{bmatrix} C_1 & D_{12} \end{bmatrix} e^{t\underline{A}}\, dt. \end{aligned}$$

The first integral involved, namely

$$I_1 := \int_0^h e^{t\underline{A}'} \begin{bmatrix} C_1' \\ D_{12}' \end{bmatrix} \begin{bmatrix} 0 & C_1 \end{bmatrix} Q(t)\, dt,$$

can be computed as follows: Define the matrix M (and P) as in

$$\begin{bmatrix} P & M \\ 0 & Q \end{bmatrix} = \exp\left\{ h \begin{bmatrix} -\underline{A}' & \begin{bmatrix} C_1' \\ D_{12}' \end{bmatrix} \begin{bmatrix} 0 & C_1 \end{bmatrix} \\ 0 & E \end{bmatrix} \right\};$$

then by Lemma 10.5.1,

$$I_1 = e^{h\underline{A}'} M. \tag{13.20}$$

For the second integral, note that

$$N(t) = \begin{bmatrix} I & 0 \end{bmatrix} e^{tH} \begin{bmatrix} 0 \\ I \end{bmatrix}.$$

So the second integral in J, namely

$$I_2 := \int_0^h e^{t\underline{A}'} \begin{bmatrix} C_1' \\ D_{12}' \end{bmatrix} \begin{bmatrix} 0 & C_1 \end{bmatrix} \begin{bmatrix} I & 0 \end{bmatrix} e^{tH}\, dt \begin{bmatrix} 0 \\ I \end{bmatrix},$$

can be computed in the same way: Define V via

$$\begin{bmatrix} P & V \\ 0 & e^{hH} \end{bmatrix} = \exp\left\{ h \begin{bmatrix} -\underline{A}' & \begin{bmatrix} C_1' \\ D_{12}' \end{bmatrix} \begin{bmatrix} 0 & C_1 & 0 \end{bmatrix} \\ 0 & H \end{bmatrix} \right\}.$$

Conformably with the blocks in H, partition

$$V = \begin{bmatrix} M & L \end{bmatrix}.$$

(Note that the first block is the same M defined earlier.) Thus

$$\begin{aligned} I_2 &= e^{h\underline{A}'} V \begin{bmatrix} 0 \\ I \end{bmatrix} \\ &= e^{h\underline{A}'} L. \end{aligned}$$

The last integral in J, defined as

$$J_\infty := \int_0^h e^{t\underline{A}'} \begin{bmatrix} C_1' \\ D_{12}' \end{bmatrix} \begin{bmatrix} C_1 & D_{12} \end{bmatrix} e^{t\underline{A}} \, dt,$$

is in fact equal to $\lim_{\gamma \to \infty} J$ and was computed in Chapter 12. Therefore

$$\begin{aligned} J &= I_1 \begin{bmatrix} Q_{11}^{-1} & 0 \\ 0 & 0 \end{bmatrix} N - I_2 + J_\infty \\ &= e^{h\underline{A}'} M \begin{bmatrix} Q_{11}^{-1} & 0 \\ 0 & 0 \end{bmatrix} N - e^{h\underline{A}'} L + J_\infty \\ &= R' M \begin{bmatrix} Q_{11}^{-1} & 0 \\ 0 & 0 \end{bmatrix} N - R' L + J_\infty. \end{aligned}$$

In the following summary, the argument h is dropped [for example, in $Q_{11}(h)$]. First compute the following matrix exponential

$$\begin{bmatrix} P & M & L \\ 0 & Q & N \\ 0 & 0 & R \end{bmatrix}$$

$$= \exp\left\{ h \begin{bmatrix} -\underline{A}' & \begin{bmatrix} C_1' \\ D_{12}' \end{bmatrix} \begin{bmatrix} 0 & C_1 \end{bmatrix} & 0 \\ 0 & E & \begin{bmatrix} C_1' \\ 0 \end{bmatrix} \begin{bmatrix} C_1 & D_{12} \end{bmatrix} \\ 0 & 0 & \underline{A} \end{bmatrix} \right\},$$

where, conformably with the further partitions on the right-hand matrix, each block matrix on the left is further partitioned into a 2×2 block matrix in the obvious way, for example,

$$\begin{bmatrix} R_{11} & R_{12} \\ R_{21} & R_{22} \end{bmatrix} = \begin{bmatrix} A_d & B_{2d} \\ 0 & I \end{bmatrix},$$

the right-hand matrix being $e^{h\underline{A}}$. Then

$$J = R' M \begin{bmatrix} Q_{11}^{-1} & 0 \\ 0 & 0 \end{bmatrix} N - R' L + J_\infty.$$

Computing B_{1d}

The matrix B_{1d} can be found by a Cholesky factorization, or square root computation, of the matrix

$$V := \underline{B}_1 \left(I - \gamma^{-2}\underline{D}_{11}^*\underline{D}_{11}\right)^{-1} \underline{B}_1^*.$$

Thus we shall focus on computing V. Recall that

$$\underline{B}_1 : \ \mathcal{K} \to \mathcal{E}, \quad \underline{B}_1 u = \int_0^h e^{(h-\tau)A} B_1 u(\tau)\, d\tau$$

$$\underline{B}_1^* : \ \mathcal{E} \to \mathcal{K}, \quad (\underline{B}_1^* x)(t) = B_1' e^{(h-t)A'} x.$$

Our first task is to compute the action of $(I - \gamma^{-2}\underline{D}_{11}^*\underline{D}_{11})^{-1}$. As in (13.18), it can be derived that $(I - \gamma^{-2}\underline{D}_{11}^*\underline{D}_{11})^{-1}$ maps u to y, where

$$\begin{aligned} y(t) = &\left[\ \gamma^{-1}B_1' \quad 0\ \right] \\ &\times \left\{ Q(t) \begin{bmatrix} -Q_{11}(h)^{-1} & 0 \\ 0 & 0 \end{bmatrix} \int_0^h Q(h-\tau) \begin{bmatrix} 0 \\ \gamma^{-1}B_1 \end{bmatrix} u(\tau)\, d\tau \right. \\ &\left. + \int_0^t Q(t-\tau) \begin{bmatrix} 0 \\ \gamma^{-1}B_1 \end{bmatrix} u(\tau)\, d\tau \right\} + u(t). \end{aligned}$$

Define the operator

$$\underline{F} = \left(I - \gamma^{-2}\underline{D}_{11}^*\underline{D}_{11}\right)^{-1} \underline{B}_1^* \ : \mathcal{E} \to \mathcal{K}.$$

Then $(\underline{F}x)(t) = f(t)x$, where

$$\begin{aligned} f(t) = &\left[\ B_1' \quad 0\ \right] \\ &\times \left\{ Q(t) \begin{bmatrix} -Q_{11}(h)^{-1} & 0 \\ 0 & 0 \end{bmatrix} \int_0^h Q(h-\tau) \begin{bmatrix} 0 \\ \gamma^{-2}B_1B_1' \end{bmatrix} e^{(h-\tau)A'}\, d\tau \right. \\ &\left. + \int_0^t Q(t-\tau) \begin{bmatrix} 0 \\ \gamma^{-2}B_1B_1' \end{bmatrix} e^{(h-\tau)A'}\, d\tau \right\} + B_1' e^{(h-t)A'}. \end{aligned}$$

The two integrals in $f(t)$ can be evaluated using the identity

$$\frac{d}{d\tau}\left\{ Q(-\tau) \begin{bmatrix} I \\ 0 \end{bmatrix} e^{-\tau A'} \right\} = -Q(-\tau) \begin{bmatrix} 0 \\ \gamma^{-2}B_1B_1' \end{bmatrix} e^{-\tau A'},$$

and after some algebra we obtain

$$f(t) = \left[\ B_1' \quad 0\ \right] Q(t) \begin{bmatrix} Q_{11}(h)^{-1} \\ 0 \end{bmatrix}. \tag{13.21}$$

Thus

$$\begin{aligned} V &= \underline{B}_1\underline{F} \\ &= \int_0^h e^{(h-t)A} \begin{bmatrix} B_1B_1' & 0 \end{bmatrix} Q(t) \begin{bmatrix} Q_{11}(h)^{-1} \\ 0 \end{bmatrix} dt. \end{aligned}$$

The integral here can be computed by the identity

$$\frac{d}{dt}\left\{\gamma^2 e^{-tA} \begin{bmatrix} 0 & I \end{bmatrix} Q(t)\right\} = e^{-tA} \begin{bmatrix} B_1B_1' & 0 \end{bmatrix} Q(t), \tag{13.22}$$

which leads to

$$\begin{aligned} V &= \gamma^2 \begin{bmatrix} 0 & I \end{bmatrix} Q(h) \begin{bmatrix} Q_{11}(h)^{-1} \\ 0 \end{bmatrix} \\ &= \gamma^2 Q_{21}(h)Q_{11}(h)^{-1}. \end{aligned}$$

The details of the derivation are left as an exercise (Exercise 13.7).

Computing A_{dd} and B_{2dd}

Since the matrices A_d and B_{2d} are readily computed, for example, they are contained in the matrix R computed earlier, to compute A_{dd} and B_{2dd} we need to compute the matrix

$$F = \begin{bmatrix} F_1 & F_2 \end{bmatrix} := \underline{B}_1\underline{D}_{11}^*(\gamma^2 - \underline{D}_{11}\underline{D}_{11}^*)^{-1} \begin{bmatrix} \underline{C}_1 & \underline{D}_{12} \end{bmatrix}.$$

Then $A_{dd} = A_d + F_1$ and $B_{2dd} = B_{2d} + F_2$. It turns out that it is easier to compute the matrix

$$\begin{aligned} F' &= \begin{bmatrix} \underline{C}_1^* \\ \underline{D}_{12}^* \end{bmatrix} \left(\gamma^2 - \underline{D}_{11}\underline{D}_{11}^*\right)^{-1} \underline{B}_1^* \\ &= \gamma^{-2}\underline{D}_{11}\left(I - \gamma^{-2}\underline{D}_{11}^*\underline{D}_{11}\right)^{-1} \underline{B}_1^*. \end{aligned}$$

From (13.21), $[(I - \gamma^{-2}\underline{D}_{11}^*\underline{D}_{11})^{-1}\underline{B}_1^*x](t) = f(t)x$, where

$$f(t) = \begin{bmatrix} B_1' & 0 \end{bmatrix} Q(t) \begin{bmatrix} Q_{11}^{-1} \\ 0 \end{bmatrix}.$$

Thus $[\underline{D}_{11}(I - \gamma^{-2}\underline{D}_{11}^*\underline{D}_{11})^{-1}\underline{B}_1^*x](t) = g(t)x$, where

$$\begin{aligned} g(t) &:= \int_0^t C_1 e^{(t-\tau)A} B_1 f(\tau)\, d\tau \\ &= C_1 e^{tA} \int_0^t e^{-\tau A} \begin{bmatrix} B_1B_1' & 0 \end{bmatrix} Q(\tau)\, d\tau \begin{bmatrix} Q_{11}(h)^{-1} \\ 0 \end{bmatrix} \\ &= \gamma^2 C_1 e^{tA} \left\{ e^{-tA} \begin{bmatrix} 0 & I \end{bmatrix} Q(t) - \begin{bmatrix} 0 & I \end{bmatrix} \right\} \begin{bmatrix} Q_{11}(h)^{-1} \\ 0 \end{bmatrix} \\ &\qquad \text{[by (13.22)]} \\ &= \gamma^2 \begin{bmatrix} 0 & C_1 \end{bmatrix} Q(t) \begin{bmatrix} Q_{11}(h)^{-1} \\ 0 \end{bmatrix}. \end{aligned}$$

Therefore

$$\begin{aligned} F' &= \frac{1}{\gamma^2}\int_0^h e^{t\underline{A}'}\begin{bmatrix} C_1' \\ D_{12}' \end{bmatrix} g(t)\,dt \\ &= \int_0^h e^{t\underline{A}'}\begin{bmatrix} C_1' \\ D_{12}' \end{bmatrix}\begin{bmatrix} 0 & C_1 \end{bmatrix} Q(t)\,dt \begin{bmatrix} Q_{11}(h)^{-1} \\ 0 \end{bmatrix}. \end{aligned}$$

The integral involved is exactly the integral I_1 which was computed in (13.20), so

$$F' = e^{h\underline{A}'} M \begin{bmatrix} Q_{11}(h)^{-1} \\ 0 \end{bmatrix},$$

where M was computed before and $e^{h\underline{A}'} = R$. Thus

$$F = \begin{bmatrix} (Q_{11}(h)^{-1})' & 0 \end{bmatrix} M'R.$$

Computational Procedure

Let us recap the steps to compute the matrices in $G_{eq,d}$:

Step 1 Start with G:

$$\hat{g}(s) = \left[\begin{array}{c|cc} A & B_1 & B_2 \\ \hline C_1 & 0 & D_{12} \\ C_2 & 0 & 0 \end{array}\right]$$

and $\gamma > \|\underline{D}_{11}\|$. ($\|\underline{D}_{11}\|$ can be computed using results in the preceding section.)

Step 2 Define the square matrix

$$\underline{A} = \begin{bmatrix} A & B_2 \\ 0 & 0 \end{bmatrix}$$

and compute as in Chapter 12

$$J_\infty = \int_0^h e^{t\underline{A}'}\begin{bmatrix} C_1 & D_{12} \end{bmatrix}'\begin{bmatrix} C_1 & D_{12} \end{bmatrix} e^{t\underline{A}}\,dt.$$

Step 3 Define

$$\begin{aligned} E &= \begin{bmatrix} -A' & -C_1'C_1 \\ B_1B_1'/\gamma^2 & A \end{bmatrix} \\ X &= \begin{bmatrix} C_1 & D_{12} \end{bmatrix}'\begin{bmatrix} 0 & C_1 \end{bmatrix} \\ Y &= \begin{bmatrix} C_1 & 0 \end{bmatrix}'\begin{bmatrix} C_1 & D_{12} \end{bmatrix}, \end{aligned}$$

and compute

$$\begin{bmatrix} P & M & L \\ 0 & Q & N \\ 0 & 0 & R \end{bmatrix} = \exp\left\{ h \begin{bmatrix} -\underline{A}' & X & 0 \\ 0 & E & Y \\ 0 & 0 & \underline{A} \end{bmatrix} \right\}.$$

Partition each block on the left-hand side into a 2×2 block matrix, for example,

$$\begin{aligned} Q &= \begin{bmatrix} Q_{11} & Q_{12} \\ Q_{21} & Q_{22} \end{bmatrix} \\ R &= \begin{bmatrix} R_{11} & R_{12} \\ 0 & I \end{bmatrix}, \end{aligned}$$

and hence $A_d = R_{11}$, $B_{2d} = R_{12}$.

Step 4 Compute

$$\begin{aligned} F &= \begin{bmatrix} F_1 & F_2 \end{bmatrix} = \begin{bmatrix} (Q_{11}^{-1})' & 0 \end{bmatrix} M'R \\ A_{dd} &= A_d + F_1 \\ B_{2dd} &= B_{2d} + F_2. \end{aligned}$$

Step 5 Compute B_{1d} (via Cholesky factorization, or square root computation) satisfying

$$B_{1d}B_{1d}' = \gamma^2 Q_{21} Q_{11}^{-1}.$$

Step 6 Compute

$$J = R'M \begin{bmatrix} Q_{11}^{-1} & 0 \\ 0 & 0 \end{bmatrix} N - R'L + J_\infty$$

and C_{1d} and D_{12d} (via Cholesky factorization, or square root computation) satisfying

$$\begin{bmatrix} C_{1d} & D_{12d} \end{bmatrix}' \begin{bmatrix} C_{1d} & D_{12d} \end{bmatrix} = J.$$

Step 7 Then

$$\hat{g}_{eq,d}(\lambda) = \left[\begin{array}{c|cc} A_{dd} & B_{1d} & B_{2dd} \\ \hline C_{1d} & 0 & D_{12d} \\ C_2 & 0 & 0 \end{array} \right].$$

Several examples using this procedure will be given in the next two sections.

13.7 $\mathcal{H}_\infty$ SD Analysis

With the results derived so far, we can now answer the first question—the analysis question—asked at the beginning of this chapter: Given G and K_d in the standard SD setup, how to compute the $\mathcal{L}_2(\mathbb{R}_+)$-induced norm of T_{zw}? This can be resolved by the following analysis procedure:

Step 1 Start with G, K_d, h:

$$\hat{g}(s) = \left[\begin{array}{c|cc} A & B_1 & B_2 \\ \hline C_1 & 0 & D_{12} \\ C_2 & 0 & 0 \end{array}\right],$$

$$\hat{k}_d(\lambda) = \left[\begin{array}{c|c} A_K & B_K \\ \hline C_K & D_K \end{array}\right].$$

Assume internal stability.

Step 2 Compute $\|\underline{D}_{11}\|$ (Section 13.5).

Step 3 For $\gamma > \|\underline{D}_{11}\|$, compute the $\mathcal{H}_\infty$ discretization

$$\hat{g}_{eq,d}(\lambda) = \left[\begin{array}{c|cc} A_{dd} & B_{1d} & B_{2dd} \\ \hline C_{1d} & 0 & D_{12d} \\ C_2 & 0 & 0 \end{array}\right]$$

Step 4 Form

$$\left[\begin{array}{c|c} A_{cld} & B_{cld} \\ \hline C_{cld} & 0 \end{array}\right] = \left[\begin{array}{cc|c} A_{dd} + B_{2dd}D_KC_2 & B_{2dd}C_K & B_{1d} \\ B_KC_2 & A_K & 0 \\ \hline C_{1d} + D_{12d}D_KC_2 & D_{12d}C_K & 0 \end{array}\right]$$

(if necessary, take a minimal realization) and the symplectic pair

$$(S_l, S_r) = \left(\begin{bmatrix} A_{cld} & 0 \\ -C'_{cld}C_{cld}/\gamma & I \end{bmatrix}, \begin{bmatrix} I & -B_{cld}B'_{cld}/\gamma \\ 0 & A'_{cld} \end{bmatrix} \right).$$

Compute

$$\gamma_{max} := \max\{\gamma : (S_l, S_r) \text{ has an eigenvalue on } \partial\mathbb{D}\}.$$

Step 5 $\|T_{zw}\| = \max\{\|\underline{D}_{11}\|, \gamma_{max}\}$.

A bisection search can be used in Steps 3 and 4 to compute γ_{max}; but attention must be paid to the selection of upper and lower bounds for γ_{max}; see Example 13.7.2 below. In most cases, $\gamma_{max} > \|\underline{D}_{11}\|$ and then in Step 5 we have simply: $\|T_{zw}\| = \gamma_{max}$.

Let us illustrate with a few examples.

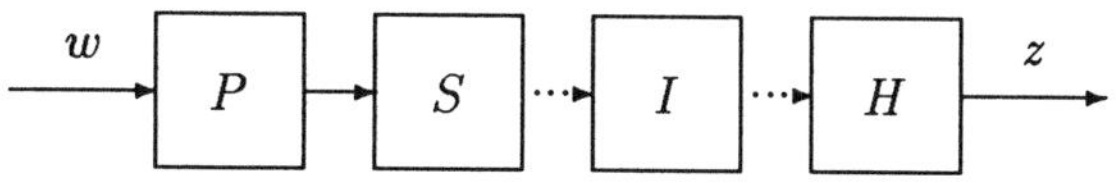

Figure 13.7: The system HSP.

Example 13.7.1 As a check of the formulas, let us recompute the norm of HSP, where $\hat{p}(s) = 1/(s+1)$; this operator is the system from w to z in the open-loop system in Figure 13.7. The norm of HSP was computed in Example 10.5.1:

$$\|HSP\| = \left[\frac{h(1+\mathrm{e}^{-h})}{2(1-\mathrm{e}^{-h})}\right]^{1/2}.$$

In particular, for $h = 2$, $\|HSP\| = 1.1459$.

It can be verified that the map $w \mapsto z$ in Figure 13.7 is exactly the map T_{zw} of the standard SD system with

$$G = \begin{bmatrix} 0 & I \\ P & 0 \end{bmatrix}, \quad K_d = I.$$

Thus we can also compute $\|HSP\|$ via the general procedure above. Note that for this example, $\|\underline{D}_{11}\| = 0$. To find γ_{max}, define $\beta(\gamma)$ to be the minimum distance from the eigenvalues of (S_l, S_r) to the unit circle. For $h = 2$, we compute $\beta(\gamma)$ for a range of γ and plot the function in Figure 13.8, from which we see that γ_{max} agrees with the norm of HSP given before, namely, the maximum γ such that $\beta = 0$ equals 1.1459.

Let us look at another simple example but with a SD feedback.

Example 13.7.2 The SD system is depicted in Figure 13.9 with $\hat{p}(s) = 1/s$ and $\hat{k}_d(\lambda) = 1$. This can be put into the standard setup with

$$\hat{g}(s) = \begin{bmatrix} \hat{p}(s) & -\hat{p}(s) \\ \hat{p}(s) & -\hat{p}(s) \end{bmatrix} = \left[\begin{array}{c|cc} 0 & 1 & -1 \\ \hline 1 & 0 & 0 \\ 1 & 0 & 0 \end{array}\right].$$

It is easily checked that the SD system is internally stable iff $0 < h < 2$. We would like to compute the $\mathcal{H}_\infty$-norm of $T_{zw} : w \mapsto z$.

First, we take, for example, $h = 1.5$, and compute $\|\underline{D}_{11}\| = 0.955$ using either of the methods in Section 13.5. To find γ_{max}, again we let $\beta(\gamma)$ denote the minimum distance from the eigenvalues of (S_l, S_r) to the unit circle. The function $\beta(\gamma)$ for γ varying from 0.995 to 2.2 is graphed in Figure 13.10. We see that $\gamma_{max} = 1.74$ and this is also $\|T_{zw}\|$.

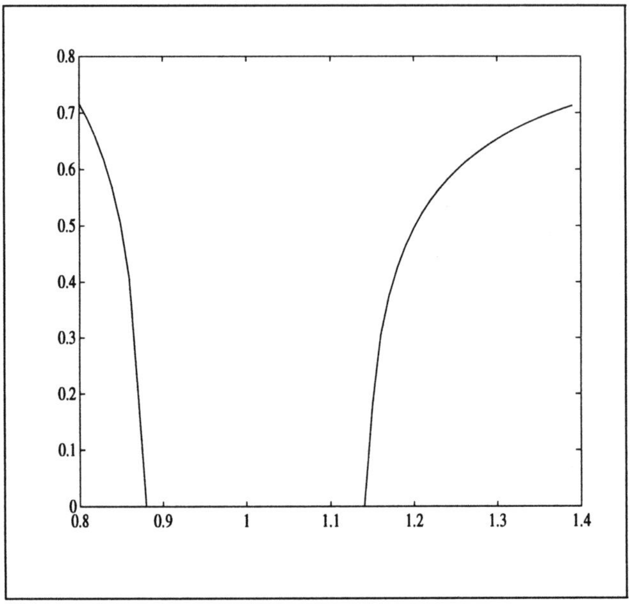

Figure 13.8: β versus γ.

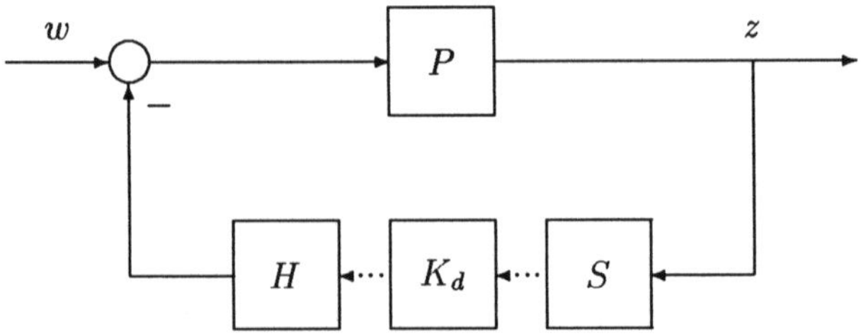

Figure 13.9: A SD system.

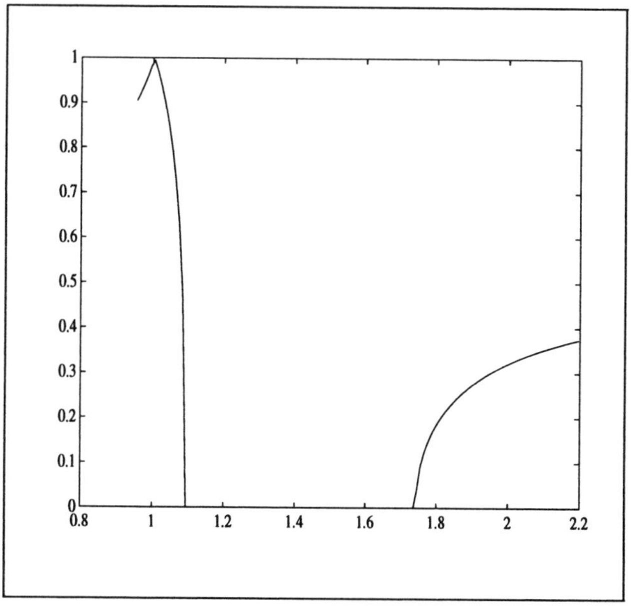

Figure 13.10: β versus γ for $h = 1.5$.

Next, we want to compute $\|T_{zw}\|$ for a range of h and see the performance degradation when h becomes large. The method we used before for computing γ_{max}, namely, computing $\beta(\gamma)$ for a fine grid of γ, gives good illustration but requires considerable computation even for a single value of h. A more efficient method is bisection search. Since γ_{max} is the maximum γ such that $\beta(\gamma) = 0$, a bisection search for γ_{max} goes as follows:

Step 1 Start with a lower bound, γ_l, and an upper bound, γ_u, with the following properties: $\gamma_l < \gamma_{max} < \gamma_u$, $\beta(\gamma_l) = 0$, and $\beta(\gamma_u) \neq 0$. Then we know that γ_{max} is the only point at which the curve $\beta(\gamma)$ breaks away from the horizontal axis. (γ_l and γ_u may be obtained by trial and error.)

Step 2 Let the middle point between γ_l and γ_u be γ_m and test $\beta(\gamma_m)$: If $\beta(\gamma_m) = 0$, improve γ_l to γ_m; if $\beta(\gamma_m) \neq 0$, improve γ_u to γ_m. In this way we narrow the possible interval for γ_{max} by a half.

Step 3 Repeat Step 2 until a satisfactory accuracy is achieved.

Using this method we can compute γ_{max} to any desired accuracy.

Now we compute $\|T_{zw}\|$ for $h \in [1, 1.9]$ via bisection search and plot the result in Figure 13.11. It is clear that $\|T_{zw}\|$ is an increasing function of h: The larger h, the worse the performance. Theoretically, $\lim_{h\to 2} \|T_{zw}\| = \infty$, because the system becomes unstable when $h \to 2$; this trend is seen in

Figure 13.11. To study the performance when $h \to 0$, we first regard the SD system as a digital implementation of the analog system with the controller $\hat{k}(s) = 1$; for this analog system, $\|T_{zw}\|$ equals the $\mathcal{H}_\infty$-norm of $1/(s+1)$, namely, 1. With this in mind, Figure 13.11 also shows that the SD system recovers the analog performance as $h \to 0$.

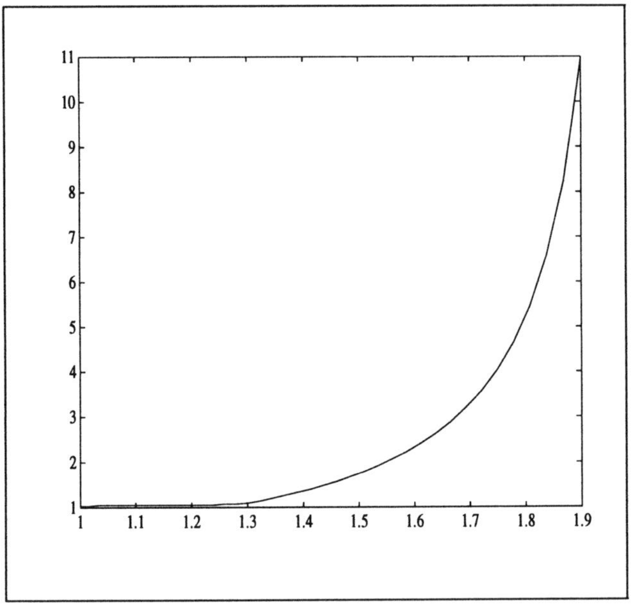

Figure 13.11: $\|T_{zw}\|$ versus h.

Example 13.7.3 We now return to the scenario of analog $\mathcal{H}_\infty$ design and digital implementation in Example 2.3.1. The analog controller was designed to get good tracking for a range of frequencies and this was achieved via minimizing the $\mathcal{H}_\infty$-norm of the system

$$T_{zw} : \begin{bmatrix} w_1 \\ w_2 \end{bmatrix} \mapsto \begin{bmatrix} z_1 \\ z_2 \end{bmatrix}$$

in Figure 2.9, redrawn here as Figure 13.12, with $\epsilon_1 = \epsilon_2 = 0.01$ and

$$\hat{p}(s) = \frac{20 - s}{(s + 0.01)(s + 20)}, \quad \hat{f}(s) = \frac{1}{(0.5/\pi)s + 1}, \quad \hat{w}(s) = \frac{1}{[(2.5/\pi)s + 1]^2}.$$

The analog controller designed using the MATLAB function *hinfsyn* is

$$\hat{k}(s) = \frac{1.4261 \times 10^5 (s + 20)(s + 6.2832)(s + 3.9436)(s + 0.01)}{(s + 631.69)(s + 159.56)(s + 39.230)(s + 1.3212)(s + 1.1876)}$$

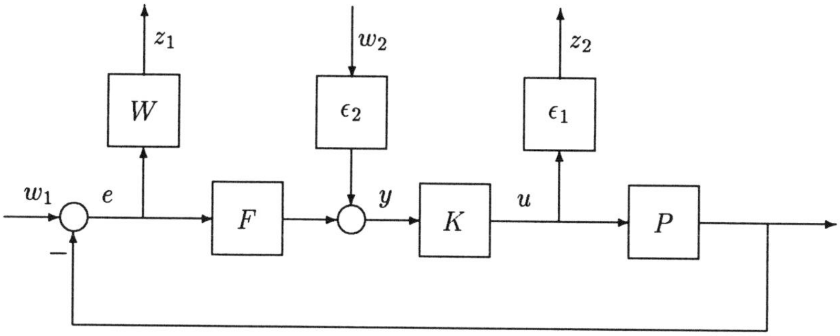

Figure 13.12: Analog feedback system.

and achieves $||T_{zw}|| = 0.0813$. When K is implemented digitally, how does $||T_{zw}||$ change with the sampling period h? (In Example 2.3.1, the performance of the digital implementation using the step-invariant transformation was analyzed *approximately*, since we could not compute $||T_{zw}||$ for the SD system then.)

Notice that the small number ϵ_2 (and ϵ_1) was introduced to regularized the analog problem; its effect on $||T_{zw}||$ of the analog system is negligible. However, for the SD systems, ϵ_2 must be 0 for otherwise $||T_{zw}||$ would be infinite. Thus in the following computation we take $\epsilon_2 = 0$.

Digital implementations of K via step-invariant transformation and bilinear transformation will be considered. First, step-invariant transformation. As in Example 11.1.1, we can determine that the resulting SD system is internally stable for $0 < h < 0.04$ (quite a small range). Using the procedures in this chapter we can compute $||T_{zw}||$ for a given h. Figure 13.13 shows $||T_{zw}||$ as a function of h, $0 < h < 0.02$ (solid curve). It is clear that the $\mathcal{H}_\infty$ performance degrades very quickly when h increases.

Next, bilinear transformation. In this case, the SD system is internally stable for $0 < h < 0.46$. Figure 13.13 gives $||T_{zw}||$ for $0 < h < 0.4$ (dash curve). Note that now the performance degrades much more slowly. Note also that $||T_{zw}||$ is not a monotonic function of h.

This example shows again that bilinear transformation is superior to the step-invariant transformation in digital implementation.

13.8 $\mathcal{H}_\infty$ SD Synthesis

The second problem posed at the beginning of this chapter is $\mathcal{H}_\infty$-optimal SD synthesis: Design a K_d to give internal stability and minimize $||T_{zw}||$.

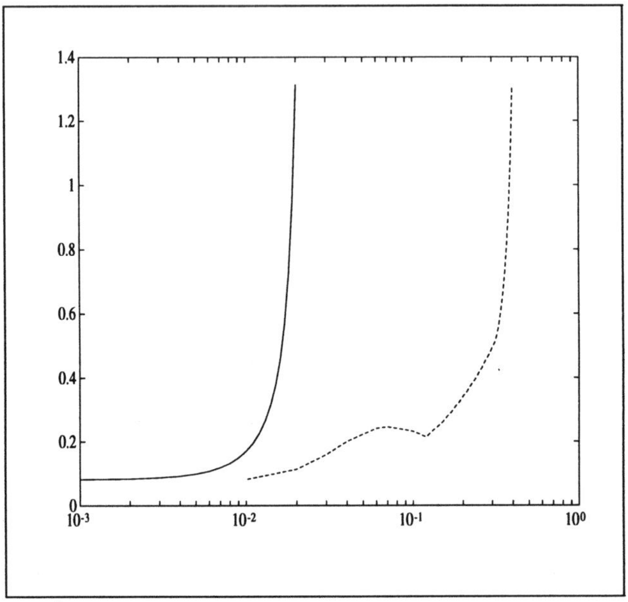

Figure 13.13: $\|T_{zw}\|$ versus h: Digital implementation via step-invariant transformation (solid) and via bilinear transformation (dash).

This again can be achieved using $\mathcal{H}_\infty$ discretization. The design procedure is summarized below:

Step 1 Start with G and h:

$$\hat{g}(s) = \left[\begin{array}{c|cc} A & B_1 & B_2 \\ \hline C_1 & 0 & D_{12} \\ C_2 & 0 & 0 \end{array}\right].$$

Step 2 Compute $\|\underline{D}_{11}\|$ (Section 13.5).

Step 3 For $\gamma > \|\underline{D}_{11}\|$, compute the $\mathcal{H}_\infty$ discretization $G_{eq,d}$:

$$\hat{g}_{eq,d}(\lambda) = \left[\begin{array}{c|cc} A_{dd} & B_{1d} & B_{2dd} \\ \hline C_{1d} & 0 & D_{12d} \\ C_2 & 0 & 0 \end{array}\right].$$

Step 4 Let $T_{eq,d}$ be the closed-loop map with discrete-time plant $G_{eq,d}$ and controller K_d. Test the solvability condition for the standard discrete-time $\mathcal{H}_\infty$ problem: There exists a $K_{eq,d}$ to achieve internal stability and $\|\hat{t}_{eq,d}\|_\infty < \gamma$. If the test fails, increase γ; otherwise, decrease γ; and go back to Step 3.

Step 5 When a satisfatory γ is found, solve the discrete-time $\mathcal{H}_\infty$ problem $\|\hat{t}_{eq,d}\|_\infty < \gamma$ for a stabilizing K_d. This K_d also achieves $\|T_{zw}\| < \gamma$.

Steps 3 and 4 can be combined together to give a bisection search for the optimal γ, γ_{opt}. Therefore, the procedure can generate a controller which gives a $\mathcal{H}_\infty$ performance which is arbitrarily close to optimality.

Several examples will be given below.

Example 13.8.1 In Example 13.7.3, we looked at discretizing an $\mathcal{H}_\infty$-optimal analog controller and saw that the performance degrades rapidly as the sampling period increases. Now we consider direct $\mathcal{H}_\infty$ SD design of the same system. The setup is depicted in Figure 13.14. Note that we have to set

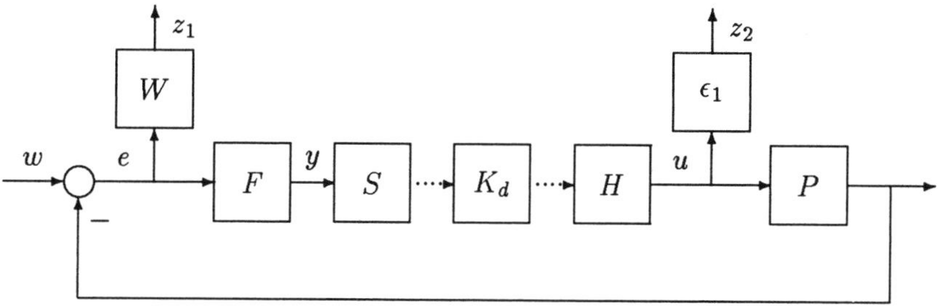

Figure 13.14: SD feedback system.

$\epsilon_2 = 0$ for the same reason stated earlier, namely, that the sampler S must be lowpass filtered.

Let T_{zw} be the SD system from w to $z = \begin{bmatrix} z_1 \\ z_2 \end{bmatrix}$ in Figure 13.14. Fix $h > 0$. Let $\gamma_{opt}(h)$ be the optimal performance, that is,

$$\gamma_{opt}(h) = \inf\{\|T_{zw}\| : \ K_d \text{ stabilizing }\}.$$

This can be computed using the preceding procedure. In order to check the discrete-time solvability condition in Step 4, we transform the problem into an analog one via bilinear transformation (see Section 7.2) and then use the MATLAB function *hinfsyn*. For $h = 0.1, 0.5$, we obtain $\gamma_{opt} = 0.178, 0.646$, respectively. The corresponding optimal controllers are also given by *hinfsyn*; transforming them back to discrete time, we get the optimal K_d. For example, for $h = 0.1$,

$$\hat{k}_{d,opt}(\lambda) = \frac{7.9354(\lambda - 7.3891)(\lambda - 1.8745)(\lambda - 1.3554)(\lambda - 1.001)(\lambda + 1.0007)}{(\lambda + 6.666)(\lambda^2 - 0.3467\lambda + 3.096)(\lambda - 1.1793)(\lambda - 1.0233)}.$$

This process is repeated in order to generate the plot of γ_{opt} versus h, as is given in Figure 13.15 (solid curve). Also shown are the plots from

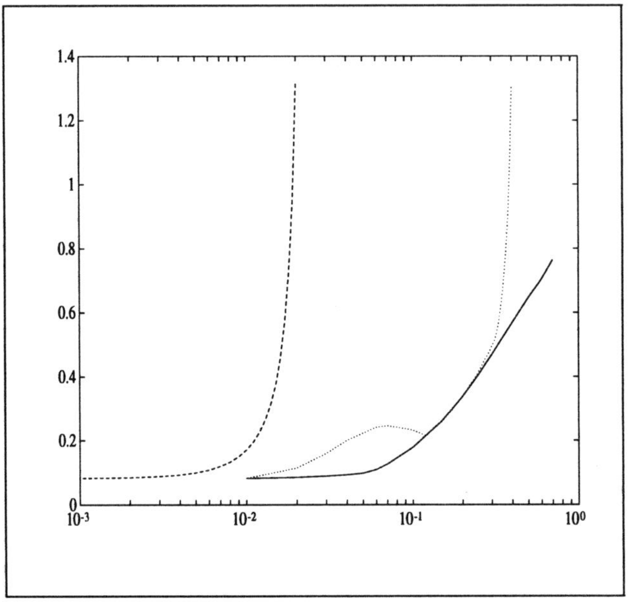

Figure 13.15: γ_{opt} versus h (solid); $\|T_{zw}\|$ versus h for digital implementation via step-invariant transformation (dash); $\|T_{zw}\|$ versus h for digital implementation via bilinear transformation (dot).

Figure 13.13 for the designs by discretizing the optimal analog controller. We see the obvious advantage of the direct SD design over the discretized analog designs: For the same sampling period, the SD design achieves much better performance than the discretized design by step-invariant transformation, and much better performance than the discretized design by bilinear transformation for $h > 0.3$. Furthermore, the SD design method always produces a stabilizing controller, whereas the discretization methods produce stabilizing controllers only for h small enough.

Another way to use the curves of performance versus h is to answer the following question: Given a pre-specified performance level γ, determine the largest sampling period, h_{max}, required to achieve that performance. From the performance-versus-h plots, we can find h_{max} for a given value of γ for the three methods: discretizing the analog design via step-invariant transformation and via bilinear transformation, and direct SD design. This is summarized in the following table:

	step-invariant trans.	bilinear trans.	direct SD
$\gamma = 0.1$	$h_{max} = 0.005$	$h_{max} = 0.015$	$h_{max} = 0.05$
$\gamma = 0.2$	$h_{max} = 0.0112$	$h_{max} = 0.04$	$h_{max} = 0.113$
$\gamma = 0.5$	$h_{max} = 0.0164$	$h_{max} = 0.3$	$h_{max} = 0.33$

Recall that the optimal analog performance is 0.0813. So, for example, a performance of $\gamma = 0.1$ in the table means allowing about 23% performance degradation from the optimal analog system; in this case, direct SD design requires a sampling rate which is 10 times slower than digital implementation via step-invariant transformation and 3.3 times slower than via bilinar transformation.

Exercises

13.1 Consider the setup

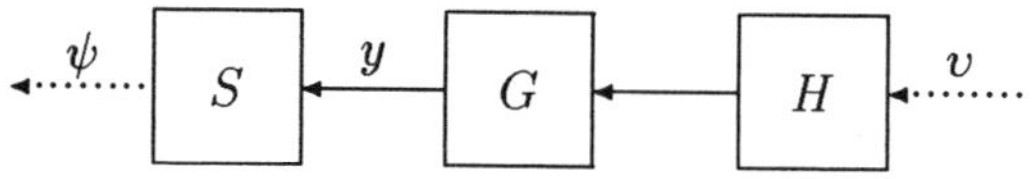

Assume G is causal, LTI, SISO, and stable. Suppose v belongs to the vector space of all discrete-time sinusoids of frequency θ, that is,

$$v \in \left\{\phi : \phi(k) = e^{jk\theta}a, \ a \in \mathbb{R}\right\}.$$

Then so does ψ (because SGH is LTI). To what vector space does y belong?

13.2 If a continuous-time signal is sampled and the resulting discrete-time signal is a sinusoid, of course this doesn't imply that the continuous-time signal is a sinusoid. What about if the continuous-time signal is first shifted by an arbitrary amount, and then sampled?

Consider the setup

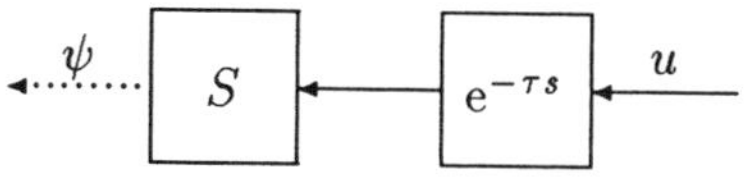

Suppose that for some input $u(t)$, $\psi(k)$ is a discrete-time sinusoid of frequency θ for every time delay $0 < \tau < h$. Show that $u(t)$ is not necessarily a continuous-time sinusoid. Hint: Try $u(t) = e^{j\theta t/h}w(t)$, with w h-periodic.

13.3 This exercise completes the proof of Theorem 13.4.1. Assume A_{cld} is stable, $\gamma > \|\underline{D}_{11}\|$, and $\|\hat{t}_{eq,d}\|_\infty < \gamma$. Show that A_{cl} is stable by the following steps:

1. Introduce an intermediate system

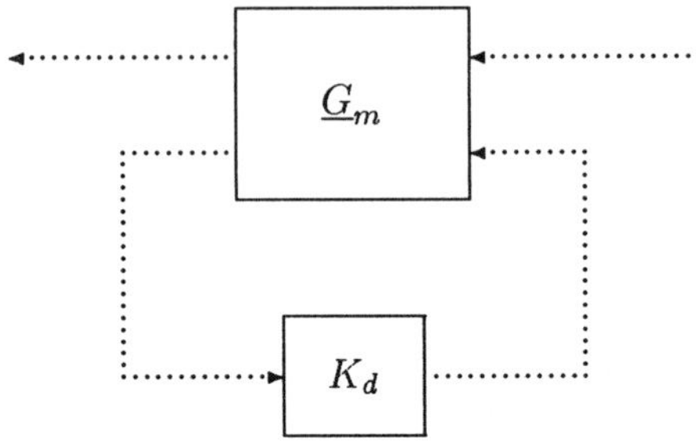

with

$$\hat{\underline{g}}_m(\lambda) = \left[\begin{array}{c|cc} A_{dd} & \gamma \underline{B}_1 \underline{R}_1 & B_{2dd} \\ \hline \gamma \underline{R}_2 \underline{C}_1 & 0 & \gamma \underline{R}_2 \underline{D}_{12} \\ C_2 & 0 & 0 \end{array}\right]$$

$$\underline{R}_1 = (\gamma^2 - \underline{D}_{11}^* \underline{D}_{11})^{-1/2}$$
$$\underline{R}_2 = (\gamma^2 - \underline{D}_{11} \underline{D}_{11}^*)^{-1/2},$$

and denote the closed-loop map by

$$\hat{\underline{t}}_m(\lambda) = \left[\begin{array}{c|c} A_{cld} & \underline{B}_{clm} \\ \hline \underline{C}_{clm} & 0 \end{array}\right].$$

Show that $\|\hat{\underline{t}}_m\|_\infty < \gamma$.

2. Show that A_{cl} is the A-matrix of $(\gamma^2 + \underline{D}_{11}^* \hat{\underline{t}}_m)^{-1}$; conclude that A_{cl} stable.

13.4 Let G denote the SISO LTI system with transfer function $1/(s+1)$. Compute the induced norm of the compression of G to $\mathcal{L}_2[0,1]$.

13.5 Show that $\underline{D}_{11}^*$ is described by the state equations in (13.10) and (13.11) with the boundary condition $x_2(h) = 0$.

13.6 Assume (A, B_1) is controllable and (C_1, A) is observable. Show that the pair in (13.14) is observable.

13.7 Verify in detail the derivation of the formulas for B_{1d} in Section 13.6.

13.8 The methods developed in this chapter provide another way to compute the $\mathcal{H}_\infty$-norm of an analog system via the eigenproblem of some symplectic pair of matrices. For the analog system given by

$$\hat{p}(s) = \left[\begin{array}{c|c} A & B \\ \hline C & 0 \end{array}\right]$$

with A stable, list the steps in computing $\|\hat{p}\|_\infty$ via this approach. Test your formulas by computing the $\mathcal{H}_\infty$-norm of

$$\hat{p}(s) = \frac{1}{s^2 + 2s + 2}.$$

Notes and References

Frequency-response functions for SD systems were introduced by Bamieh and Pearson [16], who defined them as transfer functions of lifted systems. Araki, Yamamoto, and co-workers [152], [153], [6], and [5] independently developed frequency-response functions in terms of Fourier series. This concept was also used by Dullerud and Glover [44].

The problem of computing and optimizing the $\mathcal{L}_2$-induced norm of general SD systems was treated by Hayakawa, Hara, and Yamamoto [73], Kabamba and Hara [82], [83], Toivonen [140], Bamieh and Pearson [16], Tadmor [136], Sivashankar and Khargonekar [126], and Sun *et al.* [132], [133].

Theorem 13.2.1 is from [135]. Theorem 13.4.1 is from [16], though not by symplectic pairs as here. The computations in Sections 13.5 and 13.6 are based on [16]. Computing $\|\underline{D}_{11}\|$ is related to $\mathcal{H}_\infty$ control of time-delay systems [50], [157]; Theorem 13.5.1 is due to Zhou and Khargonekar [157], where the minimality condition is assumed. The formula for B_{1d} is from [16]; the other formulas are different from and more general than those in [16] because here $D_{12} \neq 0$. It is possible to generalized the formulas to the case of $D_{11} \neq 0$; but this would considerably complicate the matter.

Bisection search was used for computing $\mathcal{H}_\infty$-norms of FDLTI systems in [20].

Appendix A

State Models

This appendix reviews state models for continuous and discrete time, and the basic notions of controllability, observability, stabilizability, and detectability.

Continuous Time

A finite-dimensional, linear, time-invariant (FDLTI), causal system has a state model of the form

$$\begin{aligned} \dot{x} &= Ax + Bu \\ y &= Cx + Du, \end{aligned}$$

where $A \in \mathbb{R}^{n\times n}$, $B \in \mathbb{R}^{n\times m}$, $C \in \mathbb{R}^{p\times n}$, and $D \in \mathbb{R}^{p\times m}$. (As usual, $\dot{x}$ denotes the time derivative of x.) If $m = p = 1$, this is a single-input, single-output (SISO) system; if $m = 1$, $p > 1$, it is single-input, multi-output (SIMO); and so on.

Frequently one wants to go back and forth between transfer functions and state models. Let us look first at going from a state model to a transfer function. In the above state model take Laplace transforms with zero initial condition:

$$\begin{aligned} s\hat{x}(s) &= A\hat{x}(s) + B\hat{u}(s) \\ \hat{y}(s) &= C\hat{x}(s) + D\hat{u}(s). \end{aligned}$$

Thus the transfer function is

$$\hat{g}(s) = D + C(s - A)^{-1}B.$$

It is a $p \times m$ matrix of rational functions.

Now the other way, from a transfer function to a state model. First, the SISO case: A state realization of

$$\hat{g}(s) = d + \frac{b_{n-1}s^{n-1} + \ldots + b_0}{s^n + a_{n-1}s^{n-1} + \ldots + a_1 s + a_0}$$

is

$$A = \begin{bmatrix} 0 & 1 & 0 & & & \\ 0 & 0 & 1 & \ddots & & \\ & & \ddots & \ddots & \ddots & \\ & & & \ddots & \ddots & 0 \\ & & & & 0 & 1 \\ -a_0 & -a_1 & & \cdots & -a_{n-2} & -a_{n-1} \end{bmatrix}, \quad B = \begin{bmatrix} 0 \\ \vdots \\ 0 \\ 1 \end{bmatrix}$$

$$C = \begin{bmatrix} b_0 & \cdots & b_{n-1} \end{bmatrix}, \qquad D = d.$$

If $\hat{g}(s)$ is a matrix, a realization can be obtained from realizations for the individual entries of $\hat{g}(s)$. For example, suppose

$$\hat{g}(s) = \begin{bmatrix} \hat{g}_1(s) \\ \hat{g}_2(s) \end{bmatrix}, \qquad \hat{g}_i(s) = \left[\begin{array}{c|c} A_i & B_i \\ \hline C_i & D_i \end{array}\right]$$

($\hat{g}_1$, $\hat{g}_2$ could be submatrices). The block diagram is

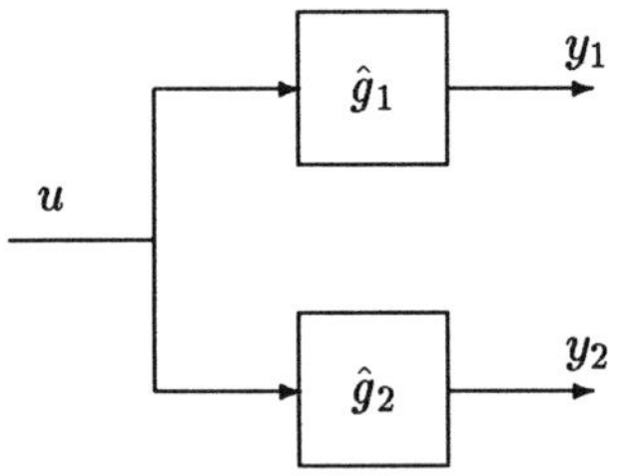

$$y = \begin{bmatrix} y_1 \\ y_2 \end{bmatrix}$$

The state equations are

$$\begin{aligned} \dot{x}_1 &= A_1 x_1 + B_1 u \\ \dot{x}_2 &= A_2 x_2 + B_2 u \\ y_1 &= C_1 x_1 + D_1 u \\ y_2 &= C_2 x_2 + D_2 u. \end{aligned}$$

If we take $x = \begin{bmatrix} x_1 \\ x_2 \end{bmatrix}$, we get

$$\hat{g}(s) = \left[\begin{array}{cc|c} A_1 & 0 & B_1 \\ 0 & A_2 & B_2 \\ \hline C_1 & 0 & D_1 \\ 0 & C_2 & D_2 \end{array}\right].$$

On the other hand, if

$$\hat{g}(s) = \begin{bmatrix} \hat{g}_1(s) & \hat{g}_2(s) \end{bmatrix}, \qquad \hat{g}_i(s) = \left[\begin{array}{c|c} A_i & B_i \\ \hline C_i & D_i \end{array}\right]$$

(again, $\hat{g}_1$, $\hat{g}_2$ could be submatrices), then the block diagram is

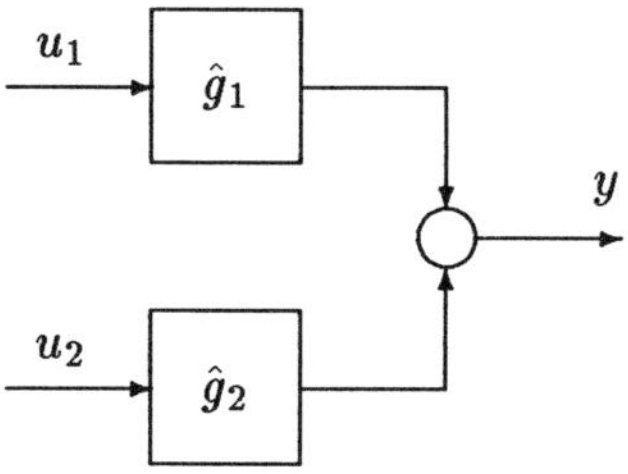

$$u = \begin{bmatrix} u_1 \\ u_2 \end{bmatrix}$$

whose state equations are

$$\begin{aligned} \dot{x}_1 &= A_1 x_1 + B_1 u_1 \\ \dot{x}_2 &= A_2 x_2 + B_2 u_2 \\ y &= C_1 x_1 + D_1 u_1 + C_2 x_2 + D_2 u_2. \end{aligned}$$

Thus

$$\hat{g}(s) = \left[\begin{array}{cc|cc} A_1 & 0 & B_1 & 0 \\ 0 & A_2 & 0 & B_2 \\ \hline C_1 & C_2 & D_1 & D_2 \end{array} \right].$$

From these two examples one can do a general matrix $\hat{g}(s)$. For example, if

$$\hat{g} = \begin{bmatrix} \hat{g}_{11} & \hat{g}_{12} \\ \hat{g}_{21} & \hat{g}_{22} \end{bmatrix}$$

one can get four realizations for the four entries, then realize the two rows,

$$\hat{g}_1 := \begin{bmatrix} \hat{g}_{11} & \hat{g}_{12} \end{bmatrix}, \qquad \hat{g}_2 := \begin{bmatrix} \hat{g}_{21} & \hat{g}_{22} \end{bmatrix},$$

and finally realize $\hat{g}$ via

$$\hat{g} = \begin{bmatrix} \hat{g}_1 \\ \hat{g}_2 \end{bmatrix}.$$

The MATLAB function *tf2ss* will take as input a SIMO transfer function, that is, a column vector; so to do a matrix one applies *tf2ss* column-wise.

Discrete Time

Now we turn to state models for discrete-time systems. We use "dot" in discrete time to denote time advance, for example, $\dot{\xi}(k) := \xi(k+1)$. The general form of a FDLTI discrete-time state model is

$$\begin{aligned} \dot{\xi} &= A\xi + Bv \\ \psi &= C\xi + Dv, \end{aligned}$$

where as before $A \in \mathbb{R}^{n \times n}$, $B \in \mathbb{R}^{n \times m}$, $C \in \mathbb{R}^{p \times n}$, and $D \in \mathbb{R}^{p \times m}$. The corresponding transfer function is

$$\hat{g}(\lambda) = D + \lambda C(I - \lambda A)^{-1} B.$$

Controllability: Continuous Time

The system

$$\dot{x} = Ax + Bu, \quad x(0) = 0 \tag{A.1}$$

is *controllable* [or the pair of matrices (A, B) is controllable] if for every target time $t_1 > 0$ and every target vector v, there is a control signal $u(t)$, $0 \leq t \leq t_1$, such that $x(t_1) = v$. Controllability is a property just of the two matrices A and B. There is a simple algebraic test:

$$(A, B) \text{ is controllable} \Longleftrightarrow$$

$$\text{rank} \begin{bmatrix} B & AB & A^2B & \ldots & A^{n-1}B \end{bmatrix} = n. \tag{A.2}$$

The latter matrix, called the *controllability matrix*, is $n \times nm$. (See the MATLAB function *ctrb*.)

The rank test (A.2) is numerically ill-conditioned. It is convenient to say that an eigenvalue λ of A is *controllable* if

$$\text{rank} \begin{bmatrix} A - \lambda & B \end{bmatrix} = n.$$

Then a more useful test is as follows: (A, B) is controllable iff each eigenvalue of A is controllable.

The most important fact about controllable systems is that their eigenvalues can be reassigned by state feedback. That is, consider applying the control signal

$$u = Fx + v, \quad v \text{ an external input}$$

to the given state model, (A.1). That is, (A, B) is transformed to $(A+BF, B)$. Then (A, B) is controllable iff for every set of desired eigenvalues, there exists a matrix F such that $A + BF$ has exactly that set of eigenvalues. Of course, the desired set of eigenvalues must have conjugate symmetry; that is, if λ is a desired eigenvalue, so too must $\bar{\lambda}$ be.

Now we move to the related notion of stabilizability: (A, B) is *stabilizable* if there exists a matrix F such that $A+BF$ is stable, that is, all its eigenvalues are in the open left half-plane. Not surprisingly, (A, B) is stabilizable iff the (unstable) eigenvalues of A in Re $s \geq 0$ are controllable.

Controllability: Discrete Time

The relevant system is

$$\dot{\xi} = A\xi + Bv. \tag{A.3}$$

The story here is almost the same as before; the only difference is that we can't expect to hit a target vector in an arbitrarily small time because time is discrete. The results are summarized as follows:

1. System (A.3), or the pair (A, B), is defined to be *controllable* if every target vector is reachable at some time by some input sequence, starting from $\xi(0) = 0$.

2. (A, B) is controllable iff
$$\operatorname{rank}\begin{bmatrix} B & AB & \dots & A^{n-1}B \end{bmatrix} = n.$$

3. (A, B) is controllable iff
$$\operatorname{rank}\begin{bmatrix} A-\lambda & B \end{bmatrix} = n \quad \text{for each eigenvalue } \lambda \text{ of } A.$$

4. (A, B) is defined to be *stabilizable* if there exists a matrix F such that the eigenvalues of $A + BF$ are all inside the open unit disk.

5. (A, B) is stabilizable iff
$$\operatorname{rank}\begin{bmatrix} A-\lambda & B \end{bmatrix} = n \quad \text{for each eigenvalue } \lambda \text{ of } A \text{ with } |\lambda| \geq 1.$$

Observability: Continuous Time

The system

$$\dot{x} = Ax, \quad y = Cx$$

[or more commonly the pair of (C, A)] is *observable* if for every $x(0)$ and $t_1 > 0$, $x(0)$ can be computed from the data $\{y(t) : 0 \leq t \leq t_1\}$. The main result is that the following five conditions are equivalent:

1. (C, A) is observable.

2. (A', C') is controllable.

3. $$\operatorname{rank}\begin{bmatrix} C \\ CA \\ \vdots \\ CA^{n-1} \end{bmatrix} = n.$$

   (This matrix is called the observability matrix.)

4. $$\operatorname{rank}\begin{bmatrix} A-\lambda \\ C \end{bmatrix} = n \quad \text{for each eigenvalue } \lambda \text{ of } A.$$

5. The eigenvalues of $A+HC$ can be arbitrarily assigned by suitable choice of H.

By analogy with controllable eigenvalues and in view of the fourth condition above, we say that an eigenvalue λ of A is *observable* if

$$\operatorname{rank}\begin{bmatrix} A-\lambda \\ C \end{bmatrix} = n.$$

If an H exists to make $A+HC$ stable, (C,A) is said to be *detectable.* The following three conditions are equivalent:

1. (C,A) is detectable.
2. (A',C') is stabilizable.
3. Every eigenvalue of A in Re $s \geq 0$ is observable.

Observability: Discrete Time

The relevant system is

$$\dot{\xi} = A\xi, \quad \psi = C\xi.$$

Again, the results parallel the continuous-time case, the major difference being that the observation interval cannot be arbitrarily small. The results are as follows:

1. The pair (C,A) is *observable* if the initial state can be computed from the output sequence

   $$\{\psi(0), \psi(1), \ldots, \psi(k_1)\}$$

   for some sufficiently large k_1.

2. (C,A) is observable iff

   $$\operatorname{rank}\begin{bmatrix} C \\ CA \\ \vdots \\ CA^{n-1} \end{bmatrix} = n.$$

3. (C,A) is observable iff

   $$\operatorname{rank}\begin{bmatrix} A-\lambda \\ C \end{bmatrix} = n \quad \text{for each eigenvalue } \lambda \text{ of } A.$$

4. (C,A) is defined to be *detectable* if there exists a matrix H such that all the eigenvalues of $A+HC$ are inside the open unit disk.

5. (C,A) is detectable iff

   $$\operatorname{rank}\begin{bmatrix} A-\lambda \\ C \end{bmatrix} = n \quad \text{for each eigenvalue } \lambda \text{ of } A \text{ with } |\lambda| \geq 1.$$

Notes and References

The material in this appendix is entirely standard; see for example [22].

Bibliography

[1] J. Ackermann. *Sampled-Data Control Systems: Analysis and Synthesis, Robust System Design.* Springer-Verlag, Berlin, 1985.

[2] H.M. Al-Rahmani and G.F. Franklin. A new optimal multirate control of linear periodic and time-invariant systems. *IEEE Trans. Auto. Control*, 35:406–415, 1990.

[3] B.D.O. Anderson. Controller design: moving from theory to practice. *IEEE Control Systems Magazine*, 13:16–25, 1993.

[4] B.D.O. Anderson and J.B. Moore. *Optimal Filtering.* Prentice-Hall, Englewood Cliffs, N.J., 1979.

[5] M. Araki, T. Hagiwara, and Y. Ito. Frequency-response of sampled-data systems II: closed-loop consideration. Technical Report, Division of Applied System Science, Kyoto University, 1992.

[6] M. Araki and Y. Ito. Frequency-response of sampled-data systems I: open-loop consideration. Technical Report, Division of Applied System Science, Kyoto University, 1992.

[7] M. Araki and K. Yamamoto. Multivariable multirate sampled-data systems: state-space description, transfer characteristics, and Nyquist criterion. *IEEE Trans. Auto. Control*, 31:145–154, 1986.

[8] K.J. Åström, P. Hagander, and J. Sternby. Zeros of sampled systems. *Automatica*, 20:31–38, 1984.

[9] K.J. Åström and B. Wittenmark. *Computer Controlled Systems: Theory and Design.* Prentice-Hall, Englewood Cliffs, N.J., 1984.

[10] M. Athans. The role and use of the stochastic linear-quadratic-gaussian problem in control system design. *IEEE Trans. Auto. Control*, 16:529–552, 1971.

[11] D.M. Auslander and C.H. Tham. *Real-Time Software for Control: Program Examples in C.* Prentice-Hall, Englewood Cliffs, N.J., 1990.

[12] R. Avedon and B.A. Francis. Digital control design via convex optimization. *ASME J. Dynamic Systems, Meas. and Control*, 115:579–586, 1993.

[13] T. Başar. Game theory and $\mathcal{H}_\infty$-optimal control: the discrete-time case. In *Bilkent Int. Conf. New Trends in Communications, Control, and Signal Processing*, 1990.

[14] T. Başar. Optimum $\mathcal{H}_\infty$ designs under sampled-data state measurements. *Systems & Control Letters*, 16:399–409, 1991.

[15] B. Bamieh, M.A. Dahleh, and J.B. Pearson. Minimization of the L^∞-induced norm for sampled-data systems. *IEEE Trans. Auto. Control*, 38:717–732, 1993.

[16] B. Bamieh and J.B. Pearson. A general framework for linear periodic systems with application to $\mathcal{H}_\infty$ sampled-data control. *IEEE Trans. Auto. Control*, 37:418–435, 1992.

[17] B. Bamieh and J.B. Pearson. The $\mathcal{H}_2$ problem for sampled-data systems. *Systems & Control Letters*, 19:1–12, 1992.

[18] B. Bamieh, J.B. Pearson, B.A. Francis, and A. Tannenbaum. A lifting technique for linear periodic systems with applications to sampled-data control. *Systems & Control Letters*, 17:79–88, 1991.

[19] H. Bercovici, C. Foias, and A. Tannenbaum. On skew Toeplitz operators. *Operator Theory: Advances and Applications*, 32:21–43, 1988.

[20] S.P. Boyd, V. Balakrishnan, and P. Kabamba. A bisection method for computing the $\mathcal{H}_\infty$ norm of a transfer matrix and related problems. *Math. Cont. Sig. Sys.*, 2:207–219, 1989.

[21] S.P. Boyd and C.H. Barrett. *Linear Controller Design: Limits of Performance.* Prentice-Hall, Englewood Cliffs, 1991.

[22] F.M. Callier and C.A. Desoer. *Linear System Theory.* Springer-Verlag, New York, 1991.

[23] T. Chen. Control of sampled-data systems. Ph.D. Thesis, Dept. of Electrical Engineering, University of Toronto, 1991.

[24] T. Chen. A simple derivation of the $\mathcal{H}_2$-optimal sampled-data controller. *Systems & Control Letters*, 20:49–56, 1993.

[25] T. Chen, A. Feintuch, and B.A. Francis. On the existence of the $\mathcal{H}_\infty$-optimal sampled-data controllers. In *Proc. IEEE CDC*, 1990.

[26] T. Chen and B.A. Francis. On the $\mathcal{L}_2$-induced norm of a sampled-data system. *Systems & Control Letters*, 15:211–219, 1990.

[27] T. Chen and B.A. Francis. $\mathcal{H}_2$-optimal sampled-data control. *IEEE Trans. Auto. Control*, AC-36:387–397, 1991.

[28] T. Chen and B.A. Francis. Input-output stability of sampled-data systems. *IEEE Trans. Auto. Control*, 36:50–58, 1991.

[29] T. Chen and B.A. Francis. Linear time-varying $\mathcal{H}_2$-optimal control of sampled-data systems. *Automatica*, 27:963–974, 1991.

[30] T. Chen and B.A. Francis. Sampled-data optimal design and robust stabilization. *ASME J. Dynamic Systems, Meas. and Control*, 114:538–543, 1992.

[31] T. Chen and L. Qiu. $\mathcal{H}_\infty$ design of general multirate sampled-data control systems. *Automatica*, 30:1139–1152, 1994.

[32] R.C. Cobden. Design and digital implementation of $\mathcal{H}_2$ and $\mathcal{H}_\infty$ optimal controllers for systems with time delays. M.A.Sc. Thesis, Dept. of Electrical Engineering, University of Toronto, 1990.

[33] J.B. Conway. *A Course in Functional Analysis*. Springer-Verlag, New York, 1985.

[34] M.A. Dahleh and J.B. Pearson. ℓ_1-optimal feedback controllers for MIMO discrete-time systems. *IEEE Trans. Auto. Control*, 32:314–322, 1987.

[35] J.H. Davis. Stability conditions derived from spectral theory: discrete system with periodic feedback. *SIAM J. Control*, 10:1–13, 1972.

[36] D.F. Delchamps. Stabilizing a linear system with quantized state feedback. *IEEE Trans. Auto. Control*, 35:916–924, 1990.

[37] R. Doraiswami. Robust control strategy for a linear time-invariant multivariable sampled-data servomechanism problem. *IEE Proc., Pt. D*, 129:383–292, 1982.

[38] P. Dorato and A.H. Levis. Optimal linear regulators: the discrete-time case. *IEEE Trans. Auto. Control*, 16:613–620, 1971.

[39] J.C. Doyle, B.A. Francis, and A.R. Tannenbaum. *Feedback Control Theory*. Macmillan, New York, 1992.

[40] J.C. Doyle, K. Glover, P. P. Khargonekar, and B.A. Francis. State-space solutions to standard $\mathcal{H}_2$ and $\mathcal{H}_\infty$ control problems. *IEEE Trans. Auto. Control*, 34:831–847, 1989.

[41] G. Dullerud. Tracking and $\mathcal{L}_1$ performance in sampled-data control systems. M.A.Sc. Thesis, Dept. of Electrical Engineering, University of Toronto, 1990.

[42] G. Dullerud and B.A. Francis. $\mathcal{L}_1$ performance in sampled-data systems. *IEEE Trans. Auto. Control*, 37:436–446, 1992.

[43] G. Dullerud and K. Glover. Analysis of structured LTI uncertainty in sampled-data systems. Report, Dept. Engineering, University of Cambridge, 1993.

[44] G. Dullerud and K. Glover. Robust stabilization of sampled-data systems to structured LTI perturbations. *IEEE Trans. Auto. Control*, 38:1497–1508, 1993.

[45] J.C. Doyle et al. *Lecture Notes in Advances in Multivariable Control.* ONR/Honeywell Workshop, Minneapolis, MN, 1984.

[46] A. Feintuch and B.A. Francis. Uniformly optimal control of linear feedback systems. *Automatica*, 21:563–574, 1985.

[47] A. Feintuch and R. Saeks. *System Theory: A Hilbert Space Approach.* Academic Press, New York, 1982.

[48] V. Feliu, J.A. Cerrada, and C. Cerrada. A method to design multirate controllers for plants sampled at a low rate. *IEEE Trans. Auto. Control*, 35:57–60, 1990.

[49] D.S. Flamm. Single-loop stability margins for multirate and periodic control systems. *IEEE Trans. Auto. Control*, 38:1232–1236, 1993.

[50] C. Foias, A. Tannenbaum, and G. Zames. On the $\mathcal{H}_\infty$-optimal sensitivity problem for systems with delays. *SIAM J. Control and Opt.*, 25:686–706, 1987.

[51] B.A. Francis. *A Course in $\mathcal{H}_\infty$ Control Theory.* Springer-Verlag, New York, 1987.

[52] B.A. Francis and T.T. Georgiou. Stability theory for linear time-invariant plants with periodic digital controllers. *IEEE Trans. Auto. Control*, 33:820–832, 1988.

[53] G.F. Franklin and A. Emami-Naeini. Design of ripple-free multivariable robust servomechanisms. *IEEE Trans. Auto. Control*, 31:661–664, 1986.

[54] G.F. Franklin, J.D. Powell, and M.L. Workman. *Digital Control of Dynamic Systems.* Addison-Wesley, 1990.

[55] H. Freeman and O. Lowenschuss. Bibliography of sampled-data control systems and z-transform applications. *IRE Trans. Auto. Control*, pages 28–30, 1958.

[56] B. Friedland. Sampled-data control systems containing periodically varying members. In *Proc. First IFAC Congress*, 1960.

[57] K. Furuta. Alternative robust servo-control system and its digital control. *Int. J. Control*, 45:183–194, 1987.

[58] J.B. Garnett. *Bounded Analytic Functions.* Academic Press, New York, 1981.

[59] I.C. Gohberg and M.G. Krein. *Introduction to the Theory of Linear Nonselfadjoint Operators.* American Mathematical Society, Providence, Rhode Island, 1969.

[60] G.C. Goodwin and A. Feuer. Linear periodic control: a frequency domain viewpoint. *Systems & Control Letters*, 19:379–390, 1992.

[61] M. Green and D. Limebeer. $\mathcal{H}_\infty$ optimal full information control for discrete-time systems. In *Proc. IEEE CDC*, pages 1769–1774, 1990.

[62] M. Green and D.J.N. Limebeer. *Linear Robust Control.* Prentice-Hall, Englewood Cliffs, N.J., 1995.

[63] M. Grimble and D. Fragopoulos. Solution of discrete $\mathcal{H}_\infty$ optimal control problems using a state-space approach. In *Proc. IEEE CDC*, pages 1775–1780, 1990.

[64] D.W. Gu, M.C. Tsai, S.D. O'Young, and I. Postlethwaite. State space formulae for discrete-time $\mathcal{H}_\infty$ optimization. *Int. J. Control*, 49:1683–1723, 1989.

[65] T. Hagiwara and M. Araki. Design of a stable feedback controller based on the multirate sampling of the plant output. *IEEE Trans. Auto. Control*, 33:812–819, 1988.

[66] T. Hagiwara and M. Araki. On preservation of strong stabilizability under sampling. *IEEE Trans. Auto. Control*, 33:1080–1082, 1988.

[67] P.R. Halmos. *A Hilbert Space Problem Book.* Second Edition, Springer-Verlag, New York, 1985.

[68] C.C. Hang, K.W. Lim, and B.W. Chong. A dual-rate adaptive digital Smith predictor. *Automatica*, 25:1–16, 1989.

[69] S. Hara, M. Nakajima, and P.T. Kabamba. Robust stabilization in digital control systems. In *Proc. MTNS*, 1991.

[70] S. Hara and H-K Sung. Sensitivity improvement by a stable controller in SISO digital control systems. *Systems & Control Letters*, 12:123–128, 1989.

[71] S. Hara and H-K Sung. Ripple-free conditions in sampled-data control systems. In *Proc. IEEE CDC*, 1991.

[72] M.L. Hautus. Controllability and stabilizability of sampled systems. *IEEE Trans. Auto. Control*, 17:528–531, 1972.

[73] Y. Hayakawa, S. Hara, and Y. Yamamoto. $\mathcal{H}_\infty$ type problem for sampled-data control systems—a solution via minimum energy characterization. *IEEE Trans. Auto. Control*, 39:2278–2284, 1994.

[74] I. Horowitz and O. Yaniv. Quantitative design for SISO nonminimum-phase unstable plants by the singular-G method. *Int. J. Control*, 46:281–294, 1987.

[75] P. Iglesias and K. Glover. State space approach to discrete time $\mathcal{H}_\infty$ control. *Int. J. Control*, 54:1031–1073, 1991.

[76] P.A. Iglesias. Input-output stability of sampled-data linear time-varying systems. Technical Report JHU/ECE-94/03, Dept. of Electrical and Computer Engineering, The Johns Hopkins University, 1994.

[77] P.A. Iglesias, D. Mustafa, and K. Glover. Discrete-time $\mathcal{H}_\infty$ controllers satisfying a minimum entropy condition. *Systems & Control Letters*, 14:275–286, 1990.

[78] Y.C. Juan and P.T. Kabamba. Optimal hold functions for sampled data regulation. *Automatica*, 27:177–181, 1991.

[79] E.I. Jury. Sampled-data systems, revisited: reflections, recollections, and reassessments. *ASME J. of Dynamic Systems, Meas., and Control*, 102:208–217, 1980.

[80] E.I. Jury and F.J. Mullin. The analysis of sampled-data control systems with a periodically time-varying sampling rates. *IRE Trans Auto. Control*, 24:15–21, 1959.

[81] P.T. Kabamba. Control of linear systems using generalized sampled-data hold functions. *IEEE Trans. Auto. Control*, 32:772–783, 1987.

[82] P.T. Kabamba and S. Hara. On computing the induced norm of a sampled data system. In *Proc. ACC*, 1990.

[83] P.T. Kabamba and S. Hara. Worst case analysis and design of sampled-data control systems. *IEEE Trans. Auto. Control*, 38:1337–1357, 1993.

[84] R. Kalman, Y.C. Ho, and K. Narendra. Controllability of linear dynamical systems. In *Contributions to Differential Equations*, volume 1. Interscience, New York, 1963.

[85] R.E. Kalman and J.E. Bertram. A unified approach to the theory of sampling systems. *J. Franklin Inst.*, 267:405–436, 1959.

[86] Y. Kannai and G. Weiss. Approximating signals by fast impulse sampling. *Math. of Control, Signals, and Systems*, 6:166–179, 1993.

[87] J.P. Keller and B.D.O. Anderson. A new approach to the discretization of continuous-time controllers. *IEEE Trans. Auto. Control*, 37:214–223, 1992.

[88] M. Khammash. Necessary and sufficient conditions for the robustness of time-varying systems with applications to sampled-data systems. *IEEE Trans. Auto. Control*, 38:49–57, 1993.

[89] P. Khargonekar and E. Sontag. On the relation between stable matrix fraction factorizations and regulable realizations of linear systems over rings. *IEEE Trans. Auto. Control*, AC-37:627–638, 1982.

[90] P.P. Khargonekar, K. Poolla, and A. Tannenbaum. Robust control of linear time-invariant plants using periodic compensation. *IEEE Trans. Auto. Control*, 30:1088–1096, 1985.

[91] P.P. Khargonekar and N. Sivashankar. $\mathcal{H}_2$ optimal control for sampled-data systems. *Systems & Control Letters*, 18:627–631, 1992.

[92] R. Kondo and K. Furuta. Sampled-data optimal control of continuous systems for quadratic criterion function taking account of delayed control action. *Int. J. Control*, 41:1051–1060, 1985.

[93] R. Kondo, S. Hara, and T. Itou. Characterization of discrete-time $\mathcal{H}_\infty$ controllers via bilinear transformation. In *Proc. IEEE CDC*, pages 1763–1768, 1990.

[94] G.M. Kranc. Input-output analysis of multirate feedback systems. *IRE Trans. Auto. Control*, 3:21–28, 1957.

[95] V. Kučera. The discrete Riccati equation of optimal control. *Kybernetica*, 8:430–447, 1972.

[96] V. Kučera. Stability of discrete linear feedback systems. In *Proc. 6th IFAC Congress*, 1975.

[97] V. Kučera. *Discrete Linear Control: The Polynomial Equation Approach.* Wiley, Prague, 1979.

[98] E.A. Larson. Fly-by-wire fight controls for the X29. *Scientific Honeyweller*, 10(1):135–143, 1989.

[99] S. Lee, S.M. Meerkov, and T. Runolfsson. Vibrational feedback control: zero placement capabilities. *IEEE Trans. Auto. Control*, 32:604–611, 1987.

[100] G.M.H. Leung. Performance analysis of sampled-data systems. M.A.Sc. Thesis, Dept. of Electrical Engineering, University of Toronto, 1990.

[101] G.M.H. Leung, T.P. Perry, and B.A. Francis. Performance analysis of sampled-data control systems. *Automatica*, 27:699–704, 1991.

[102] A.H. Levis, R.A. Schlueter, and M. Athans. On the behavior of optimal linear sampled-data regulators. *Int. J. Control*, 13(2):343–361, 1971.

[103] D. Limebeer, M. Green, and D. Walker. Discrete-time $\mathcal{H}_\infty$ control. In *Proc. IEEE CDC*, 1989.

[104] A. Linneman. On robust stability of continuous-time systems under sampled-data control. In *Proc. IEEE CDC*, 1990.

[105] A. Linnemann. $\mathcal{L}_\infty$-induced norm optimal performance in sampled-data systems. *Systems & Control Letters*, 18:265–276, 1992.

[106] K. Liu and T. Mita. Conjugation and $\mathcal{H}_\infty$ control of discrete time systems. In *Proc. IEEE CDC*, 1989.

[107] K. Liu and T. Mita. Complete solution to the standard $\mathcal{H}_\infty$ control problem of discrete-time systems. In *Proc. IEEE CDC*, pages 1786–1793, 1990.

[108] C.F. Van Loan. Computing integrals involving the matrix exponential. *IEEE Trans. Auto. Control*, 23:395–404, 1978.

[109] D.G. Meyer. A parametrization of stabilizing controllers for multirate sampled-data systems. *IEEE Trans. Auto. Control*, 35:233–236, 1990.

[110] D.G. Meyer. A solution via lifting to the general multirate LQG problem. Preprint, September, 1990.

[111] D.G. Meyer. Cost translation and a lifting approach to the multirate LQG problem. *IEEE Trans. Auto. Control*, 37:1411–1415, 1992.

[112] R.A. Meyer and C.S. Burrus. A unified analysis of multirate and periodically time-varying filters. *IEEE Trans. Circuits and Systems*, 22:162–168, 1975.

[113] M. Morari and E. Zafiriou. *Robust Process Control*. Prentice-Hall, Englewood, Cliffs, N.J., 1989.

[114] C.N. Nett, C.A. Jacobson, and M.J. Balas. A connection between state-space and doubly coprime fractional representations. *IEEE Trans. Auto. Control*, AC-39:831–832, 1984.

[115] A.W. Olbrot. Robust stabilization of uncertain systems by periodic feedback. *Int. J. Control*, 45:747–758, 1987.

[116] A.V. Oppenheim and R.W. Schafer. *Discrete-Time Signal Processing*. Prentice-Hall, Englewood Cliffs, N.J., 1989.

[117] U. Ozguner and E. J. Davison. Sampling and decentralized fixed modes. In *Proc. ACC*, 1989.

[118] T. Pappas, A.J. Laub, and N.R. Sandell Jr. On the numerical solution of the discrete-time algebraic Riccati equation. *IEEE Trans. Auto. Control*, AC-25:631–641, 1980.

[119] T.P. Perry. On the recovery of $\mathcal{H}_\infty$ bounds in sampled-data systems. M.A.Sc. Thesis, Dept. of Electrical Engineering, University of Toronto, 1989.

[120] S.E. Posner. $\mathcal{L}_p$-optimal discretization of analog filters. M.A.Sc. Thesis, Dept. of Electrical Engineering, University of Toronto, 1991.

[121] J.R. Ragazzini and G.F. Franklin. *Sampled-Data Control Systems*. McGraw-Hill, New York, 1958.

[122] M. Rosenblum and J. Rovnyak. *Hardy Classes and Operator Theory*. Oxford University Press, New York, 1985.

[123] T.B. Sheridan. *Telerobotics, Automation, and Human Supervisory Control*. MIT Press, Cambridge, Mass., 1992.

[124] N. Sivashankar and P.P. Khargonekar. Induced norms of sampled-data systems. *Automatica*, 28:1267–1272, 1992.

[125] N. Sivashankar and P.P. Khargonekar. Robust stability analysis of sampled-data systems. *IEEE Trans. Auto. Control*, 38:58–69, 1993.

[126] N. Sivashankar and P.P. Khargonekar. Characterization and computation of the $\mathcal{L}_2$ induced norm of sampled-data systems. *SIAM J. Cont. Opt.*, 32:1128–1150, 1994.

[127] R.A. Skoog and G.L. Blankenship. Generalized pulse-modulated feedback systems: norms, gains, Lipschitz constants, and stability. *IEEE Trans. Auto. Control*, 15:300–315, 1970.

[128] H.W. Smith. Optimal sampled approximations for analog filters. Technical Report, Electrical and Computer Engineering, University of Toronto, 1992.

[129] E.M. Stein and G. Weiss. *Introduction to Fourier Analysis on Euclidean Spaces*. Princeton University Press, Princeton, N.J., 1971.

[130] A.A. Stoorvogel. The $\mathcal{H}_\infty$ control problem: A state space approach. Ph.D. Thesis, Eindhoven University of Technology, The Netherlands, 1990.

[131] A.A. Stoorvogel. The discrete-time $\mathcal{H}_\infty$ control problem with measurement feedback. *SIAM J. Control and Opt.*, 29:160–184, 1991.

[132] W. Sun, K.M. Nagpal, and P.P. Khargonekar. $\mathcal{H}_\infty$ control and filtering for sampled-data systems. *IEEE Trans. Auto. Control*, 38:1162–1175, 1993.

[133] W. Sun, K.M. Nagpal, P.P. Khargonekar, and K.R. Poolla. Digital control systems: $\mathcal{H}_\infty$ controller design with a zero order hold function. In *Proc. IEEE CDC*, pages 475–480, 1992.

[134] H-K Sung and S. Hara. Properties of sensitivity and complementary sensitivity functions in single-input single-output digital control systems. *Int. J. Control*, 48:2429–2439, 1988.

[135] B. Sz.-Nagy and C. Foias. *Harmonic Analysis of Operators on Hilbert Space*. North-Holland, 1970.

[136] G. Tadmor. $\mathcal{H}_\infty$ optimal sampled-data control in continuous time systems. *Int. J. Control*, 56:99–141, 1992.

[137] P.M. Thompson, R.L. Dailey, and J.C. Doyle. New conic sectors for sampled-data feedback systems. *Systems & Control Letters*, 7:395–404, 1986.

[138] P.M. Thompson, G. Stein, and M. Athans. Conic sectors for sampled-data feedback systems. *Systems & Control Letters*, 3:77–82, 1983.

[139] H.T. Toivonen. Discretization of analog filters via $\mathcal{H}_\infty$ model matching theory. Technical Report, Dept. of Chemical Engineering, Swedish University of Åbo, Finland, 1990.

[140] H.T. Toivonen. Sampled-data control of continuous-time systems with an $\mathcal{H}_\infty$ optimality criterion. *Automatica*, 28:45–54, 1992.

[141] J.T. Tou. *Digital and Sampled-Data Control Systems*. McGraw-Hill, New York, 1959.

[142] H.L. Trentelman and A.A. Stoorvogel. Sampled-data and discrete-time $\mathcal{H}_2$ optimal control. Technical Report W-9222, Dept. of Mathematics, University of Groningen, 1992.

[143] S. Urikura and A. Nagata. Ripple-free deadbeat control for sampled-data systems. *IEEE Trans. Auto. Control*, 32:474–482, 1987.

[144] P.P. Vaidyanathan. *Multirate Systems and Filter Banks.* Prentice-Hall, Englewood Cliffs, N.J., 1993.

[145] M. Vidyasagar. *Nonlinear Systems Analysis, 2nd ed.* Prentice-Hall, New Jersey, 1992.

[146] J.L. Willems. Time-varying feedback for the stabilization of fixed modes in decentralized control systems. *Automatica*, 25:127–131, 1989.

[147] C.S. Williams. *Designing Digital Filters.* Prentice-Hall, Englewood Cliffs, N.J., 1986.

[148] D. Williamson and K. Kadiman. Optimal finite wordlength linear quadratic regulation. *IEEE Trans. Auto. Control*, 34:1218–28, 1989.

[149] H.K. Wimmer. Normal forms of symplectic pencils and the discrete-time algebraic Riccati equation. *Linear Algebra and its Applications*, 147:411–440, 1991.

[150] W.M. Wonham. *Linear Multivariable Control: A Geometric Approach.* 3rd ed., Springer-Verlag, New York, 1985.

[151] Y. Yamamoto. A new approach to sampled-data control systems – a function space method. In *Proc. IEEE CDC*, 1990.

[152] Y. Yamamoto. Frequency response and its computation for sampled-data systems. Technical Report, Division of Applied System Science, Kyoto University, 1992.

[153] Y. Yamamoto. On the state-space and frequency domain characterization of $\mathcal{H}^\infty$-norm of sampled-data systems. Technical Report, Division of Applied System Science, Kyoto University, 1992.

[154] Y. Yamamoto. A function space approach to sampled-data control systems and tracking problems. *IEEE Trans. Auto. Control*, 39:703–713, 1994.

[155] D.C. Youla, H.A. Jabr, and J. J. Bongiorno Jr. Modern Wiener-Hopf design of optimal controllers: part 2. *IEEE Trans. Auto. Control*, AC-21:319–338, 1976.

[156] K. Zhou, J. Doyle, and K. Glover. *Robust and Optimal Control.* In preparation, 1994.

[157] K. Zhou and P.P. Khargonekar. On the weighted sensitivity minimization problem for delay systems. *Systems & Control Letters*, 8:307–312, 1987.

Index